Analytical Methods in Chemical Analysis

Also of interest

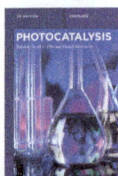

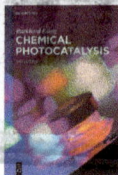

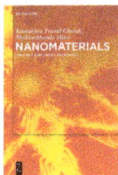

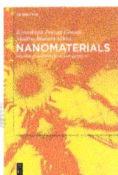

Analytical Methods in Chemical Analysis

An Introduction

Edited by
Shikha Kaushik and Banty Kumar

DE GRUYTER

Editors
Dr. Shikha Kaushik
Department of Chemistry
Rajdhani College
University of Delhi
New Delhi 110015, India
kaushik.shikha@gmail.com
shikha.kaushik@rajdhani.du.ac.in

Dr. Banty Kumar
Department of Chemistry
Rajdhani College
University of Delhi
New Delhi 110015, India
bantykumar881@gmail.com

ISBN 978-3-11-079480-9
e-ISBN (PDF) 978-3-11-079481-6
e-ISBN (EPUB) 978-3-11-079490-8

Library of Congress Control Number: 2023931050

Bibliographic information published by the Deutsche Nationalbibliothek
The Deutsche Nationalbibliothek lists this publication in the Deutsche Nationalbibliografie;
detailed bibliographic data are available on the internet at http://dnb.dnb.de.

© 2023 Walter de Gruyter GmbH, Berlin/Boston
Cover image: vchal/iStock/Getty Images
Typesetting: Integra Software Services Pvt. Ltd.
Printing and binding: CPI books GmbH, Leck

www.degruyter.com

Preface

Languages are true analytical methods . . . .

Antoine Lavoisier

Analytical Chemistry is an important and applied experimental field of science that has gained immense importance in recent years because of its practical utility. It is the branch of science that employs different instruments and methods for collection, separation, identification, and quantification of various organic, inorganic, and biological molecules. Analytical Chemistry is based not only on chemistry but also on other disciplines such as biology, physics, pharmaceutical, and many fields of technology.

The book entitled *Analytical Methods in Chemical Analysis* provides information pertinent to the important techniques and methods employed in analytical chemistry for the analysis of various organic and inorganic compounds. The book is organized into six sections, which cover the basic concepts of qualitative and quantitative analysis, spectrochemical methods of analysis, and also thermal and electroanalytical methods. Qualitative analysis identifies analytes, while quantitative analysis determines the concentration or numerical amount of the molecules under study. This book also exposes students to the different laws of spectroscopy and various electronic transitions that occur in the different regions of the electromagnetic spectra. The main objective of this work is to develop an understanding and make learners familiar with the basic analytical methods employed in the chemical analysis of various compounds. The authors have made significant efforts to make this book a valuable piece of information for the students and fellow teachers. The contents of the book will be useful for students of Chemistry, Analytical Chemistry, and Industrial Chemistry.

Our heartfelt thanks go to Prof. Rajesh Giri (Principal, Rajdhani College, University of Delhi) for his constant support and encouragement. His immeasurable contribution and unwavering support have always been of great value to us. The authors wish to express their warm and sincere thanks to Prof. Shrikant Kukreti (Senior Professor, Department of Chemistry, University of Delhi) for his kind guidance and valuable suggestions. His thoughtful encouragement and continual support have always kept us motivated. We are delighted to extend appreciation for the cooperation we received from Late Prof. Ashok Kumar Prasad (Senior Professor, Department of Chemistry, University of Delhi) for his invaluable help and support. We also express our heartfelt appreciation to all authors. It was the mere cooperation and spirit of the authors that made this endeavor possible. A big thanks go to dear students and worthy colleagues at Rajdhani College who helped in one or the other way. Our warm and most earnest thanks go to our families for their kind patience and utmost cooperation during the entire work span. The authors would love to mention a special thanks to our dear Siyonaa, Sarth, Advvay, and Master Yatharth for their smiling faces and tender love which kept us going. Last, but not least, we would like to acknowledge with gratitude the support of the publishing team at esteemed De Gruyter.

https://doi.org/10.1515/9783110794816-202

All possible efforts have been made so that there would not be any errors in the book; however, no work is perfect. We shall be pleased to receive suggestions from the readers for improvement of subsequent editions.

Shikha Kaushik
Banty Kumar

Contents

Section IV: **Thermoanalytical Methods**

Section V: **Electroanalytical Methods**

Section VI: **Separation methods and chromatography**

List of authors

Neeta Azad
Department of Chemistry, Atma Ram Sanatan
Dharma College, University of Delhi, Dhaula
Kuan, Delhi 110021, India
neetaazad@arsd.du.ac.in
Chapter 4

Jyoti Devi
Department of Chemistry, SRM University
Delhi-NCR, Sonepat, Haryana 131029, India
jyotikaushik916@gmail.com
Chapter 12

Meenakshi Gupta
Department of Chemistry, Atma Ram Sanatan
Dharma College, University of Delhi, Dhaula
Kuan, Delhi 110021, India
meenaxi1978@gmail.com
Chapters 10, 11

Mukesh
Keshav Mahavidyalaya, Near Sainik Vihar, Zone
H-4/5, Co-operative Group Housing Societies
Pitampura, Delhi 110034, India
mukeshgupta.chem@gmail.com
Chapters 6, 7

Jaspreet Kaur
Hindu College, University Enclave, North
Campus, Delhi 110007, India
jaspreet.kutaal@gmail.com
Chapters 6, 7

Vinod Khatri
Govt. College for Women, Murthal, Sonipat
131027, India
khatri.vinod49@gmail.com
Chapter 2

Mohan Kumar
Shri Varhsney College, Achal Tal, Railway Road
Aligarh 202001, U.P., India
kumar.mohan46@gmail.com
Chapter 13

Prashant Kumar
Department of Chemistry, SRM University
Delhi-NCR, Sonepat, Haryana 131029, India
prashant.kumar@srmuniversity.ac.in
Chapter 12

Ritika Nagpal
Department of Chemistry, SRM University
Delhi-NCR, Sonepat, Haryana 131029, India
ritika.nagpal@srmuniversity.ac.in
Chapter 12

Ritu Payal
Department of Chemistry, Rajdhani College
University of Delhi, Mahatma Gandhi Marg
Delhi 110015, India
ritupayal@rajdhani.du.ac.in
Chapter 8

Poonam Pipil
Department of Chemistry, Rajdhani College,
University of Delhi, Mahatma Gandhi Marg,
New Delhi 110015 India
pipilpoonam90@gmail.com
Chapters 1, 3

Mukesh Kumar Saini
Department of Chemistry, Rajdhani College,
University of Delhi, Mahatma Gandhi Marg,
New Delhi 110015 India
gudha.mukesh@gmail.com
Chapters 1, 3

Manju Saroj
Department of Chemistry, Gargi College
University of Delhi, Delhi, India
Manjuksaroj@gmail.com
Chapter 8

Chanchal Singh
D. S. N. (PG) College, CSJM University
A. B. Nagar, Unnao 209801, India
premichanchal@gmail.com
Chapter 5

https://doi.org/10.1515/9783110794816-204

Prerna Singh
Department of Chemistry,
Zakir Husain Delhi College,
University of Delhi, J. L. N. Marg
New Delhi 110002, India
prernasingh05@gmail.com
Chapter 14

Shramila Yadav
Department of Chemistry,
Rajdhani College, University of Delhi,
Mahatma Gandhi Marg
New Delhi 110015, India
syadav@rajdhani.du.ac.in
Chapters 9, 11

About the editors

Dr. Shikha Kaushik is working as an Assistant Professor at Rajdhani College, University of Delhi. She received her Bachelor's degree in Chemistry, Master's degree in Organic Chemistry, and Ph.D. from the University of Delhi. She has been teaching graduate students of various courses for more than 12 years. Dr. Kaushik was awarded Research Fellowship in Science for Meritorious Students (RFSMS) by University Grants Commission, India, for pursuing her research work. She has published a number of research papers in peer-reviewed journals of international repute.

Dr. Banty Kumar is teaching as an Associate Professor at Rajdhani College, University of Delhi. He completed his undergraduate studies in Chemistry, postgraduation in Organic Chemistry from Hansraj College, and Ph.D. from the University of Delhi. He has been teaching Organic Chemistry to B.Sc. (Hons), B.Sc. (PS) Chemistry, and B.Sc. (APS) Industrial Chemistry students since 2004. He is also actively involved in research and has published a number of research articles in international journals.

https://doi.org/10.1515/9783110794816-205

Abbreviations

A_λ	absorbance at a particular wavelength (λ)
Å	angstrom (1 Å = 0.1 nm)
aq.	aqueous
G	conductance
I	current
EMR	electromagnetic radiations
EMF	electromotive force
eV	electron volt
K	equilibrium constant
EW	equivalent weight
F	Faraday
g	grams
Hz	hertz
h	hours
Js	joule seconds
J	joules
λ	lambda (used for wavelength)
L	litres
m/s	meter per second
m	meters
μg	micrograms
μg dm^{-3}	micrograms per cubic decimeter
μl	microlitre
mg dL^{-1}	milligrams per deciliter
mL	millilitre
mins	minutes
M	molar
ε	molar extinction coefficient (epsilon)
MW	molecular weight
nm	nanometer
N	normal
OD	optical density
ppb	parts per billion
ppm	parts per million
Pm	picometer
V	potential
R	resistance
p	resistivity
sec	seconds
T	transmittance
UV	ultraviolet
Vis.	visible
vol.	volume
v/v	volume by volume
w/v	weight by volume
w/w	weight by weight

https://doi.org/10.1515/9783110794816-206

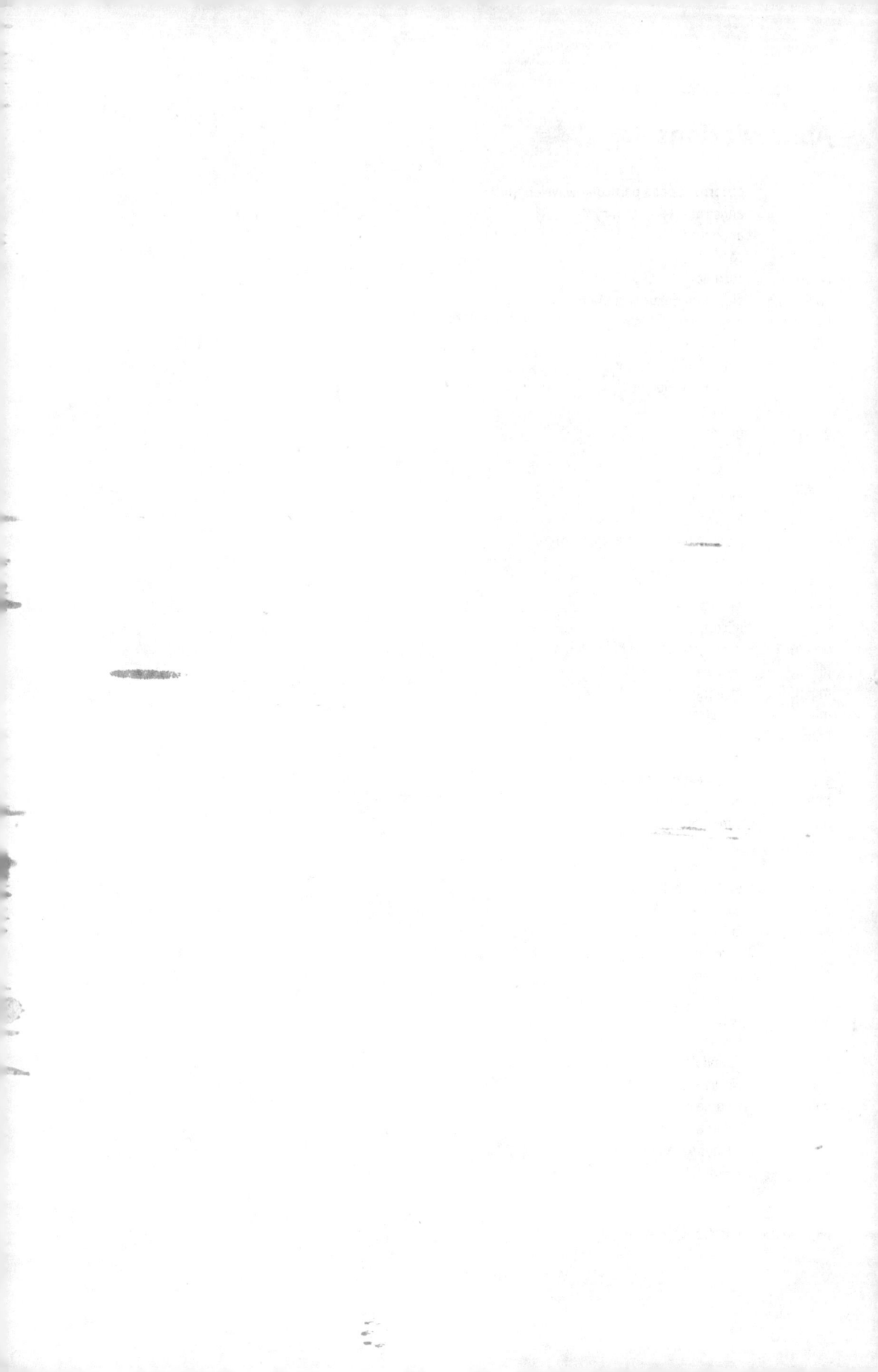

Section I: **Analytical chemistry**

Poonam Pipil and Mukesh Kumar Saini

Chapter 1
Introduction to analytical chemistry

Abstract: Analytical chemistry is an important and applied experimental field of science that has gained immense importance in recent years because of its practical utility. It is the branch of science that employs different instruments and methods for the collection, separation, identification, and quantification of various organic, inorganic, and biological molecules. Analytical chemistry is based not only on chemistry but also on other disciplines such as biology, physics, pharmaceutical, and many fields of technology. This chapter provides information pertinent to basic and important concepts of chemical characterization, that is, qualitative and quantitative analysis, different analytical techniques, and instruments employed for analysis along with their advantages and disadvantages. Analytical techniques such as chromatographic, titrimetric, spectroscopic, electrophoretic, diffraction, and electrochemical methods are used for the quantitative and qualitative analysis of compounds. Analysis of various organic, inorganic, or biochemical compounds is carried out on a specific instrument, and these instruments work on a specific principle. Data obtained from their instrumental techniques help in determining the chemical composition, quantity, and structure of various unknown molecules. Recent developments in automation of analytical instruments have helped the scientific community across the globe in numerous ways and also accelerated the advancement of analytical chemistry techniques.

1.1 Analytical chemistry

Analytical chemistry is basically a measurement science comprising powerful tools and methods that are useful in all fields of sciences. It is a branch of chemistry which deals with the methods and techniques employed for the separation, identification, and determination of substances present in a sample. This interdisciplinary branch of science plays an important role in research areas of biochemistry, clinical, agricultural, environmental, pharmaceutical, geology, space science, forensic, manufacturing, and many others (Figure 1.1). Its scope is very broad and embraces a wide range of manual, chemical, and instrumental techniques. Both quantitative and qualitative information are required in an analysis. Qualitative analysis gives the information about the identity of the constituent species (the elements or compounds) in a sample whereas quantitative analysis emphasizes on how much of a constituent species *or analytes* are present in a sample? Analytes are the substances or components of a sample that are to be determined. It has wide applicability in medical, environmental, industry, and all other sciences for example: It is used (i) in the analysis of blood or

https://doi.org/10.1515/9783110794816-001

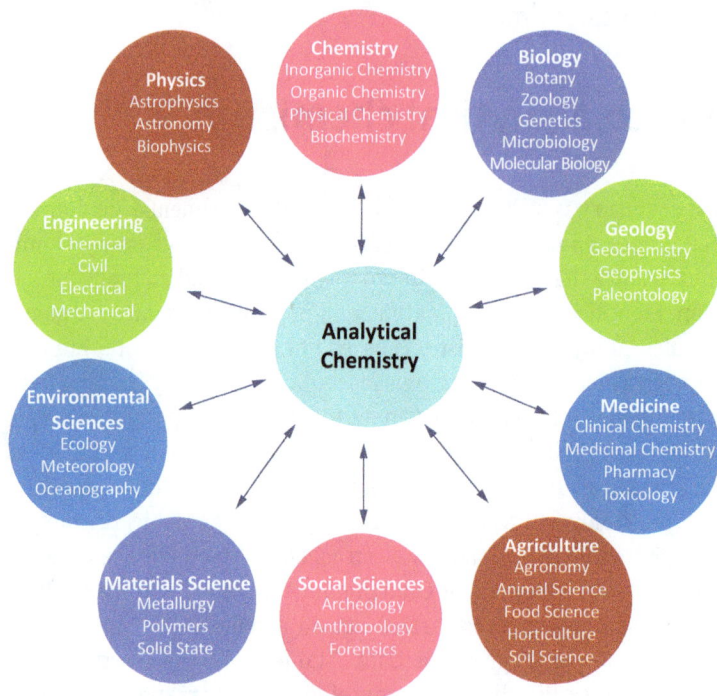

Figure 1.1: The relationship between analytical chemistry and the other sciences.

urine, (ii) for determination of heavy metals in soil or water, (iii) for comparison of DNA codes, and (iv) in the analysis of vitamin contents in food sample.

One of the main objectives of analytical chemistry is to improve scientifically corroborated methods that offer the qualitative and quantitative assessment of samples with a definite accuracy in a stipulated time. The methods for measuring the content of an element in a sample and methods for achieving an analytical goal can be separated into qualitative analysis (identification methods) and quantitative analysis, respectively. The detection method is generally the initial stage in the investigation of an unknown sample, and it also incorporates the analysis of its qualitative structure in order to classify the substance (e.g., as a mineral, dirt, and alloy).

The qualitative analytical methods have been used in the analysis for the determination of

i.   The constituents of a substance of the material under analysis, for example, identification of the grade of an alloy.
ii. The presence or absence of certain elements in the material.
iii. The total content of ions, elements, or the simplest combinations which constitute the studied product.
iv. The species of the mixtures such as sulfur, carbon, and phosphorus in the analysis of finished products such as iron and steel.

However, in various cases, determination of elements in a product is not sufficient for the estimation of its chemical properties. An idea of molecular composition is also necessary. Molecular composition means the form of compounds in which the components are present in the material and the content of these compounds. For example, in order to evaluate the properties of an alloy, it is not sufficient to know the total content of just carbon and sulfur, but it is also necessary to know the form in which they are present in the alloy.

Basically, analytical chemistry is a quantitative science. For determining the concentration of an analyte species in a solution, monitoring the rate of a reaction, estimating an equilibrium constant, and elucidating the structure of an unknown compound, it is all about measurements/recording data and performing calculations. Before proceeding further into the classification of analytical methods, let us briefly review some standard terms and definitions which are of great use in analytical chemistry.

1.2 Units of measurement

Suppose, you are in a laboratory and getting the instructions "Pipette out 20 of your standard solution to a 250 volumetric flask, and add 20 water to it." How will you do this? Clearly, these instructions are incomplete because units of measurement are not specified. Now, compare this with a complete instruction: "Pipette out 20 mL of your standard solution to a 250 mL volumetric flask, and add 20 mL water to it." You can easily follow this instruction as units are specified.

Measurement usually denotes the amount of something, and a measurement unit is a number that expresses that quantity of that unit. Sometimes, different units may be used to express the same physical measurement. For example, the mass of a sample can be expressed in kilogram (kg), ounces (oz), or pounds (lb). To remove this ambiguity, a common set of fundamental units, known as SI unit *(after the Système International d'Unités)* is followed by the scientific community (Table 1.1).

Table 1.1: Commonly used physical quantities and SI units.

Physical quantity	SI units	Symbol
Time	Second	s
Mass	Kilogram	kg
Length	Meter	M
Amount of substance	Mole	mol
Temperature	Kelvin	K
Electric current	Ampere	A
Luminous intensity	Candela	cd

1.2.1 Units for expressing concentration

The concentration of a solution is very important for studying various reactions because it helps in determining the kinetics and mechanism of that particular reaction. Concentration is a measurement unit that indicates the amount of solute present in a known amount of solution, and some of the important units of concentration are discussed below. It is also very important to know the mole concept to express the concentrations of solutions.

i.   **Moles:** The term "mole," symbol mol, is the SI unit of amount of a substance, and one mole contains exactly 6.022×10^{23} (Avogadro's number) elementary entities (atoms, molecules, or ions). For example, in the reaction

$$Ag^+ + Cl^- \rightarrow AgCl$$

we know one silver ion reacts with one chloride ion. So, 1 mol of silver ion weighs 107.87 g and reacts with 1 mol of chloride ion which weighs 35.453 g. The mole is extensively used in all branches of chemistry to express the quantity of reactants and products in chemical reactions.

ii. **Molarity (M):** This is the most widely used method that describes the concentration of a solution. A 1 M solution is defined as the solution that contains 1 mol of substance in 1 L of a solution. It is generally expressed as moles per liter or millimoles per milliliters:

$$\text{Molarity, } M = \frac{\text{Moles of solute}}{\text{Volume of solution (L)}}$$

iii. **Normality (N):** Normality (N) is another unit of concentration that is used by chemists, and it is defined as the gram equivalent weight of a solute per liter of solution. A 1 N solution can be prepared by dissolving one equivalent of solute in 1 L of solution.

$$\text{Normality, } N = \frac{\text{No. of gram equivalents of solute}}{\text{Volume of solution (L)}}$$

iv. **Formality (F):** The concentrations of solutions of ionic salts that do not exist as molecules in the solutions are given as formal. It is defined as the number of moles of solute per liter of solution (F) regardless of chemical form. Formality is numerically similar to molarity for a substance if it dissolves without dissociating into ions; for example, the molar concentration of a solution of glucose is the same as its formal concentration. However, for NaCl (sodium chloride) that ionizes in the solution, the molarity and formality are different.

v.   **Molality (m):** Molality is another important concentration term, used in addition to molarity and normality. It is defined as the number of moles of solute dissolved in 1 kg of the solvent.

$$\text{Molality, } m = \frac{\text{Moles of solute}}{\text{Mass of solvent (kg)}}$$

vi. **Percent (%) solution:** Sometimes, the concentration of molecules is expressed as units of solute per 100 units of sample. It can be represented either in terms of weight-to-volume percent (% w/v), weight-to-weight percent (% w/w), or volume-to-volume percent (% v/v). For example, a solution of 10% w/v NaCl (sodium chloride) contains 10 g of solute (NaCl) per 100 mL of solution.

Trace concentrations are usually represented by smaller units such as parts per million (ppm) or parts per billion (ppb). They are used to express the concentration contaminants in soil or water sample. Table 1.2 shows the concentration relationship for ppb and ppm.

Table 1.2: Units for expressing trace concentrations.

Unit	Weight/volume (w/v)	Weight/weight (w/w)	Volume/volume (v/v)
Parts per million (ppm)	mg/L	mg/kg	µL/L
	µg/mL	µg/g	nL/mL
Parts per billion (ppb)	µg/L	µg/kg	nL/L

So, we have seen that different weight criteria are used to express results, and it may also happen that sometimes results (concentration of analyte) are reported as different chemical species rather than the form in which it exists. For example, the hardness of water is reported in mg/L (milligrams per liter) as $CaCO_3$ (calcium carbonate), and Fe (iron) is measured in the form of Fe_2O_3 (iron (III) oxide) and reported as % Fe.

Quantitative analysis of small amount of substances and small volume of solution also requires special equipment such as microbalance and microburettes. In qualitative macroanalysis and semi-microanalysis, effects can be observed with the naked eye but instruments such as microscope and magnifying glass are required for micro and ultra-microanalysis.

1.3 Classification of analytical methods

Analytical methods can be classified into two categories:
i. **Classical methods of analysis:** These methods usually rely on the chemical reaction between the analyte (the compound under analysis) and the reagent that is added to the analyte. This analysis is also called "wet chemical analysis" and

consists of those methods that use no electronic and mechanical instruments other than a balance. Volumetric and gravimetric techniques fall under this category.

ii. **Instrumental methods of analysis:** This method of analysis uses an instrument for the measurement of a physical property of a particular analyte. These can be further categorized into different categories, depending on the type of instrument employed (Table 1.3). Some of the instrumental techniques used for analysis are conductometric, spectroscopic, thermal, and electroanalytical methods.

Table 1.3: Measured physical properties and analytical methods.

Physical property	Analytical method used for measurement of physical property
Volume	Volumetric
Mass	Gravimetric
Electrical current, electrical conductance, electrical potential, etc.	Conductometry, potentiometry, voltammetry, amperometry, polarography, and coulometry (electroanalytical methods)
Absorption of electromagnetic radiation	Atomic absorption, colorimetry, spectrophotometry (X-ray, UV–visible (UV–vis), IR (infrared), NMR (nuclear magnetic resonance), and ESR (electron spin resonance)
Emission of electromagnetic radiation	Emission spectroscopy, fluorescence, flame photometry, and radiochemical methods
Scattering of radiation	Turbidimetry, nephelometry, and Raman spectroscopy
Refraction of radiation	Interferometry, refractometry
Rotation	Polarimetry, optical rotatory dispersion
Diffraction	X-ray, electron diffraction methods
Thermal properties	Thermal conductivity and enthalpy methods
Mass to charge ratio	Mass spectrometry

Although in recent years, modern analytical chemistry is dominated by sophisticated instruments, but the roots of some of the principles used in modern instruments emerged from traditional techniques, and some of these are still used today. In fact, chemical steps form an integral part of the instrumental analysis; for example, the sample preparation, change in oxidation state, removal of excess reagent, pH, and concentration adjustments are some of the chemical steps, among many, which are required in instrumental methods.

1.3.1 Classical methods of analysis

As discussed earlier, qualitative analysis signifies the presence or absence of a particular compound in the sample under analysis but not the quantity. Numerous qualitative chemical tests and flame test can be performed to confirm the presence of certain ions or elements in an analyte. Although, with modern instrumentation, these tests are rarely used but they are quite important and form the backbone of undergraduate analytical chemistry laboratory and fieldwork.

Classical methods of analysis, often termed as "wet chemical methods," involve the sample preparation by taking the analyte and dissolving it in a suitable solvent. Acid–base titrations, oxidation–reduction titrations, and gravimetric determinations are examples of wet chemical analyses. Quantitative analysis measures the quantity of a particular constituent present in a substance, and it can be measured by volume (volumetric analysis) or mass (gravimetric analysis). Volumetric analysis measures the volume of reagent having unknown concentration (analyte) by titrating against a solution of known concentration using suitable indicators, whereas gravimetric methods determine the mass of an analyte present by weighing the sample before and/or after transformation.

The quantitative relationship between two reacting solution is quite interesting to understand. Titrimetric or volumetric analyses are the most useful and accurate analytical techniques employed for determining the concentration of an analyte, and they constitute an important topic in the curriculum at the school and college level. Before we describe the different types of volumetric titrations and principle behind gravimetric analysis, we must discuss some of the important terms which are used during a titration, and then volumetric relationships can be used to calculate the quantitative information about the titrated analyte.

i.   **Standard solution:** The solution of know concentration is termed as standard solution. A standard solution is prepared by weighing an accurate amount of a highly pure, and stable compound, known as primary standard, and dissolving it in a specific volume of solvent in a volumetric flask.

ii.   **Standardization:** It is a process of determining the exact concentration of the solution by titrating it against a weighed quantity of primary standard.

iii. **Titrant:** This solution is taken in the burette, and a standard solution is used as a titrant in titration experiments.

iv.   **Analyte:** An analyte or a component (in clinical biochemistry) or a chemical species is a substance that is being analyzed or under chemical analysis.

v.   **Titrand:** A substance whose concentration is to be determined by titration is referred to as a titrand.

vi.   **End point:** The experimental point at which the reaction is observed to be complete is called the end point. The completion of reaction is a marked change in some property of the solution. This change could be the color change of the solution (by an indicator) or a change in some electrical or other physical property of the solution.

vii. **Equivalence point:** The equivalence point (or stoichiometric point) is the point at which the number of equivalents of analyte exactly equals the number of equivalents of the titrant added.

viii. **Indicator:** An indicator is a reagent that is employed to determine the end point in a titration. A color change during a titration is usually brought about by the addition of an indicator, and the color of indicator depends on the properties of the solution.

(a) **Volumetric methods:** When one of the solutions is taken in a conical flask, and the other is added carefully from the burette till the endpoint is obtained is called titration. A solution with a known concentration is prepared as one of the components in a titration and is referred to as the titrant. The solution is then titrated with an unknown solution using a suitable indicator to determine the end point. The strength of the unknown solution can be determined by using the simple mathematical equation:

$$(C_1 V_1 = C_2 V_2)$$

This equation describes the concentration (C) and volume (V) of two solutions: unknown (analyte) and known (titrant, standard). There are four different forms of volumetric titrations; namely acid–base titration, oxidation–reduction titration, precipitation titration, and complexometric titration.

– **Acid–base titration:** An acid–base titration involves the determination of concentration of an acid or base by neutralizing it with a standard solution of a base and an acid. The end point indicates the completion of the reaction. An example in this category is titration of hydrochloric acid (HCl, strong acid) with sodium hydroxide (NaOH, strong base):

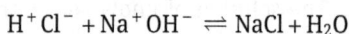

$$H^+ Cl^- + Na^+ OH^- \rightleftharpoons NaCl + H_2O$$

– **Reduction–oxidation titrations:** Another important class of titrations is oxidation–reduction or "redox" titrations. These titrations involve redox reactions between the titrant and the analyte to determine the concentration of a given analyte. These types of titrations require redox indicators and can also be carried out using potentiometer. The reaction between potassium permanganate and oxalic acid is an example of redox titration:

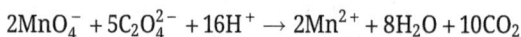

$$2MnO_4^- + 5C_2O_4^{2-} + 16H^+ \rightarrow 2Mn^{2+} + 8H_2O + 10CO_2$$

– **Precipitation titrations:** Precipitation titrations are those titrimetric method of analysis in which the analyte and titrant react to form a precipitate, and the titration continues till the whole of the analyte is consumed. These titrations are not nearly so numerous as acid–base or redox titrations and are limited to mostly those reactions that involve precipitation of silver ions with anions such as thiocyanates and halogens:

$$AgNO_3 + NaCl \rightarrow AgCl + NaNO_3$$

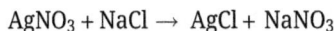

- **Complexometric titration:** These titrations are employed for the determination of different metal ions in solution. In complexometric titrations, the titrant (a complexing agent) forms a colored complex with the analyte, a metal ion. The most useful chelating agents used in these titrations is ethylenediaminetetraacetic acid (EDTA), and the titration of Mg(II) (magnesium) and Ca(II) (calcium) with EDTA is an example of complexometric titrations.

(b) **Gravimetric analysis:** It is one of the most important quantitative methods employed for accurately determining the mass of a substance. This method involves the selective precipitation of analyte from an aqueous solution. The precipitate is separated by filtration from rest of the solution, dried, and is accurately weighed. From the weight of the precipitate and its chemical composition, the weight of the analyte can be determined. A successful gravimetric analysis consists of some important steps that are needed to be followed to obtain a pure and filtrable precipitate for weighing, which are as follows:

  i.   Preparation of the solution
  ii.   Precipitation
  iii. Digestion
  iv.   Filtration
  v.   Washing
  vi.   Drying or igniting
  vii. Weighing
  viii. Calculation

There are four types of gravimetric methods of analysis, namely

i. **Precipitation gravimetry:** A gravimetric method that uses a precipitation reaction to separate ions from solution, for example, the direct determination of chloride ions (Cl^-) by precipitating AgCl (silver chloride).

ii. **Electrogravimetry:** In this method, the analyte is deposited as a solid film on the cathode or anode in an electrochemical cell. Example in this category includes the oxidation and deposition of Pb^{2+} as PbO_2 on a Pt anode.

iii. **Volatilization gravimetry:** This method employs the chemical or thermal energy to remove the target ion from the sample's solution. For determining the moisture content of food, thermal energy vaporized the H_2O (water molecules).

iv. **Particulate gravimetry:** A particulate gravimetry is a method in which the mass of a particular analyte is determined following its separation from the sample matrix by extraction or filtration. The suspended solids can be determined by the method of particulate gravimetry.

Classical methods of analysis require highly skilled and attentive analyst, if accuracy and precise results are to be obtained. The procedure employed is also time-consuming, and the demands of today's high-throughput forensics, environmental, quality control, and pharmaceutical labs do not allow the use of such time-consuming methods for routine analysis. Also, for such types of analyses, nondestructive analytical methods are preferred, and in that case, wet chemical analysis will not serve the purpose. Although these classical analytical methods are still in use in specialized areas of analysis, some of these have been shifted to automated instruments.

1.3.2 Instrumental methods of analysis

Today, most of analyses are performed with specially designed electronic instruments, such as spectrophotometers, that are attached to computers for data collection and analysis. These instruments either make use of some physical property of matter or interaction of electromagnetic radiation and matter for characterization of the sample under analysis. For example, electroanalytical methods employed the measurement of electrical properties such as current, potential, resistance, and quantity of electrical charge, whereas spectroscopic methods measure the interaction of electromagnetic radiations with analyte atoms or molecules. Other miscellaneous methods of analysis include the measurement of rate of radioactive decay, heat of reaction, and optical activity. The classification of basic instrumental methods of chemical analysis and the underlying principle is summarized in Table 1.4.

Table 1.4: Classification of basic instrumental methods of chemical analysis and the principle involved.

Method	Principle involved in the method	Quantity or parameter measured
(A) Electrochemical methods	Variation in system's electrical properties	Mass of the deposited substance
(a) Electrogravimetry	Deposition of matter on an electrode under electrolysis conditions	
(b) Coulometry	Deposition of matter on an electrode under electrolysis conditions	Quantity of electricity
(c) Potentiometry	Change in system's electrode potential during the course of reaction	Electrode potential

Table 1.4 (continued)

Method	Principle involved in the method	Quantity or parameter measured
(d) Conductometry	Change in the electric conductivity of a solution during the course of reaction	Electrical conductivity, electric resistance
(e) Polarography	Electrode polarization	Current, voltage
(B) Optical methods	Interaction of matter with electromagnetic radiation	
(a) Emission spectroscopy	Appearance of emission spectra because of high temperatures	Intensity and location of spectral lines
(b) Luminescence analysis	Emission of electromagnetic energy by matter caused by UV radiation	Intensity of emission
(c) X-ray spectroscopy	Emission of electromagnetic energy from atoms under the impact of X-rays	Intensity and location of spectral lines
(d) Raman spectroscopy	Absorption of monochromatic radiation by matter followed by emission of new radiation different from the absorbed one	Intensity and location of spectral lines
(e) Atomic absorption spectroscopy	Absorption by atoms in the source of excitation of monochromatic radiation	Intensity of absorption
(f) Absorption spectrophotometry (photometry, colorimetry, photoelectron-colorimetry)	Absorption of poly- and monochromatic radiation energy by molecules and ions present in solution	Optical density of the solution
(g) Turbidimetry	Absorption and scattering of a radiation by nonhomogeneous or turbid media	Optical density of the medium
(h) Nephelometry	Scattering and reflection of a light beam by nonhomogeneous media or colloidal solution	Optical density of the medium
(i) Refractometry	Refraction of radiation by matter	Refractive index
(C) Radiometric method	Transformation of a stable element's isotopes to radioisotopes upon radiation of matter by nuclear particles	Radioactivity
(a) Isotopic dilution	Modification in specific activity of the compounds labeled with a radio isotope	Radioactivity

Table 1.4 (continued)

Method	Principle involved in the method	Quantity or parameter measured
(D) **Mass spectrometry**	Ionization of atoms, ions, and molecules by combined action of electric and magnetic fields	Position and intensity of signals in mass spectra
(E) **NMR**	Nuclear magnetism (resonance absorption of electromagnetic radiation by matter in a constant magnetic field)	Position and intensity of signals in NMR spectrum

Both methods of analysis, classical and instrumental, are similar in many aspects, for instance, in proper sampling, preparation of sample, assessment of accuracy and precision, and maintaining of proper record.

1.4 Steps in analysis

1.4.1 Goal setting

We must have some target of the analysis and should know how to access it through suitable method at the lowest cost and fastest time while maintaining good precision and accuracy.

1.4.2 Selecting a method of analysis

The suitable method of analysis depends on several factors such as:

i.   **Cost:** There are some methods like volumetric analysis and gravimetry method which are inexpensive but most of the instrumental methods of analysis are expensive. So, depending upon the level of precision, appropriate method may be selected.

ii. **Availability of equipments and materials:** The equipments and the materials of the selected method of analysis should be available in the laboratory.

iii. **Required accuracy and precision:** Generally, the accurate and precise methods of analysis are found to be expensive. Therefore, depending on the analysis, the appropriate equipment may be chosen. For example, for the estimation of salt content in sea water, there is no requirement of selecting an expensive method because a simple mistake will not cause big harm but if we wish to analyze content of heavy metal like mercury in sea water, the accurate and precise method must be selected because mercury is poisonous and even a small error can cause big harm.

iv. **Sensitivity:** The selected method must be able to find out even the minute amount of an analyte in the sample. For example, if the concentration of the analyte in the sample is around 0.00001 M then the chosen method must be able to determine 0.00001 M. Therefore, instruments should be of high sensitivity.

v. **Robustness:** Since the methods are subjected to different chemical and physical conditions, we should choose the method that is almost free from any kind of interference. This method is considered to be a robust method and can be used for the determination of analyte in all sorts of samples.

vi. **Type of interferences:** The method selected for the analysis should be such where the level of interference from the existing impurities should be minimum. The interference from the impurities may vary depending upon the method; in some method, it may be observed while in others it is not observed.

vii. **The complexity of the sample** and the number of components significantly influence the choice of method.

viii. **The number of samples to be analyzed:** If the number of samples to be analyzed is high, then automatic and fast instruments may be chosen.

1.4.3 Sampling

Sampling is the most difficult step in an analysis process and it is the source of greatest error. The sample is taken from different sources like water or soil or air. The sample to be prepared must represent the average chemical composition of material to be analyzed. The number of samples taken from the source depends on the homogeneity of material, for example, in air which is homogeneous, few samples are required, whereas liquid such as water is less homogeneous, so larger number of samples should be taken while with the solid which is commonly most nonhomogeneous such as soil or rock very larger number of samples are required and so on. The sampling is very crucial step, and final results of an analysis will greatly depend upon the reliability of the sampling step.

i. **Drying the sample:** Any loss or gain of water can significantly affect the chemical composition of solids. The solid samples generally contain different amounts of adsorbed water, and this factor is difficult to control; therefore, we should remove moisture before weighing the sample by drying it in the oven. If the analyte is found to be thermally unstable, then desiccator can be used for drying. Alternatively, the moisture content present in the sample can be determined by separate analytical procedure. On the other hand, liquid samples are subject to solvent evaporation for drying. Some special measures like sample manipulation and measurement in an inert atmosphere are done to preserve the quality of the sample.

ii. **Replicate samples:** The replicate sampling or replica are done which are of approximately same size and are done at the same time and in the same way. Replication can improve the quality of the results and reliability. All measurements obtained on replicates are averaged, and different statistical tests are done on the results to confirm reliability.

iii. **Sample dissolution:** The chemical reactions are generally slow in the solid state as compared to solution; therefore for better results the sample must be dissolved in suitable medium for analysis. The solvent should dissolve the entire sample, including the analyte, completely. To dissolve the sample, the water is tried first and then diluted acids, concentrated acids, different mixtures of acids solutions, and so on. The fusion can also be done where an acidic sample is melted with a base such as Na_2CO_3 (sodium carbonate) and a base is melted with an acid. Similarly, an oxidant is melted with a reductant material. Organic compounds are dissolved in noninterfering organic solvents. For the estimation of inorganic elements in organic samples, the organic matter is removed by dry burning or wet burning processes using hot oxidizing acids such as HNO_3 (nitric acid) or $HClO_4$ (perchloric acid) which convert the organic matter to water vapor and carbon dioxide. Through this, the interference of organic matter in the analysis is removed.

iv. **Storage of sample solution:** During the storage, it must be ensured that sample is not contaminated, and its loss or disintegration must be avoided. The container used for storing must be made of noninterfering material so that the sample does not interact with it, and it must be air tight to prevent air exposure. The sample must be protected from sun and heat because they may have impact on the sample composition.

1.4.4 Interferences

An interference is caused by substance which can affect the analysis. For example, cobalt ion (Co^{2+}) is detected by the addition of thiocyanate ion (SCN^-) which reacts with the cobalt to give a red-color complex, but other ion such as Fe^{3+} (iron) also give red-color complex. Therefore, Fe^{3+} causes interference in the detection of Co^{2+} ion and it must be removed before the analysis. The separation of interfering ion is done by extraction, precipitation, or any other means. Interferences can also be removed by pH control and by masking agents. Specific method of analysis is a special method for a single substance where no interference is observed with other ions. These special methods of analysis are practically difficult to perform. All methods of analysis either respond to a limited substances and the possibility of their interferences is too high. Techniques or reactions that work for only one analyte are said to be *specific* whereas techniques or reactions that apply for only a few analytes are termed as *selective.*

1.4.5 Measurement

The concentration of analyte is measured in several aliquots. The main purpose of repeated measurements is to find out the uncertainty in the analysis and to provide some knowledge about the gross error which may be present in the analysis of a single aliquot. The uncertainty of a measurement is equally important as the measurement itself because it confirms the reliability of measurement. The different types of analytical methods are tested on the same sample to see whether they are providing same results. The effect of the sampling procedure may also be tested by constructing and analyzing several different bulk samples.

1.4.6 Calculating and evaluating the results

Computing analyte concentrations are based on the raw experimental data collected in the measurement step, the characteristics of the measurement instruments, and the stoichiometry of the analytical reaction. The results must be confirmed statistically otherwise they are not of much use.

1.4.7 Evaluating results by estimating their reliability

Analytical results are incomplete without an estimate of their reliability.

1.5 Advantages of instrumental methods

As discussed earlier, instrumental methods rely on machines, and with the availability of instruments for carrying out different experiments, concerns about time required for analysis are no more a problem. Modern instrumental methods have provided many distinct benefits to the scientific community; however, they have their own limitations such as they are expensive. They also required trained personnel for their handling and regular maintenance.

Advantages of instrumental method of analysis are:
i.   Sample requirement – These analyses required a small amount of sample in comparison with other laboratory methods.
ii. Time saving – These methods are quick, and the analysis time has been reduced from hours to days or even minutes.
iii. Accuracy – Sufficient reliability and improved accuracy is obtained in results by eliminating errors introduced due to personal bias.

iv. Sensitivity – They are highly sensitive as they can detect and analyze sample at micro- and even at femto-mole level. Such sensitivities could not be possible with conventional classical approach.

There is no doubt that modern instrumental methods offer unlimited advantages and have made analysis easier but at the same time, we should not overlook the basic analytical operational skills. Such basic skills ensure high reliability of results if performed with utmost attention and in a careful manner.

1.6 Role of analytical chemistry in scientific areas

Chemical science is the constituent part of analytical chemistry and it plays a great role in the discovery of many laws of nature, for example, law of conservation of mass, law of constant composition, law of equivalent weight, law of mass action, the periodic law, and law of multiple properties. These methods are utilized for theoretical substantiation and developments. For example, law of constant composition allows an analytical chemist the isolation of a desired element in the pure form. This law is combined with the laws of conservation of mass and gravitation which forms the basis of gravimetry. The law of equivalent weights supports the volumetric or titrimetric analysis.

The Mendeleev's periodic law forms, theoretically, the basis of analytical classification of ions and facilitates the development of new methods to detect and determine the elements and their compounds. The position of an element in the periodic table predetermines the analytical properties of an element, its ions, and compounds in many aspects. The law of mass action and its importance for equilibrium systems are of great practical and theoretical importance for analytical chemistry. Electrochemical laws form the theoretical foundation of electrogravimetry, polarography, potentiometry, and other electrochemical methods of analysis.

Analytical chemistry is also closely related to other sciences and primarily to mathematics. Analytical results are generally expressed graphically as mathematical functions and are used to estimate the analytical methods's accuracy of results and to reveal errors, if any.

Physical methods of analysis in analytical chemistry are expressed by diverse applications which are based on inner electronic process and nuclear levels process, for example, mass spectroscopy and X-ray. The interaction of analytical chemistry and its applications in the technologies and instrumentation makes is particularly remarkable. Analytical methods have been progressively improving by using various instrumental methods. Moreover, an increase in sensitivity of analytical determination allows more precise determinations as well as processing of peculiar properties of materials.

Literature is rich in studies illustrating the role of analytical chemistry in medicine, whether in practical application, or in research. It helps in the diagnosis of diseases through clinical chemistry, for instance, in analysis of enzymes, bacteria, blood gases, and other medical samples. The development of latest sensitive instrumental techniques helps in detecting various abnormal and normal components of body fluids. Analytical chemistry aids in understanding the disease through its molecular makeup via pathology. Clinical tests are not limited to general determinations only, but also helpful in the diagnosis of specific metabolic diseases. In pharmaceutical studies, it is essential to find the properties and therapeutic value of a drug before it is approved and made available to the general public. Various methods of chemical and instrumental analysis, statistics, and computers help analytical chemists in determining the properties of drugs and made it essential for medical research and pharmaceutical development. And, there would be no forensic science – the sciences employed for the purposes of law, without analytical chemistry.

The role of analytical chemistry is very important in industries as it plays a crucial role in the chemical analysis of the raw materials, the intermediate compounds, and finished products as well as helps in analytical control of the process. The product of nonferrous metallurgy is more complex and varied that includes mass of more than 70 metals, semiconductors, solid alloys, and electrode materials. As a result, analytical tasks are extremely necessary, and it is important for a metallurgical engineer to have a clear understanding of current methods of analyzing raw materials and products as well as the methods for controlling the technological processes. Technical control in industry, that is, a means of evaluating the product quality and the technological process of preventing defects in products and making the quality of product meet the standard and technical specification, is quite important, which can be achieved by employing various physicochemical and analytical methods. Analytical chemistry plays a significant role in geochemistry as well. It is used in the exploration of mineral deposits as well as in the processing of material.

The above-cited examples clearly illustrate the great significance and importance of analytical methods and analytical chemistry in scientific areas. Scientists working in various research areas such as medicine, clinical biochemistry, environmental, space science, forensics, and geology need better tools for analyzing materials, and this can be achieved with the help of analytical chemistry.

References

[1]   Christian, G.D., Dasgupta, P.K., and Schug, K.A. Analytical chemistry. John Wiley & Sons; 2013. Hoboken, New Jersey, US.
[2]   Fifield, F.W., and Kealey, D. Principles and practice of analytical chemistry. Hoboken: Blackwell Science; 2000, p. 200.
[3]   Hage, D.S., and Carr, J.D. Analytical chemistry and quantitative analysis. Boston: Prentice Hall; 2011.
[4]   Harvey, D. Modern analytical chemistry. New York: McGraw-Hill; vol. 1, 2000.
[5]   Skoog, D.A., West, D.M., Holler, F.J., and Crouch, S.R. Fundamentals of analytical chemistry. Cengage learning; 2013. United States of America.

Section II: **Qualitative and quantitative aspects of analysis**

Vinod Khatri

Chapter 2
Quantitative aspects of analysis

Abstract: The quantitative analysis is the branch of analytical chemistry where we determine the quantity of the substance. It indicates the amount of each substance in a sample. The amount may be expressed in terms of mass, concentration, or relative abundance of one or all components of a sample. The quantitative analysis can be done by classical methods (gravimetric and volumetric analysis) and instrumental methods (electroanalytical, spectroscopic, and chromatographic analysis). The main steps performed during the quantitative analysis include sampling, field sample pretreatment, laboratory treatment, laboratory assay, calculation, and result. In this chapter, we have discussed the quantitative aspects of analysis.

2.1 Introduction

Quantitative analysis is the determination of absolute or relative abundance of one, several, or all components present in the given sample. This analytical tool enables us to calculate the quantity of materials in a sample.

The sample used in either quantitative or qualitative analysis is called as analyte, and most of the techniques and methods employed in quantitative as well qualitative analysis are common. The first step of every analysis involves the identification of number of constituents in analyte by using quantitative analysis which is followed by qualitative analysis. There are several methods present to determine the composition of sample like volumetric analysis (titrimetry), gravimetric analysis, and instrumental analysis. The instrumental analysis includes the electroanalytical method, spectroscopy method, and chromatographic analysis (Figure 2.1).

The basis of volumetric and gravimetric analysis is the quantitative execution of chemical reaction. In volumetric analysis, substance to be analyzed is allowed to react with standard solution of reagent. The volume of solution of reagent needed to complete the reaction is determined. Gravimetric analysis involves the precipitation of insoluble salt of a compound from a solution. For example, barium is precipitated as insoluble barium sulfate which is then filtered, washed, dried, and weighed for the estimation of barium quantitatively. The accuracy of gravimetric analysis is more than the volumetric analysis.

https://doi.org/10.1515/9783110794816-002

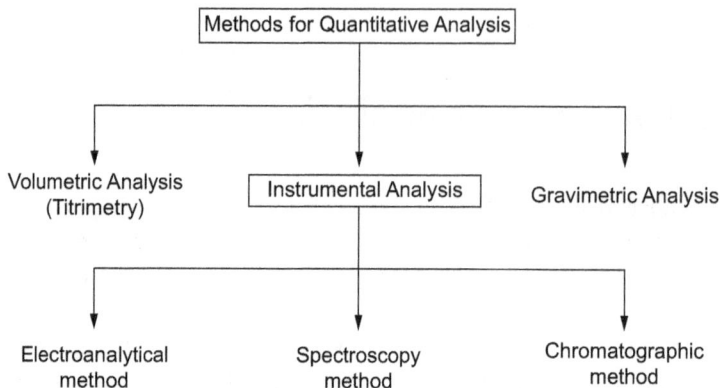

Figure 2.1: Flow chart of methods for quantitative analysis.

2.2 Fundamental units of measurement

There are different units of measurement in analytical chemistry and to understand their importance imagine that you find the instruction in laboratory procedure. "Transfer 2 of your compound to a 250 volumetric flask, and do the dilution." How will you do this? As we can see, these instructions are confusing since the units of measurement are missing. Now read the complete instruction: "Transfer 2 g of your sample to a 250 mL volumetric flask, and do the required dilution." This instruction can be easily followed. There are different units which are used to express any parameter and many units may be used to express the same physical measurement. For example, the mass of a compound weighing 2 g may be expressed as 2,000 mg. For avoiding the confusion, common set of fundamental units are used, some of which are given in Table 2.1. These units are called SI units. All different measurements are defined in relation to these fundamental SI units. For example, the quantity of heat produced during any chemical reaction is determined in joules (J). Table 2.2 provides a list of other important SI units and some non-SI units, and Table 2.3 summarizes some of the common prefixes used for exponential notation.

Analytical chemists work with measurements that may range from very large to very small. Take the example of a mole which contains 602,213,670,000,000,000,000,000 particles, and analytical techniques are sufficient enough to detect as little as 0.000000000000001 g of a compound. To keep it simple, these measurements are expressed through scientific notation; now a mole is written as 6.0221367×10^{23} particles. Sometimes measurements are expressed without using the exponential term, changing it with a prefix, for example, a mass of 1×10^{-9} g is the same as 1 ng.

Table 2.1: Fundamental SI units.

Measurement	Unit	Symbol
Mass	Kilogram	kg
Volume	Litre	L
Distance	Meter	m
Temperature	Kelvin	K
Time	Second	s
Current	Ampere	A
Amount of substance	Mole	mol

Table 2.2: Other SI and non-SI units.

Measurement	Unit	Symbol	Equivalent SI units
Length	Angstrom	Å	$1\,\text{Å} = 1 \times 10^{-10}\,\text{m}$
Force	Newton	N	$1\,\text{N} = 1\,\text{m kg/s}^2$
Pressure	Pascal	Pa	$1\,\text{Pa} = 1\,\text{N/m}^2$
	Atmosphere	atm	101,325 Pa
Energy, heat	Joule	J	$1\,\text{J} = 1\,\text{N m}$
Power	Watt	W	$1\,\text{W} = 1\,\text{J/s}$
Charge	Coulomb	C	$1\text{C} = 1\text{A s}$
Potential	Volt	V	$1\,\text{V} = 1\,\text{W/A}$
Temperature	Degree Celsius	°C	°C = K-273.15

Table 2.3: Common prefixes for exponential notation.

Exponential	Prefix	Symbol
10^{12}	Tera	T
10^{9}	Giga	G
10^{6}	Mega	M
10^{3}	Kilo	K
10^{-1}	Deci	d
10^{-2}	Centi	c
10^{-3}	Milli	m
10^{-6}	Micro	μ
10^{-9}	Nano	n
10^{-12}	Pico	p
10^{-15}	Femto	f
10^{-18}	Atto	a

2.3 Steps involved in quantitative analysis

The main steps involved in quantitative analysis are as follows:
i.   Sampling
ii. Sample preparation
iii. Analysis of sample
iv. Interpretation of results

Each step of quantitative analysis must be executed correctly in order to obtain the correct results. All the steps involved in the quantitative analysis are explained in the flow chart as illustrated in Figure 2.2.

2.3.1 Sampling

A small portion of bulk sample is taken in order to analyze. The homogeneity of bulk sample decides the number of samples needed for analysis. If the sample is homogeneous

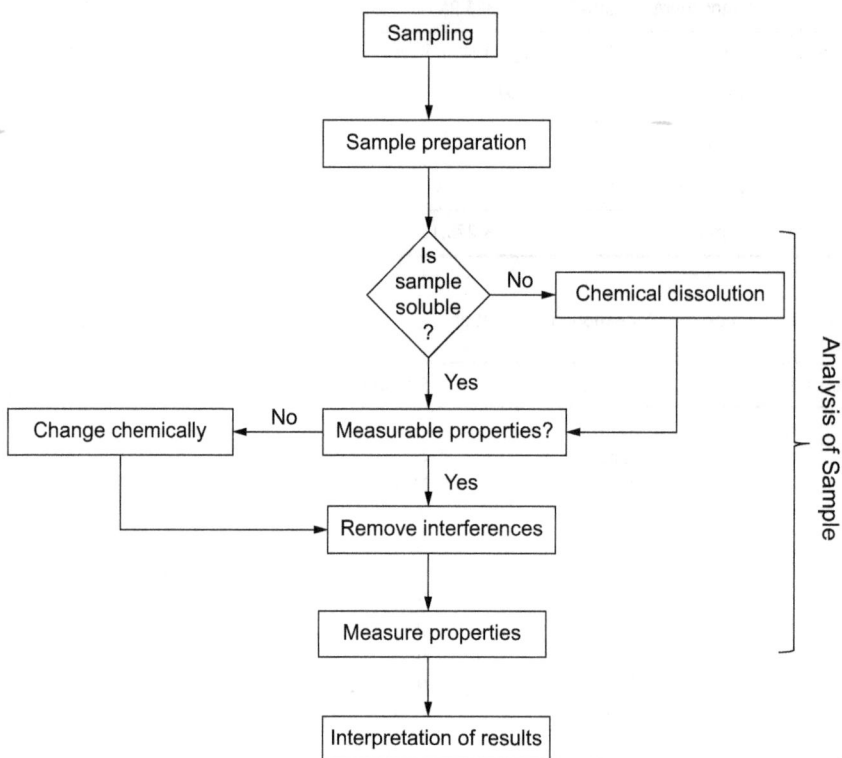

Figure 2.2: Flow chart of steps involved in quantitative analysis.

then only one sample is enough for accurate analysis but if sample is heterogeneous then more number of samples are required. For example, if we wish to find out the chemical pollutant of a lake, then we need to take the sample from the center as well as the periphery of the lake.

2.3.2 Sample preparation

After sampling, it is mandatory to treat the sample physically or chemically to ensure that the concentration or component will not change which may include grounding, mixing, and storing. Multiple samples collected from heterogeneous bulk sample may be mixed to get the one homogeneous sample. For example, water samples taken from the different part of lake must be sealed in a container to avoid the change in the oxygen concentration in sample. Normally all samples are prepared in triplicate to obtain accurate results.

2.3.3 Analysis of sample

A smaller homogeneous sample from bulk must be dissolved properly in the solvent so that no part of analyte remains undetected. After dissolving the sample, chemical species which may interfere with the analysis should be removed. There are various methods to eliminate metal ions from the sample. For example, if lead is present in your sample and lead can interference with your analysis, then we can remove the lead by passing H_2S (hydrogen sulfide) gas through the solution of sample. Lead can be precipitated out from the sample as lead sulfide (PbS); hence, we can remove this by simple filtration step. After removal of interference, measure the concentration of analyte by preparing the various aliquots of sample. Measurement of any property of sample is directly proportional to the concentration of that sample.

2.3.4 Interpretation of results

Analytical results are considered complete when their reliability has been estimated. After finishing the assay, results are presented in a meaningful manner. Results are represented by two values in which first value indicates the correctness of result and second value indicates the amount of random error in the analysis.

2.4 Calibration of glassware and instruments

Calibration of glassware and instrument is very important to ensure the accurate results of an analysis. Calibration of glassware mainly includes the volumetric flask, pipettes, and burettes. All volumetric glassware should be free from water before being calibrated. Glassware is calibrated using a liquid of known specific density and an analytical balance. The procedure is used to determine the mass of liquid hold by the glassware which further can be converted into the volume by using the density (d).

2.4.1 Calibration of volumetric flask

The calibration of volumetric flask can be done by the following steps:
i.   Clean the volumetric flask with water and then rinse the flask two to three times with small amount of acetone. Allow the residual acetone to evaporate from the flask.
ii.   Determine the mass of dry flask and record this mass on the notebook. Always make sure that you tare the analytical balance before measuring the mass.
iii. Fill the volumetric flask up to the fill mark (Figure 2.3) on the neck of volumetric flask with deionized water followed by the measurement of mass of the flask. Record the mass of filled volumetric flask in the notebook.
iv.   Remove the deionized water from the flask. Repeat steps i–iii until you got the triplicate measurement of mass of water in the flask.
v.   Calculate the volume of water contained in the flask.

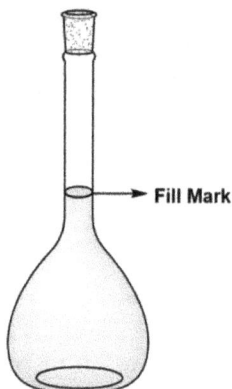

Figure 2.3: A typical volumetric flask.

Table 2.4: Calibration of volumetric flask.

S. no.	Mass of empty volumetric flask (g)	Mass of water filled volumetric flask (g)	Mass of water (g)	Calculated volume of water (mL) at 25 °C ($d = 0.997$ g/L)
1.	35.1311	85.1632	50.0321	50.1826
2.	35.1311	85.1605	50.0294	50.1799
3.	35.1311	85.1611	50.0300	50.1805

From Table 2.4, calculate the mean and sample deviation for each flask.

2.4.2 Calibration of pipette

The calibration of pipette can be done by the following steps:
i.   Clean and rinse the pipette with deionized water.
ii. Measure the mass of suitable size of beaker (depending on the volume of flask) on the analytical balance. Record the mass of beaker in the notebook.
iii. Fill the pipette with the deionized water up to the fill mark of pipette (Figure 2.4). Wipe out excess of water with tissue paper and then dispense the water in the beaker.
iv. Measure the mass of filled beaker and water in the beaker by subtracting the mass of beaker. Record the mass of water in the notebook. Repeat this procedure in triplicate.
v.   Calculate the volume of water dispensed by the pipette.

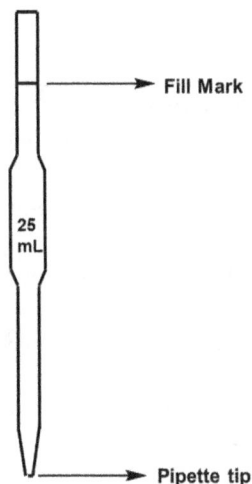

Figure 2.4: A typical pipette.

Table 2.5: Calibration of pipette.

S. no.	Mass of dispensed water (g) from pipette	Calculated volume of water (mL) at 25 °C ($d = 0.997$ g/L)
1.	10.012	10.042
2.	10.009	10.039
3.	10.006	10.036

From the calculated volume of Table 2.5, calculate the mean and sample deviation from each pipette.

Figure 2.5: A typical burette.

2.4.3 Calibration of burette

The calibration of burette can be done by the following steps:
i.   Select a suitable size burette, clean it, and rinse with only deionized water.
ii.   Determine the mass of empty beaker of suitable size with analytical balance and record it on the notebook.
iii. Fill the burette (Figure 2.5) with deionized water and dispense a portion of water depending upon the size of burette, say 10 mL for 50 mL burette.
iv.   Determine the mass of dispensed water from the burette using analytical balance and note down in the notebook.
v.   Repeat steps (iii) and (iv) two to three times and note down the mass of water dispensed from burette in the notebook.
vi.   Refill the burette with deionized water and repeat the protocol from step (ii) to (v) to obtain three calibrations.
vii. Determine the volume of water dispensed by dividing the mass of dispensed water with density of water at that temperature (Table 2.6).

Table 2.6: Calibration of burette.

S. no.	Volume dispensed from burette (mL)	Mass of dispensed water (g) from burette	Calculated volume of water (mL) at 25 °C ($d = 0.997$ g/L)
1.	5.13	5.05	5.06
2.	10.12	10.00	10.03
3.	15.01	14.92	14.96

viii. Determine the volume difference by subtracting the calculated volume of water from the dispensed volume of water from the burette.

ix. Prepare the graph by plotting the difference in volume (y-axis) versus volume of water dispensed from the burette (x-axis) to obtain the single calibration value.

2.5 Estimation of geometrical isomers and keto–enol tautomers

The quantitative estimation of geometrical isomers and keto–enol tautomers can be performed with the help of ^{1}H-NMR (nuclear magnetic resonance) spectroscopy. ^{1}H-NMR can determine the relative amount of geometrical isomers/tautomers in the mixture.

For slow tautomerism, two sets of protons will appear in the ^{1}H-NMR for keto and enol form. Thus, it is possible to determine the keto–enol ratio by comparing the relative intensities of NMR peaks of keto and enol form. For example, the spectrum of acetylacetone clearly shows NMR peaks for keto and enol forms.

Keto form   **enol form**

Similarly, in geometrical isomerism, we can determine the amount of E-isomer and Z-isomer form of a mixture using the intensity of peak of corresponding protons. In Figure 2.6, two peaks observed correspond to the three protons of –CH$_3$ (methyl) group of E- and Z-isomers. The one peak of E-isomer is at 2.34 ppm and has the integration of 1.20 protons and another peak of Z-isomer at 2.48 ppm has integration of 1.82 protons. From this spectrum, percentage of E-isomer was found to be $\frac{1.20}{3} \times 100 = 40\%$ and percentage of Z-isomer is $\frac{1.82}{3} \times 100 = 60\%$.

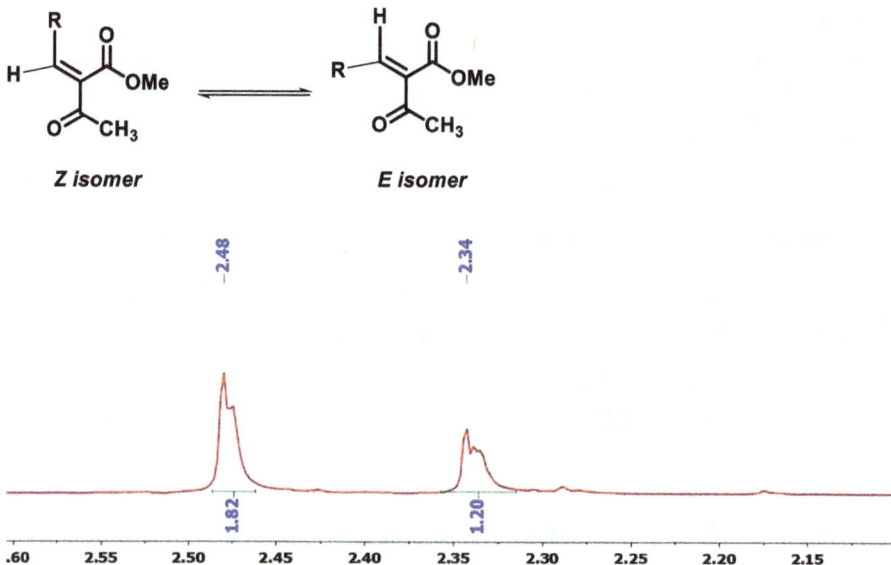

Figure 2.6: Partial ^{1}H NMR spectra of *E*- and *Z*-isomers.

2.6 Determination of composition of metal complexes using job's method of continuous variation and mole ratio method

Absorption spectroscopy is the powerful tool for determining the formula of metal complexes formed by reaction between the metal and ligand. It is an important technique since absorbance determination can be done easily without disturbing the equilibrium between the metal and the ligands. The only requirement for determination of composition is that either reactant or complex formed should absorb light. There are three common methods for determining the stoichiometry of metal complexes:

i.   Job's method of continuous variation
ii.   Mole ratio method
iii. Slope ratio method

By these methods, we can easily determine the composition and formation constant without isolating the pure complex.

2.6.1 Job's method of continuous variation

Job's method of continuous variation is used extensively in chemistry and biochemistry to study molecular associations. This is also called method of continuous variation (MCV). MCV provides qualitative and quantitative insights into the stoichiometries of different complexes. This method has varied applications such as studying selectivity in metal removal or extraction by chelation and is mainly used to study the composition of wide range of coordination compounds. In this method, stock solution of equimolar amount of metal ion and ligand is mixed in such a way that total molar concentration of metal and ligand is maintained constant but the molar ratio is changed (Table 2.7). After mixing the equimolar amount of metal ion and ligand, the absorbance of different compositions is measured that correspond to the complex formation.

Table 2.7: Method for preparation of solution for Job's plot.

C_M (0.1 mM)	C_L (0.1 mM)	C_{ML} ($C_M + C_L$) 0.1 mM
0 mL	1 mL	1 mL
0.1 mL	0.9 mL	1 mL
0.2 mL	0.8 mL	1 mL
0.3 mL	0.7 mL	1 mL
0.4 mL	0.6 mL	1 mL
0.5 mL	0.5 mL	1 mL
0.6 mL	0.4 mL	1 mL
0.7 mL	0.3 mL	1 mL
0.8 mL	0.2 mL	1 mL
0.9 mL	0.1 mL	1 mL
1 mL	0 mL	1 mL

C_M = concentration of metal,
C_L = concentration of ligand, and
C_{ML} = concentration of metal-ligand complex

For example, metal M is mixed with ligand L and the complex ML_n is formed after mixing. The absorbance value corresponds to ML_n and is measured for a series of samples and plotted against the mole fractions of ligands (X_L) or metal ion. The average number of ligands (n) bound to one metal can be calculated from the plot. The nature of curve obtained in graph provides the idea of relative stoichiometry of ligand and metal ions. Even without determining the molar response of AB, the shape of the curve provides qualitative insight into K_{eq} too. The exact stoichiometry of ligand and metal can be determined from the maxima of Job's plot. In Figure 2.7(A), maxima of the graph is at $X = 0.5$ in which mole fraction of metal ion is 0.5 and mole fraction of ligand is also 0.5. Hence, the curve obtained is for the complex ML. In Figure 2.7(B), a maximum absorbance of graph is at $X = 0.67$, in which mole fraction of ligand is 0.67 and mole fraction of metal ion is 0.33. Hence, the curve obtained is for the complex ML_2.

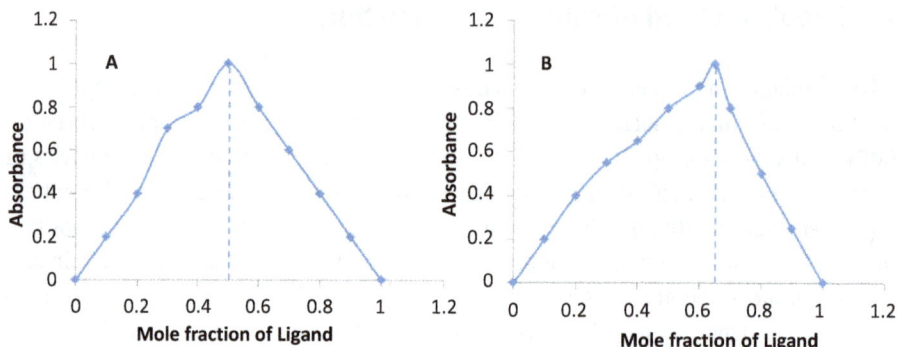

Figure 2.7: Job's plot for the formation of metal–ligand complexes (A) ML, metal:ligand (1:1) and (B) ML_2, metal:ligand (1:2).

2.6.1.1 Limitation of job's method

i.   This method is pertinent to one type of equilibria, that is, only one type of complex formation in the solution.
ii.   This method is only applicable for stable complexes but not for very weak complexes.
iii. If the complex is not absorbing the UV–visible light, spectrophotometer cannot be used for the determination of this method.
iv.   pH and ionic strength must be maintained constant during the course of experiment.

2.6.2 Mole ratio method

The mole ratio method is an alternative to the Job's plot for determining the stoichiometry of metal and ligand in the complex formed. This method is applicable for strong complexes only. In this method, mole of metal is held constant while the moles of ligand are increased. The series of solutions are prepared in which the mole ratio may vary from 0.1 to 10 or 20. Keeping one of the reactants constant can make preparation of solution simple and offer the scope for using the reactant which is present in the limited amount. The absorbance corresponds to the formation of complex and is recorded and plotted against the ratio of mole of ligand to metal. The break in the curve corresponds to the stoichiometry of the complex formed (Figure 2.8). The stoichiometry of the complex can be determined by extrapolating the straight-line part of the graph. The point at which these two lines intersect corresponds to the ratio of metal to ligand. The break in the curve is sharp if complex has high ionic strength. The absence of change of slope indicates that complex of that particular ratio does not form. This exercise is useful for weakly dissociated complex but if the dissociation constant is too large then the curve becomes smooth which makes locating the stoichiometric point difficult. The certain region around the curvature can be used effectively to determine the dissociation constant of the complex.

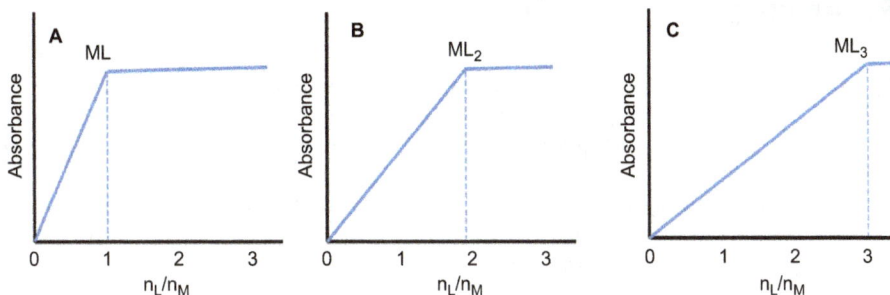

Figure 2.8: Mole ratio plot for the formation of metal–ligand complexes: (A) ML, metal:ligand (1:1), (B) ML_2, metal:ligand (1:2), and (C) ML_3, metal:ligand (1:3).

This method is more advantageous than continuous variation method if the ratio of the reactants in the complex is high. This method is also very useful for differentiating the different types of complexes such as ML_5 versus ML_6.

We can also obtain the mole ratio plot for the stepwise formation of complexes like ML, ML_2, and ML_3. In the stepwise formation of complexes, the mole ratio plot has more than one break in the curve depending upon the number of complexes formed (Figure 2.9). The improvement in this method is observed by the use of dual wavelength spectrophotometry.

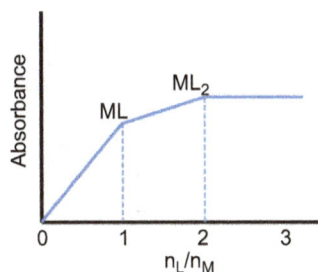

Figure 2.9: Mole ratio plot for stepwise formation of metal–ligand complex.

2.6.3 Slope ratio method

Slope ratio method is applicable for weak complexes which have large dissociation constant, where a single complex is formed. This method can be applied to comparatively few cases. In this method, it is assumed that (a) the complex formation reaction can be forced to completion by adding large excess of either reactant metal ion or ligand and (b) Beer's law is followed during the complex formation.

References

[1]   Leonard, M.A. 1990. Vogel's textbook of quantitative chemical analysis, 5th edn, Revised by Jeffery, G.H., Bassett, J., Mendham, J., Denney, R.C. Longman: Harlow; 1989, p. 877.

[2]   Nan, Z., and Chun-Xiang, H. Improved mole ratio method by dual-wavelength spectrophotometry. Analyst. 1993; 118(8): 1077–1080.

[3]   Ononuju, A.O., Ogwuegbu, M.O.C., Ayuk, A.A., and Onu, U.L. Spectroscopic determination of the stoichiometries and stability constants of iron (III) and aluminium (III) complexes of salicylic acid. Journal of Chemical Society of Nigeria. 2016; 41(2).

Mukesh Kumar Saini and Poonam Pipil

Chapter 3
Data handling in qualitative and quantitative analysis

Abstract: This chapter describes various aspects of quantitative analysis such as sampling, evaluation of analytical data, errors, accuracy, and precision, and methods of data expression. Normal law of distribution of indeterminate errors, statistical test of data, F-, Q-, and t-tests, rejection of data, and confidence intervals will also be discussed.

The knowledge of sampling, for example, sample size, number of sample analysis, and method of sampling, is needed to design a correct analysis method. Accuracy and precision are necessary steps which should be taken into consideration while collecting data. After this, different statistical tests, that is, t-test, Q-test, and F-test are performed to express the uncertainty in measured values for each step of the analysis. By doing this, we can find out the limitations of analytical method and also understand the significance of data set so that we can minimize error in the analyzed result.

3.1 Sampling and evaluation of analytical data

Sampling is the most important step in the quantitative analysis, and the importance and accuracy of measurements can be limited by the sampling process. If sampling is not performed correctly, it becomes the weak link in the chain of analysis. Sampling means collecting a smaller material that is usually representative of the true bulk material in a homogeneous state. A representative sample that gives the true value and distribution of the analysis is the original bulk material. This statement is especially true when the material to be analyzed is large, and in homogeneous liquid, such as a lake, or a nonhomogeneous solid, such as an ore, clay, or a piece of animal tissue. The first step in the sampling process is to identify the items selected for analysis, often referred to as the sample population unit. Then, the second step is to obtain a sample that is representative of the whole. This sample is called the gross sample. Its size may vary from small to large depending on the type of bulk material. The representative gross sample is then reduced to a size sufficiently small to make it homogenous. Once the sample is homogenized, an aliquot or a small part thereof can be analyzed. This aliquot is called the laboratory sample.

https://doi.org/10.1515/9783110794816-003

The various important factors of sampling which may affect the analysis of material are:

i.   **Collection and storage of samples:** Samples are collected using some sort of collection device and put into a container. These equipment and containers can often contaminate the sample. For example, stainless steel needles used to collect blood or serum samples may contain traces of metals. Metal spatulas, drill bits, scissors, glass pipettes, filter paper, plastic, and rubber tubing can add unwanted organic and inorganic contaminants to samples. To avoid metal contamination such as chromium, iron, and nickel, one can use tongs and tweezers with platinum or gold tips.

ii. **Size of samples:** Based on the size of sample, analysis can be categorized into four categories: (i) macroanalysis, (ii) semi-microanalysis, (iii) microscopic analysis, and (iv) ultra-microanalysis. When the sample volume is >0.1 g, the analysis is called macroanalysis, and if the sample volume is between 0.01 and 0.1 g, it is termed as semi-microanalysis. The third type of analysis is microscopic analysis in which the sample volume varies from 10^{-4} to 10^{-2} g, and the fourth analysis is ultra-microanalysis, where the sample volume is <10^{-4} g. In addition to this, depending on the concentration of the analyte to be analyzed; the analytes are classified as (i) major (1–100%), (ii) minor (100 ppm (parts per million) to 1%), (iii) trace (1 ppb (parts per billion) to 100 ppm), and (iv) ultra-trace (<1 ppb). The sample should be taken in sufficient amount for all analyses so that analysis can be performed in duplicate or triplicate, if required. In case, if, only a small amount of sample is available, as, may be the case with forensic samples from the crime scene or in the case of rocks brought back from the Moon, the analyst must do the best possible job with what is being provided.

iii. **Physical state of samples:** Some of the problems associated with obtaining gross samples of solids, liquids, and gases are as follows:

The nonuniformity of solid materials, variation in particle size, and within-particle variation make sampling of solids more difficult than other materials. The easiest but usually most unreliable method of sampling solids is the grab sample, in which a sample is taken at random and considered to be representative. Grab sample will be satisfactory only if the material from which it is derived is homogeneous. When the sample is heterogeneous, it is best to take 1/50 to 1/100 of the total bulk for the gross sample. Only then we will be able to get the most reliable results. If the particle size of the sample is large then the gross sample should be larger in size. A systematic sampling can be done to obtain aliquots representing all parts of the bulk.

Liquid samples are homogeneous compared to solid samples and easily we can collect a representative sample by a simple grab sampling. It only needs to be stirred normally to obtain a homogeneous mixture. However, it may also happen when stirring is not required as in the case of analysis of the oily layer floating on the water, the oil layer can be removed with the sucker.

For sampling of liquid samples, we have to consider few things, such as a representative sample must be collected at remote locations from the source of contamination. For example, river water sample should be collected away from the river bank, floating foams, oils, and effluents from industrial and municipal wastewater treatment sites. Samples from different water bodies such as rivers, lakes, and ponds may require sampling at different distances or depths on a diagonal rather than directly below the shore. Such samples can be analyzed individually or mixed to obtain an average overall sample. Particles of solid matter from the liquid samples must be removed by filtration or centrifugation. But, in some cases, suspended solid particles are not removed while sampling, that is, samples of orange juice and liquid antacid. In case of immiscible liquids, such as oil spills in sea water and cream on milk samples need to be emulsified before the representative sample is collected. It is also important to note the collection times of some biological specimens, that a 24-h urine sample must be collected rather than a single spot sample. The composition of the blood sample may vary before and after a meal.

Gas samples are generally assumed to be homogeneous and the sample can be collected rapidly (also called a grab sample) or over a long period of time in an evacuated container. A stainless-steel canister or an inert polyvinyl fluoride bag is typically used for sampling and the container must be contaminant-free. Balloons and plastic bags can also be used, but they can mix volatile organic compounds (present in bag) into the gas sample. Sampling of organic gases can be done using activated charcoal because the organic vapor in the air is adsorbed onto the charcoal while most of the air passes through it. The process of taking an analyzer out of the gas phase is called "scrubbing" and the major advantage of this process is that it reduces the physical size of the sample. Sampling of toxic flammable or corrosive gases should be handled with great care using appropriate protection devices. The solids in the gas sample need to be removed by filtration and the required filter material must be of suitable pore size and contamination free.

Sample pretreatment is a preparatory work performed prior to analysis while the samples arrive in an analytical laboratory. Samples can vary in size, physical states, moisture content, and major constituents to ultra-trace levels.

The method for sample pretreatment is as follows:
The effect of variable moisture content can be eliminated by drying the samples at ~100–120 °C. From this, water content in the sample can be calculated by weighing the sample before and after drying. Before carrying out analysis, some known matrix component should be removed or subtracted, which may interfere with the measurement analysis. Analytes can be divided into different groups based on some common characteristics such as filtration, centrifugation, evaporation, distillation, and solid or solvent phase extraction. Performing these treatment methods may also increase the concentration of desired analyte, if they are present in small amounts.

This is then followed by the preparation of a laboratory sample, usually by treating the analyte with appropriate inorganic or organic reagents. In some analysis, laboratory sample is directly used for the analysis, for example, in surface analysis. However, many samples such as geological and metallurgical samples need to be first decomposed to free up analytes and to facilitate specific reactions in solution. The optimum concentration of analyte in the laboratory sample can be achieved by dilution with appropriate solvent or by concentration. The solution stability of the laboratory sample is ensured with respect to certain parameters such as pH, ionic strength, and solvent composition before finally being analyzed. It should be kept in mind that sampling errors cannot be corrected by evaluating a reagent blank because sampling errors are completely independent of analysis error.

3.2 Errors, accuracy, and precision

When we perform a chemical analysis, some uncertainty is associated with every measurement, which is called experimental error. There is absolutely no way to measure the "true" value of anything; we can only carefully apply a technique to obtain results, we have to rely on. If the results of any chemical analysis obtained by different methods agree with one another, then we can conclude that results are accurate, which means they are near the true or accepted value. Experimental results can be expressed with a high or a low degree of certainty, but never with complete certainty.

3.2.1 Accuracy and precision

Precision is nothing but reproducibility of a result that tells us repetition of one method of measurement several times. Precision expresses the closeness of two or more measurements to each other whereas accuracy describes how close a measured value is to the true or accepted value, and it can be obtained by an experienced person using a well-tested experimental method. The concept of precision and accuracy can be explained by target practice exercise. As shown in Figure 3.1(a), all the six bullets hit in the bull's eye (close or at the target), which shows that the target hit is precise and accurate, whereas Figure 3.1(b) depicts that the all six bullets hit precisely but away from bull's eye. This shows that the target hit is only precise but inaccurate, and this may be due to the nonalignment of gun. This suggests that accuracy require good precision but good precision is not the guarantee of accuracy. Figure 3.1(c) displays that the target hit is neither precise nor accurate.

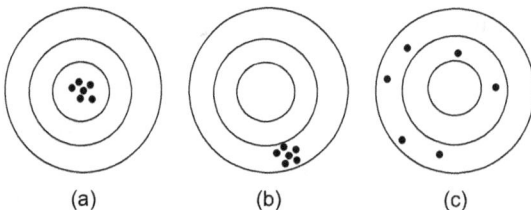

(a)   (b)   (c)

Figure 3.1: Explanation of accuracy and precision using target practice exercise.

3.2.2 Experimental errors

There are two types of experimental errors: (i) systematic and (ii) random.

3.2.2.1 Systematic error (also called determinate error)

This is a consistent error that can be detected and corrected. It may arise from a flaw in instrument or the design of an experimental method. A key feature of systematic error is that it is reproducible when we repeat the experiment in exactly the same manner as done previously. In principle, systematic error can be discovered and corrected, although this may not be easy. Systematic error has a definite value, and in some of the measurement, it may always be positive and in others, always negative. As the name determinate error suggest that it can be determined experimentally and then either avoided or corrected.

3.2.2.1.1 Sources of systematic errors

Instrumental error

Instrumental error may be caused by incorrect standardization, incorrect instrument alignment, incorrect wavelength settings, incorrect reading of values, and uncalibration of instrument. For example, a pH meter is standardized using buffer of pH 7.0 but the actual pH of buffer is 6.92. Then all pH readings will be 0.08 pH unit high. The actual reading of a sample would be pH 5.52, if the pH meter shows the reading as 5.60. This systematic error could be discovered by using a second buffer of known pH to test the meter. Instrumental error can be reduced by frequent calibration of the instrument prior to performing the experiment.

Error in method

Error in method arises from the analytical method itself. For example, in volumetric analysis titration, error is the difference between end point and equivalence point. Incompleteness of the reaction and side reactions are all the possible source of method errors. In the same way, coprecipitation of undesired entities and dissolution of desired precipitate in gravimetric analysis are the errors of the method that vary the results and hence make the results either higher or lower than the accepted value. It is difficult to identify and correct the errors in the method. Such errors can be reduced by calibration and selecting an alternative method.

Error in reagents

Such types of error arise from impurities in reagents used in measurement. For example, poor quality of water is used for solution preparations in various volumetric analyses. This type of error can be eliminated by using double-distilled water and conducting reagent blank determinations. Another major source of error in reagent is contamination. There are various source of contamination; for instance, personal care items, such as hand creams, shampoos, powders, cosmetics, and perfumes, as they contain significant amount of chemicals. Powdered gloves may contain a variety of trace elements and should not be used by analysts for performing trace element determinations.

Personal errors

Personal errors come from personal judgment during taking measurement. The person performing the measurement causes these errors. For example, end point judgment in a titrimetric analysis, when the analyst is insensitive to color changes at the end point, excess reagent is used. The other personal error may be the result of inexperience, insufficient training, or being "in a hurry." A person may use the instrument incorrectly, perhaps by placing the sample in the instrument incorrectly each time or setting the instrument to the wrong conditions for measurement. Some other personal errors are (i) carelessness, which is not as common as is generally believed; (ii) transcription errors, that is, copying the wrong information into a laboratory notebook, or spreadsheet, or onto a label; and (iii) calculation errors. Such types of errors can be corrected by providing proper training to the analyst, paying attention to even minute details during performing the experiment, and it also comes with the experience.

3.2.2.2 Random error (also called indeterminate error)

These types of error arise from uncontrolled variables in the measurement. As the name suggests, indeterminate error means that are unable to determine that is why it cannot be reduced by calibration. It might be reduced by a better experimental method or by increasing the number of measurements. Random error has an equal chance of being positive or negative.

Random errors may be caused due to incorrect readings of balances, pH meter, spectrophotometer, and scales such as rulers or dials by the same analyst under the same conditions. For example, a weighing balance that is capable of measuring only up to three digits after decimal then we cannot distinguish between two samples with weighing of 1.0153 and 1.0147 g. In one case, the measured weight is high; in the other case, it is low.

Random errors will follow a Gaussian or normal distribution when only indeterminate errors were involved in measurements; then, we can use mathematical laws of probability to get the most probable result of a series of measurements.

3.3 Methods of expression

Mathematical expression of accuracy is done by (i) absolute error and (ii) relative error.

3.3.1 Absolute error

For a single measurement, absolute error is simply expressed by the difference between true value and measured value. But when more than one measurement is taken, then the absolute error is expressed by the difference between mean measured value and true value. Unit of absolute error is same as the unit of measurement. For example, the result of a gravimetric analysis is 1.9210 g/L and the true value for that particular is measurement 1.8210 g/L, then the absolute error is 0.1000 g/L.

For single measurement, absolute error ($E_{absolute}$) can be calculated as

$$E_{absolute} = x_i - x_t$$

where x_i is the measured value and x_t is the true value.

For more than one measurement, $E_{absolute}$ is

$$E_{absolute} = \bar{x} - x_t$$

where $\bar{x}$ is the mean value.

3.3.2 Relative error

Usually relative error is expressed as percentage (%) and we can obtain it by dividing absolute error by the true value and multiplying by 100. The value of absolute error calculated earlier can be expressed as the relative error ($E_{relative}$):

$$\frac{0.1000}{1.8210} \times 100\% = 5.5\%$$

$$E_{relative} = \frac{x_i - x_t}{x_t} \times 100\% \quad \text{or} \quad E_{relative} = \frac{\bar{x} - x_t}{x_t} \times 100\%$$

Relative error is also expressed as parts per thousand (ppt)

$$E_{relative} = \frac{x_i - x_t}{x_t} \times 1,000 \text{ ppt} \quad \text{or} \quad E_{relative} = \frac{\bar{x} - x_t}{x_t} \times 1,000 \text{ ppt}$$

Precision can be expressed by various mathematical ways such as absolute deviation, mean deviation, relative deviation, percent relative deviation, standard deviation, standard deviation of the mean, and percent relative standard deviation. Before proceeding further, let us understand these standard terms which are used to express precision:

i. **Absolute deviation ($d_{absolute}$):** It can be expressed as difference between measured value and mean of the measured values.

$$d_{absolute} = |x_i - \bar{x}|$$

where x_i is the measured value and $\bar{x}$ mean of the measurements.

ii. **Mean or average deviation (d_{mean}):** It can be calculated as mean of differences over N measurements.

$$d_{mean} = \frac{\Sigma |x_i - \bar{x}|}{N}$$

iii. **Relative deviation ($D_{relative}$):** It can be expressed as absolute deviation divided by the mean value of the measurements.

$$D_{relative} = \frac{d_{absolute}}{\bar{x}}$$

iv. **Percent relative deviation (% $D_{relative}$):** It can be calculated from relative deviation equation simply multiplying by 100.

$$\% \, D_{relative} = \frac{d_{absolute}}{\bar{x}} \times 100\%$$

v. **Standard deviation:** For an infinite set of measurements, standard deviation can be expressed by the following equation:

$$\sigma_{sd} = \sqrt{\frac{\sum_{i=1}^{i=N}(x_i - \mu)^2}{N}} \quad (\text{when } N > 30)$$

where σ_{sd} = standard deviation, x_i is the individual measurement value, μ is the mean of the infinite measurements, and N is the number of measurements (infinite).

However, standard deviation is represented as estimated standard deviation when measured data sets are less than 30. It is denoted by s instead of σ. Here, population mean (μ) is replaced with mean of the measurements ($\bar{x}$) and the N is replaced with $N - 1$, which is defined as the number of independent deviations ($x_i - \bar{x}$):

$$s = \sqrt{\frac{\sum_{i=1}^{i=N}(x_i - \bar{x})^2}{N-1}} \quad (\text{when } N < 30)$$

where $N - 1$ is the number of degrees of freedom.

vi. **Standard deviation of the mean (s_{mean}):** It is sometimes introduced as the standard error and associated with mean of N measurement data sets.

$$s_{mean} = \frac{s}{\sqrt{N}}$$

vii. **Percent relative standard deviation:** It is just the standard deviation expressed as a fraction of the mean which is given as the percentage of the mean, which is often called the coefficient of variation.

$$\text{Percent relative standard deviation} = \frac{s}{\bar{x}} \times 100 = \% \, \text{RSD}$$

viii. **Variance:** It is expressed by s^2 (for small number of measurements) or σ^2 (for a large number of measurements). From the expression, it is clear that variance is the square of standard deviation (s) or population standard deviation (σ).

$$\text{Variance} = s^2 = \frac{\sum_{i=1}^{i=N}(x_i - \bar{x})^2}{N-1} \quad \text{(for small number of measurements)}$$

$$\text{Variance} = \sigma^2 = \frac{\sum_{i=1}^{i=N}(x_i - \mu)^2}{N} \quad \text{(for large number of measurements)}$$

Variance is used for comparison of precisions of two measured data sets (*F*-test). It can also be used for the determination of random error propagation in various calculation steps of analysis because variance is an additive statistic while standard deviation is not an additive property.

3.4 Normal law of distribution of indeterminate errors

In general, analytical results follow a Gaussian or normal distribution when only random errors are included in the measurement. If an analytical measurement is repeated many times, the obtained analytical results tend to cluster symmetrically about the mean value as shown in Figure 3.2. We can get a perfect smooth curve of the results, which is called a Gaussian curve, when large population measurements are made. The frequency density of the measurement can be expressed by the following equation of the normal distribution:

$$\text{Frequency density } f(x) = \frac{1}{\sigma\sqrt{2\pi}} e^{-\frac{(x-\mu)^2}{2\sigma^2}}$$

where σ is the population standard deviation, x is the measured value, and μ is the true (accepted) mean value or population mean value, which can be obtained by statistical calculations when there is no determinate or systematic error. In this equation, $(x - \mu)$ is expressed as the measure of indeterminate or random error. From the normal distribution of analytical results (Figure 3.2), we can observe that (i) small indeterminate (random) errors are seen more often than large errors and (ii) there is a symmetric distribution of positive and negative indeterminate errors about the maximum. Good precision of analytical measurement can be seen from the Gaussian curve if there is little variation (positive or negative) in the measured standard deviation from the accepted mean standard deviation. The shape of a Gaussian curve changes as the population standard deviation changes.

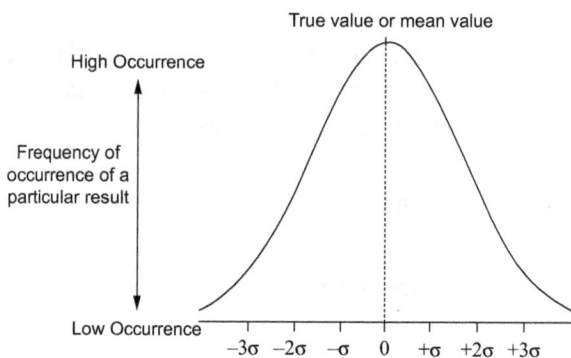

Figure 3.2: Normal indeterminate or random error curve.

3.5 Statistical test of data, *F*-, *Q*-, and *t*-tests, rejection of data, and confidence intervals

The result of a measured data set obtained from new instrumental method is often compared with result obtained from a standard method (more accepted method) using different statistical tests such as *Q*-test, *F*-test, and *t*-test to check the significance of result as well as for acceptance of data points obtained from new instrumental method. These statistical tests involve a comparison between a calculated value and a tabulated value in a confidence interval or probability level to check the suspected outliers, which is one or more data point considerably higher or lower than the remainder and appears to be outside the range in a set of measurements – *Q*-test.

Precisions of two measured data sets collected by two persons in the same laboratory or two different methods of measurement for the same analyte or results from two different laboratories – *F*-test.

Experimental means of two measured data sets or to compare the experimental mean of one measured data set with a known or standard value – *t*-test.

3.5.1 Q-test

Q-test is used to find out an "outlier" in a replicate set of measurements. Considerably, outlier is either the highest or lowest value in the measured data set. An outlier may originate due to systematic or determinate errors such as inaccurate burette reading, use of the wrong size pipette, loss of mass while taking weight measurement, incorrect dilution, or allowing the sample to boil to dryness when it should not have been done. The following relation is used to calculate the measured value of Q and then it is compared with the tabulated Q-value in a confidence interval or probability level (usually 90 or 95%). If Q-measured value is less than the Q-tabulated, then it is concluded that there is no significant difference between the out of line value and the other values, and the suspect value is considered for further data analysis. However, if Q-measured is greater than Q-tabulated, then the suspect value is considered as an outlier and is rejected, hence will not be used in the further calculations:

$$Q_{measured} = \frac{|\text{suspect value} - \text{closest value}|}{|\text{highest value} - \text{lowest value}|}$$

For example, four replicate measured data values were collected for the quantification of arsenic in ground water sample, say, 0.401, 0.403, 0.410, and 0.380 µg/dm^3.

In these values, 0.380 µg/dm^3 is a suspected value and lowest value. According to this, closest value is 0.401 µg/dm^3 and the highest value is 0.410 µg/dm^3. So,

$$Q_{measured} = \frac{|0.380 - 0.401|}{|0.410 - 0.380|}$$

$$Q_{measured} = 0.700$$

Table 3.1: Critical values of Q at the 95% ($P = 0.05$) level for a two-tailed test.

Number of measurements	4	5	6	7	8
Critical value	0.832	0.718	0.622	0.571	0.525

Q-measured value 0.700 is compared with Q-tabulated for four data measurements 0.832 (Table 3.1) is less than Q-tabulated, so it is concluded that 0.380 mg/dm^3 is not an outlier at the 95% confidence level and should be considered for further analysis.

Further three more replicate value collected for the same measurement and obtained values are 0.400, 0.411, and 0.413 mg/dm^3. Again, 0.380 µg/dm^3 is a suspected value and also lowest value, now the closest value is 0.400 mg/dm^3 and the highest value is 0.413 µg/dm^3.

$$Q_{measured} = \frac{|0.380 - 0.400|}{|0.413 - 0.380|}$$

$$Q_{measured} = 0.606$$

Q-measured value 0.606 is compared with Q-tabulated for seven data measurements 0.571 (Table 3.1) is higher than Q-tabulated, so it is concluded that 0.380 mg/dm^3 is an outlier at the 95% confidence level and should not be considered for further analysis.

Exercise: Find out the outlier from the given measurement values for a pesticide analysis of a river water measured values are 5,600, 5,640, 5,700, 5,690, and 5,810 μg/dm^3 also suggest that outlier should retained or rejected at the 95% confidence level (see the table for critical value of Q)?

3.5.2 F-test

F-test is a statistic expressed by ratio of variances s_1^2/s_2^2 (for small number of measurements) or σ_1^2/σ_2^2 (for large number of measurements) of two measured data sets collected from different analysis methods. The value of F is greater than or equal to one ($F \geq 1$) because numerator in the formula is always larger variance.

The calculated value of F ($F_{measured}$) is compared with the tabulated value ($F_{tabulated}$) in a confidence interval or probability level (usually 90% or 95%). According to null hypothesis, if $F_{measured}$ is small than the $F_{tabulated}$, then it is concluded that measured data sets obtained from different analysis methods have good precision because there is no significant difference between variances of these data sets. But, when $F_{measured}$ value is larger than the $F_{tabulated}$ value, then it is concluded that measured data sets obtained from different analysis methods have significant difference in precision as well as in variances.

Consider the following example:

Table 3.2: Critical values of F at the 95% confidence level.

$v_1 \rightarrow$ v_2 $\downarrow$	2	3	4	5	6	7	8	9	10
2	19.00	19.17	19.24	19.30	19.32	19.35	19.38	19.40	19.41
3	9.55	9.29	9.12	9.01	8.94	8.89	8.85	8.81	8.79
4	6.94	6.59	6.40	6.26	6.16	6.09	6.04	6.00	5.96
5	5.79	5.41	5.19	5.06	4.95	4.88	4.82	4.77	4.74
6	5.14	4.76	4.53	4.39	4.29	4.21	4.15	4.10	4.06
7	4.74	4.35	4.12	3.97	3.87	3.80	3.73	3.68	3.64
8	4.46	4.07	3.84	3.69	3.58	3.50	3.45	3.39	3.35

Table 3.2 (continued)

$v_1 \rightarrow$ v_2 $\downarrow$	2	3	4	5	6	7	8	9	10
9	4.26	3.86	3.63	3.48	3.37	3.29	3.23	3.19	3.14
10	4.10	3.71	3.48	3.33	3.22	3.14	3.07	3.02	2.99

Here v_1 and v_2 are the number of degrees of freedom of numerator and denominator, respectively.

A new method is proposed for the quantification of carbonate ions in a waste water analysis and variance of measured data set obtained from this new method is compared with variance of data set from the standard method (Example 3.1).

Example 3.1: Comparison of the variance obtained from new method and standard method

	New method	**Standard method**
Carbonate ion	82	85
(mg/dm^3)	80	83
	78	86
	85	84
	87	81
	81	79
	84	
	79	
Mean ($\bar{x}$)	82	83

$$s^2 = \frac{\sum_{i=1}^{i=N}(x_i - \bar{x})^2}{N-1}$$

$$\frac{\left[(0)^2 + (-2)^2 + (-4)^2 + (3)^2 \right.}{\left. + (5)^2 + (-1)^2 + (2)^2 + (-3)^2 \right]}{(8-1)}$$

$$\frac{\left[(2)^2 + (0)^2 + (3)^2 + (1)^2 \right.}{\left. + (-2)^2 + (-4)^2 \right]}{(6-1)}$$

$$= \left(\frac{68}{7}\right)$$

$$= \left(\frac{34}{5}\right)$$

$$= 9.71$$

$$= 6.8$$

$$= s_1^2 \text{ (numerator)}$$

$$= s_2^2 \text{ (denominator)}$$

$F_{measured} = \dfrac{(s_1^2)}{(s_2^2)}$	1.43
$F_{tabulated}$	4.88 (where number of degrees of freedom of the new method is 7 and number of degrees of freedom of the standard method is 5 for the 95% confidence level)

As the result shows that $F_{measured}$ (1.43) is small than the $F_{tabulated}$ (4.88) (Table 3.2), then it is concluded that measured data sets obtained from new method have good precision with standard method and the analysis by new method is acceptable.

3.5.3 t-Test

t-Test is also called Student's t because this test is developed by W. S. Gossett (a student). Gossett published his work on t-statistics with a pseudonym "student" in the year 1908.

For the following different situations, t-test is used in a different manner.

First case: When a same quantity is measured repeatedly using a measurement method (test-measurement), and mean value and standard deviation are calculated for it. We suppose to get different mean value and standard deviation for each time. We also have mean value for the same quantity from an accepted standard measurement method. Then, we can compare the mean value and standard deviation of test-measurement with accepted standard method. On the basis of comparison, we can conclude whether the test-measurement agrees with standard accepted method or not.

For example, hydrogen content in a sample of coal is analyzed by a test-measurement and values are found to be 4.22, 4.29, 4.23, and 4.30 wt%. For the same sample, 4.19 wt% hydrogen was found with a standard accepted method. Then the mean value and standard deviation calculated for the test measurement at 95% confidence level is compared with the value obtained from the standard accepted method:

$$\bar{X} = \frac{4.22 + 4.23 + 4.29 + 4.30}{4} = 4.26$$

$$S = \sqrt{\frac{\sum_{i=1}^{i=N}(x_i - \bar{x})^2}{N-1}} = \sqrt{\frac{(4.22-4.26)^2 + (4.23-4.26)^2 + (4.29-4.26)^2 + (4.30-4.26)^2}{4-1}} = 0.041$$

$$\text{At 95\% confidence level} = \bar{X} \pm \frac{t_s}{\sqrt{N}} = 4.26 \pm \frac{(3.182)\times(0.041)}{\sqrt{4}} = 4.26 \pm 0.065$$

$$\text{At 95\% confidence level} = 4.195 - 4.325 \, \text{wt\%}$$

The value of hydrogen content obtained from the standard method is 4.19 weight % which is just outside the range of measured weight %. Therefore, it is concluded that result observed from test-measurement agrees only 5% with result of acceptable standard method.

Second case: When a same quantity is measured many times using two different methods, and mean value and standard deviation are calculated. Then, using t-test we can check whether the results obtained by different methods agree with each other or not.

Table 3.3: Critical values of Student's t-test at 90%, 95%, and 99% confidence levels.

Confidence level → Degrees of freedom↓	90%	95%	99%
1	6.314	12.706	63.656
2	2.920	4.303	9.925
3	2.353	3.182	5.841
4	2.132	2.776	4.604
5	2.015	2.571	4.032
6	1.943	2.447	3.707
7	1.895	2.365	3.500
8	1.860	2.306	3.355
9	1.833	2.262	3.250
10	1.812	2.228	3.169
∞	1.645	1.960	2.576

In this case, experimental means of two different methods (Method A and Method B) are compared by the following equation:

$$t_{measured} = \frac{|\bar{X}_A - \bar{X}_B|}{s_{pooled}} \sqrt{\frac{NM}{N+M}}$$

where $|\bar{X}_A - \bar{X}_B|$ is the mode of absolute difference of mean values of both the method, s_{pooled} is the pooled standard deviation of measured values of both measurement methods, and N and M are the number of measurements for Method A and Method B, respectively:

$$s_{pooled} = \sqrt{\frac{\sum_{i=1}^{i=N}(x_i - \bar{X}_A)^2 + \sum_{i=1}^{i=M}(x_i - \bar{X}_B)^2}{N+M-1}} = \sqrt{\frac{s^2{}_A(N-1) + s^2{}_B(M-1)}{N+M-2}}$$

In this case, $t_{measured}$ is compared with $t_{tabulated}$ at a particular confidence level for $N + M - 2$ degrees of freedom. If $t_{measured}$ is larger than the $t_{tabulated}$, then it is concluded that results of two different methods do not agree with each other and considered to be different. If $t_{measured}$ is small than the $t_{tabulated}$, then it is concluded that results obtained from two different methods agree with each other and we can consider results of both the methods.

The gravimetric analysis data of iron obtained from two different methods are compared at 95% confidence level and agreement between them is checked, and the same is exemplified below (Example 3.2).

Example 3.2: Gravimetric analysis data obtained from two different methods:

Statistics	Method A	Method B
Iron %	10.10	9.20
	8.65	8.89
	10.50	9.70
	9.40	9.00
	9.99	9.40
	9.25	
$\bar{x}$	$\bar{x}_A$ = 9.65 (number of measurements 6)	$\bar{x}_B$ = 9.24 (number of measurements 5)
$(x_i - \bar{x})$	0.45	−0.04
	−1.00	−0.35
	0.85	0.46
	−0.25	−0.24
	0.34	0.16
	−0.40	
$(x_i - \bar{x})^2$	0.2025	0.0016
	1.00	0.1225
	0.7225	0.2116
	0.0625	0.0576
	0.1156	0.0256
	0.1600	
$\sum (x_i - \bar{x})^2$	2.263	0.419

$$S_{pooled} = \sqrt{\frac{\sum_{i=1}^{i=N} (x_i - \bar{x}_A)^2 + \sum_{i=1}^{i=M} (x_i - \bar{x}_B)^2}{N + M - 2}} = \sqrt{\frac{2.263 + 0.419}{6 + 5 - 2}} = 0.546$$

$$t_{measured} = \frac{|9.65 - 9.24|}{0.546} \sqrt{\frac{6 \times 5}{6 + 5}} = 1.241$$

where $t_{measured}$ value (1.241) is compared with $t_{tabulated}$ value (2.262) from Table 3.3 for 9 degrees of freedom at 95% confidence level which is less than the $t_{tabulated}$, so this is concluded that results of both the methods agree with each other and there is no significant difference in the results obtained by these two methods.

Third case: The use of paired t-test for comparing differences between individual measurements of two different methods

$$\text{Paired } t_{measured} = \frac{\bar{X}_{difference}}{S_{difference}} \sqrt{N}$$

where $\bar{X}_{difference}$ is the mean value of individual differences of measurements of two different methods, $s_{difference}$ is the standard deviation of individual differences of measurements of two different methods, and N is the number of measurements:

$$s_{difference} = \sqrt{\frac{\sum_{i=1}^{i=N}(X_{di} - \bar{X}_d)^2}{N-1}}$$

where X_d is the individual difference of measurements of two different methods, $\bar{X}_d$ is the mean value of individual difference of measurements of two different methods, and N is the number of measurements.

For instance, urea nitrogen analysis is performed in different samples of blood using a new test method, and obtained results are compared with the results obtained from the standard accepted method for the same samples. Then, paired t-test at 95% confidence level is used to check the acceptance of results obtained from new test method, as elaborated below in Example 3.3.

Example 3.3: Comparison of results obtained from two different methods for urea nitrogen analysis

Blood samples	Blood urea nitrogen analysis using new test method (mg/dL)	Blood urea nitrogen analysis using standard method (mg/dL)	Individual measurement difference (X_d)	$(X_{di} - \bar{X}_d)$	$(X_{di} - \bar{X}_d)^2$
1	13.7	12.9	0.8	0.52	0.2704
2	11.2	11.5	−0.3	−0.58	0.3364
3	18.5	17.9	0.6	0.32	0.1024
4	9.6	9.7	−0.1	−0.38	0.1444
5	12.5	12.1	0.4	0.12	0.0144
6	12.2	11.9	0.3	0.02	0.0004

$$\bar{X}_d = \frac{\sum X_d}{N} = \frac{1.7}{6} = 0.28 \qquad \sum(X_{di}-\bar{X}_d)^2 = 0.87$$

$$s_{difference} = \sqrt{\frac{\sum_{i=1}^{i=N}(X_{di} - \bar{X}_d)^2}{N-1}} = \sqrt{\frac{0.87}{6-1}} = 0.417$$

$$\text{Paired } t_{measured} = \frac{0.28}{0.417}\sqrt{6} = 1.644$$

$t_{measured}$ 1.644 is less than the $t_{tabulated}$ 2.571 (Table 3.3) at the 95% confidence level for five degrees of freedom, so it is concluded that blood urea nitrogen analysis results obtained from new method is acceptable at 95% confidence level and there is no statistical difference between the results of two methods at 95% confidence level.

When a reference value (μ) is given, then we can use the following equation of Student's t to calculate $t_{measured}$

$$t_{measured} = \frac{(\bar{X} - \mu)}{s} \sqrt{N}$$

where s is the standard deviation, $\bar{X}$ is the mean value, and N is the number of measurements.

If the $t_{measured}$ is less than $t_{tabulated}$ at a confidence level (usually 95%), then the measurement can be considered statistically not different from the reference value (μ).

3.5.4 Confidence intervals

When we perform a quantitative analysis, then most of the time it is difficult to determine a true or accepted mean value (μ) along with a population standard deviation (σ) from a small set of measurements because there is always a possibility of small difference between true mean value and the calculated mean value ($\bar{X}$). We can get this only from a large set of measurements, and it is time consuming. Then the estimation of true mean value is done by giving a range of measured values on either side of arithmetic mean ($\bar{X}$) and within this range true or accepted mean value (μ) is expected to lie with a certain level of probability, which is known as confidence interval. This is also called as confidence limit or confidence level and is typically expressed as a percentage.

With the help of following relation, we can calculate the confidence limits using standard deviation of the mean and a statistical factor t in a probability level, assuming that no systematic errors are present.

$$\text{Confidence limit for true mean value} = \bar{X} \pm \frac{ts}{\sqrt{N}}$$

where s is the standard deviation for small set of measurements, N is the number of measurements, and t is the Student's t that has varying values for different number of degrees of freedom at desired confidence level (Table 3.3).

We have the following measurements of dichlorodiphenyltrichloroethane (DDT) pesticide analysis in river water samples using gas chromatography (Example 3.4). For which confidence (in the range), we can report the results at the 95% confidence level for this.

Example 3.4: Measurement of DDT pesticide in river water samples

River water samples	Measurements of DDT pesticide (in parts per billion)	$(x_i - \bar{x})$	$(x_i - \bar{x})^2$
A	2.5	0.0	0.0
B	2.6	0.1	0.1
C	1.9	−0.6	0.36
D	3.0	0.5	0.25
E	2.7	0.3	0.09
	$\bar{x} = 2.5$		$\sum (x_i - \bar{x})^2 = 0.8$

$$s = \sqrt{\frac{\sum_{i=1}^{i=N} (x_i - \bar{x})^2}{N - 1}} = \sqrt{\frac{0.8}{5 - 1}} = 0.447$$

$$\text{Confidence limit at 95\% confidence level} = \bar{x} \pm \frac{t_s}{\sqrt{N}} = 2.5 \pm \frac{(2.776) \times (0.447)}{\sqrt{5}} = 2.5 \pm 0.6$$

On the basis of above calculation, we can say that true mean value or accepted mean value of the analysis will lie in the range of 1.9–3.1 ppb at 95% confidence level.

References

[1]   Christian, G.D., Dasgupta, P.K., and Schug, K.A. Analytical chemistry. John Wiley & Sons; 2013. Hoboken, New Jersey, US.

[2]   Daniel-Harris, C. 2010 Quantitative chemical analysis, 8th edn. New York: W. H. Freeman and Company; 2010.

[3]   Harvey, D. Modern analytical chemistry. New York: McGraw-Hill; Vol. 1, 2000.

[4]   Kealey, D., and Haines, P.J. 2004. BIOS instant notes in analytical chemistry. Taylor & Francis; 2002.

[5]   Li, N., and Hefferren, J.J. Quantitative chemical analysis. World Scientific Publishing Company; 2013.

[6]   Robinson, J.W., Frame, E.M.S., Frame, G.M., Eileen, M., and Skelly, F. Undergraduate instrumental analysis. 2005.

[7]   Rouessac, F. and Rouessac, A. Chemical analysis: modern instrumentation methods and techniques. John Wiley & Sons; 2022 United States of America.

[8]   Skoog, D.A., West, D.M., Holler, F.J., and Crouch, S.R. Fundamentals of analytical chemistry, 9th edn. Cengage learning; 2013 United States of America.

Section III: **Spectrochemical methods of analysis**

Neeta Azad

Chapter 4
UV–visible spectroscopy

Abstract: Spectroscopy is one of the important techniques used for the structural analysis of compounds, and it finds its application in all fields of science like Chemistry, Biochemistry, Biology, and Medicines. Spectroscopy deals with the interaction of matter with electromagnetic radiations (EMR). EMR can be categorized into different regions on the basis of their frequencies/wavelengths, and these are gamma rays (γ, highest frequency/lowest wavelength), X-rays, ultraviolet–visible (UV–vis), infrared (IR), microwaves, and radio waves (lowest frequency/longest wavelength). The characteristics of radiations associated with different EM regions are completely different due to their frequencies, as frequency is the measures of energy associated with the EMR. This chapter focuses on UV–visible region of the EM spectrum which falls in wavelength region from 190 to 800 nm. The interaction of matter with radiations in UV–visible region shows wide applicability in our daily life; for example, characterization of colors is done on the basis of their wavelength. Everything around us looks colorful because of visible light. In chemical analysis of compounds, the different color of solutions refers to different concentrations. Similarly, in case of conjugated organic compounds, electronic transitions between various energy states are responsible for their UV–visible spectra. This phenomenon can be studied using Beer–Lambert's law, which signifies the relation between absorbance of radiations in UV–visible region and concentration of solution under study.

The prime concern of this chapter is to present concepts of UV–visible spectroscopy in a very simple manner, which can be easily understood by undergraduate students. EM spectrum along with the importance of UV–visible region has been discussed in detail with simple examples so that students can correlate the concepts with their daily life. The highlights of this chapter are properties of EMR, origin of spectra, interaction of radiations with matter, and fundamental laws of spectroscopy. After reading this chapter, questions like how absorption and emission of EMR give rise to emission and absorption spectroscopy, and how these two are different, can be answered. Along with the explanation of role of transmittance and absorbance in Beer–Lambert's law, its mathematical treatment has also been elaborated. Important terms and symbols used in absorption measurements and basic principles of instrumentation (choice of source, monochromator, and detector) for single and double beam instrument have also been taken into consideration so that one can apply these concepts while working in laboratory. What are the important instrumental techniques of recording UV–visible spectrum and how to analyze the recorded spectrum? What kinds of compounds are active in UV and visible range? All these concepts are summed up in this chapter.

https://doi.org/10.1515/9783110794816-004

4.1 Introduction

Spectroscopy is a branch of science which deals with the interaction of electromagnetic radiations (EMR) with matter. When EMRs interact with matter, various phenomena like absorption, reflection, dispersion, scattering, and emission are possible. Depending upon these phenomena and nature of EMR, spectroscopy can be classified into various categories like absorption spectroscopy, emission spectroscopy, and Raman spectroscopy. White light is a combination of seven different wave lengths, and these wavelengths correspond to seven colors: violet, indigo, blue, green, yellow, orange, and red. Yellow color of sunflowers, green color of leaves, and everything around us that looks colorful is due to visible light in action. The progress of a chemical reaction can also be monitored on the basis of different colors of reactant(s) and product(s). This color change during the course of any chemical/biochemical reaction is due to electronic transitions between various electronic energy levels. The energy differences between various electronic energy levels fall in UV–visible range. The transition of electrons is associated with different chemical processes, and it may involve either of these processes, fluorescence, phosphorescence, and dissociation of molecules.

4.2 Electromagnetic radiations

EMR are the form of energy which can travel through space and have electric and magnetic fields associated with them. The oscillation of electric and magnetic fields is perpendicular to each other as well as to the direction of propagation of radiation. EMR does not require any medium to propagate and can travel through vacuum. A wide range of frequencies/wavelengths fall under electromagnetic spectrum. These are gamma rays, X-rays, ultraviolet rays (UV), visible rays, infrared rays (IR), microwave, and radio waves. These different EMR carries different energy and are used for different purpose. For example, X-rays are used for medical imaging, gamma rays are used in cancer treatment, UV rays help in disinfection of food and materials, etc., visible rays are the reason we can see the world, infrared rays help in finding structure of molecules, microwaves have application in cooking appliances, and radio waves are used for communication. This way EMRs play an important role in our day-to-day life. The wavelength, frequency, and energy associated with these EMR are mentioned in Table 4.1.

Table 4.1: The region of electromagnetic spectrum.

EM region	Wavelength	Frequency (Hz)
Gamma rays	<1 pm	10^{20}–10^{24}
X–rays	1 nm to 1 pm	10^{17}–10^{20}
UV–visible	750 nm to 1 nm	10^{14}–10^{17}
Infrared (IR)	25 µm to 750 nm	10^{13}–10^{14}
Microwaves	1 mm to 25 µm	10^{11}–10^{13}
Radio waves	>1 mm	$<10^{11}$

The energy, E, of an EMR can be calculated using the following equation:

$$E = h\nu \tag{4.1}$$

$$E = \frac{hc}{\lambda} \tag{4.2}$$

where h is Planck's constant ($h = 6.626 \times 10^{-34}$ Js), ν is the frequency of radiations (hertz or s^{-1}), λ is the wavelength (m), and c is the speed of light (3×10^8 m/s).

Some of the examples on energy relations are discussed further.

4.2.1 Practice problems

Example 4.1: Sun emits UV rays of frequency 1×10^{15} s^{-1}. Will this energy of radiations be sufficient to excite an electron from ground state to first excited state? The difference between the two electronic energy levels is 1 eV (1 eV = 1.602×10^{-19} J). Justify your answer.

Solution: The energy difference between two electronic energy levels, E equals the energy required to jump electron from ground state to the first excited state.

Therefore, $E = 1\text{eV} = 1.602 \times 10^{-19}$ J

Energy of UV rays, $E' = h\nu$

$\quad\quad\quad = (6.626 \times 10^{-34}$ Js$) (1 \times 10^{15}$ s$^{-1}) = 6.626 \times 10^{-19}$ J

$\quad\quad\quad E' > E$

The energy of UV radiations is more than the energy required for an electron to jump from ground state to first excited state. Hence, the electronic transition will happen in the presence of UV rays.

Example 4.2: X-rays of wavelength 10 nm falls on a metal having work function 2 eV. Will the electron eject out from the metal surface? Justify your answer.

Solution: Wavelength of X-rays, $\lambda = 10$ nm

$\quad\quad\quad = 10 \times 10^{-9}$ m

$$= 10^{-8} \text{ m}$$

Energy of X-rays, $\quad E = \dfrac{hc}{\lambda} = \dfrac{(6.626 \times 10^{-34} \text{ Js}) (3 \times 10^{8} \text{m/s})}{(10^{-8}\text{m})} = 19.878 \times 10^{-18} \text{ J}$

Energy required to eject electron out from metal surface, E' = work function of metal.
Therefore, $\qquad\qquad\qquad\qquad\qquad\qquad E' = 2 \text{ eV}$

$$E' = 2 \times 1.602 \times 10^{-19} \text{ J}$$

$$= 3.204 \times 10^{-19} \text{ J}$$

Here, $E > E'$. Hence, the electron will be ejected out from its metal surface when X-rays fall on it.

4.3 Ultraviolet–visible (UV–vis) region

White light coming from sun or other sources is the combination of seven different colors. These seven colors collectively appear white and fall in the visible region of EM spectrum. Visible region of electromagnetic spectrum lies between wavelength range from 400–800 nm. The energy of these different wavelengths in this region lies between 1.55 eV to 3.10 eV. This energy is just sufficient to cause electronic transitions in unsaturated organic compounds. More details about electronic transitions have been discussed in later section.

This white light when dispersed through a prism, various wavelengths are visible in the form of different colors (Table 4.2), and in the absence of any color (wavelength), another color (wavelength) will dominate. Hence, a different color of light is visible to our eyes rather than white light. For example, when white light falls on a grass it appears green because wavelength corresponding to green color is being reflected by grass, and rest of the wavelengths are being absorbed. It means the complementary color of green is absorbed by the grass. In simple terms, a color which dominates in the absence of another color is called its complementary color. If red color is taken out of spectrum, white light will convert to green light. Similarly, if yellow color is separated from spectrum, the color of light will appear violet. The best examples of this are whitening of the clothes. To keep them white, clothes are washed in blue- or purple-colored solutions, called cloth whiteners. This is because gradually with time, the tendency of cloth to reflect all wavelengths diminish, and it starts absorbing blue/violet color, and its complementary color yellow/orange is visible to our eyes. To correct this phenomenon, clothes are dipped into violet-colored solution to bring its tendency back so that it again reflects all colors and appear white.

UV region is not visible to eyes but the energy of this region is sufficient to cause electronic transitions. The wavelengths, frequencies, energies of UV–visible region, and complementary colors of all wavelengths of visible region are mentioned in Table 4.2.

Table 4.2: Color, wavelength, energy, and complementary colors of UV–visible region.

Color	Wavelength (nm)	Frequency (10^{12} Hz)	Energy of a photon (eV)	Complementary color
Red	750–625	400–480	1.65–1.98	Blue-green
Orange	625–590	480–510	1.98–2.10	Blue
Yellow	590–565	510–530	2.10–2.19	Blue-violet
Green	565–500	530–600	2.19–2.48	Red-violet
Blue	500–450	600–670	2.48–2.75	Orange
Violet	450–400	670–790	2.75–3.26	Yellow-green
Ultraviolet	<400	>790	>3.26	No color

Example 4.3: An aqueous solution of potassium dichromate ($K_2Cr_2O_7$) appears orange in color. When we record its absorbance using spectrophotometer, the maximum absorbance (λ_{max}) comes out to be around 470 nm, which corresponds to blue color, and not 620 nm (corresponding to orange color). Give reason.

Solution: The aqueous solution of $K_2Cr_2O_7$ appears orange in color because wavelength corresponding to orange color is being reflected by the solution. It means its complementary color, that is, blue is being absorbed by the solution. Hence, the absorbance we record for this solution using spectrophotometer is the absorbance corresponding to blue color. This is the reason, why λmax for this solution comes out to be 470 nm rather than 620 nm.

4.4 Basic principle of instrumentation

The working of spectrophotometer in UV–visible range is shown in Figure 4.1. The source of white light is kept at one end of spectrophotometer whereas detector is placed at the other end. The light source is usually a tungsten lamp (visible region) or deuterium/hydrogen lamp (for UV region). After source of light, entrance slit is present whose function is to direct the light in direction of dispersion device. The dispersion device splits white light into seven colors. Another slit, termed as exit slit, is positioned after dispersion device which again directs the dispersed wavelength toward the sample. After passing through the sample, the transmitted light falls on detector, and its intensity is measured. Absorbance of light is calculated from its intensity using Beer–Lambert's law.

4.4.1 Single-beam and double-beam spectrophotometer

The double-beam spectrophotometer is different from single-beam spectrophotometer in its light splitting part. A single-beam spectrophotometer has only one beam of light that passes through the sample whereas in double-beam spectrophotometer, the incident light after passing through exit slit splits into two parts of same intensity by the reflectors. One part of incident beam passes through reference solution, and the other part passes through the sample solution. Both are received by the detector, and the result is displayed in digital form on the screen of display unit.

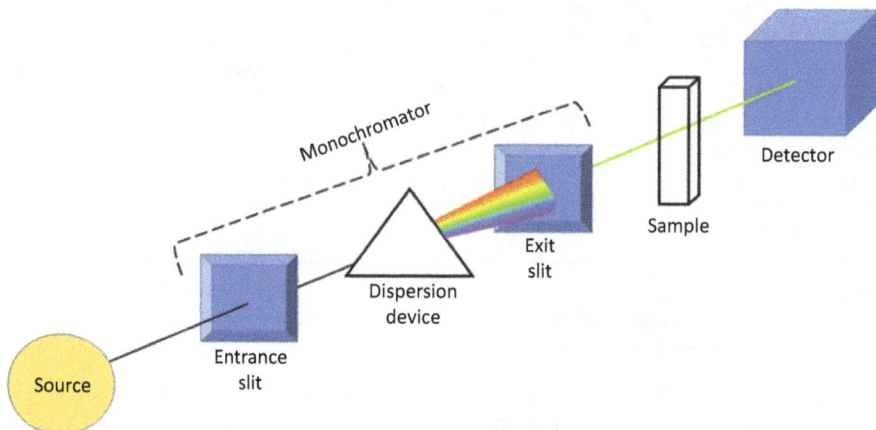

Figure 4.1: Internal setup of an UV–visible spectrophotometer.

4.5 Beer–Lambert's law, absorbance, and transmittance

The intensities of the solutions showing transitions in UV–visible region can be measured using Beer–Lambert's law, which is the combined form of Beer's law and Lambert's law.

Lambert's law: Lambert's law, named after Johann Heinrich Lambert, states that any radiation travelling through a medium absorbing the radiation, the rate of decrease of intensity with respect to the distance is proportional to the incident intensity. If "I_0" is the intensity of incident radiation, "I" is the incident intensity of transmitting radiation at any point, travelling through a path length "x" and dI/dx is the rate of decrease of intensity, then mathematically,

$$-\frac{dI}{dx} \, \alpha \, I$$

$$\frac{dI}{dx} = -k' \, I$$

$$\int_{I_0}^{I} \frac{dI}{I} = \int_{0}^{x} -k'dx$$

$$\ln \frac{I}{I_0} = -k'x$$

$$\frac{I}{I_0} = e^{-k'x}$$

$$I = I_0 e^{-k'x} \tag{4.3}$$

When the natural logarithm is changed to common logarithm, base "e" will be changed to base "10" and the value of constant k' will change to a new value "k." Here, $k = k'/2.303$. Substituting these values in eq. (4.3), then

$$I = I_0 10^{-kx} \tag{4.4}$$

Equation (4.4) represents Lambert's law. But in the case of colored solutions, it was found that the intensity of the transmitted radiations also depends upon the concentration of solution. This was studied by German mathematician and chemist August Beer and hence called Beer's law. The combination of the two is called Beer–Lambert's law, "k" in eq. (4.4) is replaced by "εC" to give

$$I = I_0 10^{-\varepsilon Cx} \tag{4.5}$$

The constant ε in eq. (4.5) is known as *molar extinction coefficient or molar absorptivity coefficient*.

The logarithmic form of eq. (4.5) is

$$\ln \frac{I}{I_0} = -\varepsilon Cx = A$$

$$\ln \frac{I_0}{I} = \varepsilon Cx = A$$

where A is the absorbance of the solution.

Hence,

$$A = \ln \frac{I_0}{I} \tag{4.6}$$

$$A = \varepsilon Cx$$

It implies, $C \, \alpha \, A$, that is, concentration is proportional to the absorbance,

$$\varepsilon = \frac{A}{Cx} = \text{constant} \qquad (4.7)$$

If $C = 1$ M (molar) or mol/L and $x = 1$ cm, then,

$$\varepsilon = \frac{A}{1 \times 1} = A$$

Hence, molar absorptivity coefficient can be defined as the absorbance of a solution with concentration 1 M, having path length 1 cm. The value of molar absorptivity coefficient is constant for a particular solution, and it depends upon the nature of the solution. The units for ε are M^{-1} cm^{-1} or $(mol/L)^{-1}$ cm^{-1}.

Transmittance is defined as the fraction of light that transmitted through the solution via given path length x.

Mathematically,

$$T = \frac{I}{I_0} \qquad (4.8)$$

where T is transmittance, I_0 is the intensity of incident light, and I is the intensity of transmitted light.

It is used to measure the drop in the intensity of transmitted light through a solution and is generally expresses as percentage (%T) and can be calculated as

Percentage transmittance,

$$\% T = \left(\frac{I}{I_0} \times 100 \right) \qquad (4.9)$$

Transmittance can vary over a range of 0–100%. Both absorbance and transmittance are unitless quantities.

4.5.1 Relation between absorbance and transmittance

From eqs. (4.6) and (4.8), we have

$$A = \ln\frac{I_0}{I} \text{ and } T = \frac{I}{I_0}$$

Hence,

$$A = \ln\left(\frac{1}{T}\right) \text{ or } A = -\ln(T) \qquad (4.10)$$

Example 4.4: What will be the transmittance of a solution if 40% light transmits through it?

Solution: Let the intensity of incident light $= I_0$
Intensity of transmitted light $= I$

$$I = \frac{40}{100} \times I_0 = 0.4I_0$$

$$T = \frac{I}{I_0} = \frac{0.4I_0}{I_0} = 0.4$$

$$T = 0.4$$

Example 4.5: A colored solution shows maximum absorbance of 0.92 at 470 nm. What will be the concentration of solution if its molar absorptivity coefficient is 0.412 $(\text{mol/L})^{-1}$ cm^{-1} and width of cuvette is 1 cm.

Solution: Given absorbance, $A = 0.92$

$$\varepsilon = 0.412\ (\text{mol/L})^{-1}\text{cm}^{-1}$$

$$x = 1\,\text{cm}$$

$$A = \varepsilon C x$$

$$C = \frac{A}{\varepsilon x} = \frac{0.92}{0.412 \times 1} = 2.233\,\text{mol/L}$$

Example 4.6: Calculate percentage transmittance (% T) of the solution in problem number (4.5).

Solution: In problem (4.5), $\qquad A = 0.92$

$$A = -\ln(T)$$

Therefore, $\qquad\qquad T = e^{-A} = e^{-0.92} = 0.3985$

Hence,

$$\% T = 0.3985 \times 100 = 39.85\%$$

4.6 Applications of UV–visible spectroscopy

There are various applications of UV–visible spectroscopy but for a chemist the most important ones have been listed below:

1. **Determination of concentration of unknown solutions:** This is one of the widest applications of UV–visible spectroscopy. The concentration of any unknown solution absorbing in UV–visible region can be determined using either spectrophotometer or colorimeter (for colored solutions only).

2. **To study electronic transitions in organic compounds:** Unsaturated organic compounds show electronic transition in ultraviolet region and are helpful in knowing how substitution of different functional groups can affect the spectrum of a particular compound. This finds a wide applicability in pharmaceutical industry.

3. **Fluorescence and phosphorescence:** Luminescence is divided into two categories: fluorescence and phosphorescence, observed in some compounds due to electronic transitions, is well explained by UV–visible spectroscopy.

4.6.1 Experimental procedure to determine the concentration of unknown solution absorbing in UV–visible region

The concentration of an unknown solution can be determined by using a spectrophotometer. It is a very sensitive instrument and should be kept in a place which is free from dust and humidity as dust and humidity may affect source of UV–visible radiations (lamp) mounted inside the device. Some of the key points should be kept in mind before recording the spectrum, which are as follows:

i. **Selection of solvent:** The choice of the solvent should be made very carefully. It should not absorb in the region under study. The solvent should not react with the compound under study and must be colorless.

ii. **Moderate concentration:** The concentration of the solution should not be too high. If it is so, dilute it to make it moderate before recording the spectrum. The absorbance of highly concentrated solutions may fall out of range. Similarly, very dilute solutions may also not give significant value of absorbance.

iii. **Clean apparatus:** The cuvette should be properly cleaned and dried before taking solution into it. Always wipe the exposed surfaces of the cuvettes using a soft tissue paper before placing them in the cuvette holder. Do not handle them from the optical faces.

iv. **Spectrophotometer:** Spectrophotometer/colorimeter should be switched on at least 30 min before running the sample to stabilize the lamp and the detector. If you start recording the spectrum immediately after switching the instrument on, readings may deviate from actual values.

v. **Reference solution:** Solvent should be used as reference (Blank) solution for zero settings to avoid any kind of errors due to absorbance shown by the solvent.

vi. **Solution in cuvettes:** Cuvettes should always be more than half-filled, and the level of reference as well as the solutions under study should be same.

vii. **Freshly prepared solutions:** All the solutions should be freshly prepared. It may happen that with time, some compounds may decompose or their composition may very due to various reasons which can lead to erroneous results.

viii. **Interference from external light:** The shutter of sample chamber of spectrophotometer should be closed before recording absorbance of the samples to avoid the absorbance of external light.

The concentration of any unknown solution absorbing in UV-visible region can be determined using spectrophotometer. Unsaturated organic compounds such as alkenes, alkynes, carbonyl compounds, conjugated systems, nucleic acids, proteins, and inorganic compounds, for example, potassium dichromate ($K_2Cr_2O_7$), potassium permanganate ($KMnO_4$), and copper sulfate ($CuSO_4$) absorb in this region. UV–visible spectroscopy can be employed to:

i.   Determine λ_{max} (wavelength at which the solution shows maximum absorbance).
ii. Verify Beer–Lambert's law and to determine molar extinction coefficient (ε).
iii. Determine the concentration of unknown solution.

4.6.1.1 Determination of λ_{max}

The procedure for determination of λ_{max} involves the following steps:
1. Switch on the instrument and prepare a set of clean and dry-labeled test tubes.
2. Prepare solutions of different concentrations from the standard solution (as shown in Figure 4.2). A standard solution is the solution of known concentration and can be prepared by dissolving a known amount of compound of interest in a suitable solvent.
3. Prepare at least 10 solutions of different concentrations to get sufficient data points to verify Beer–Lambert's law.
4. Select a solution out of prepared solutions of different concentrations to find out wavelength (λ_{max}). Fill the cuvette with approximately ¾ with the sample solution and place it in the sample compartment. Pour that solution in cuvette.

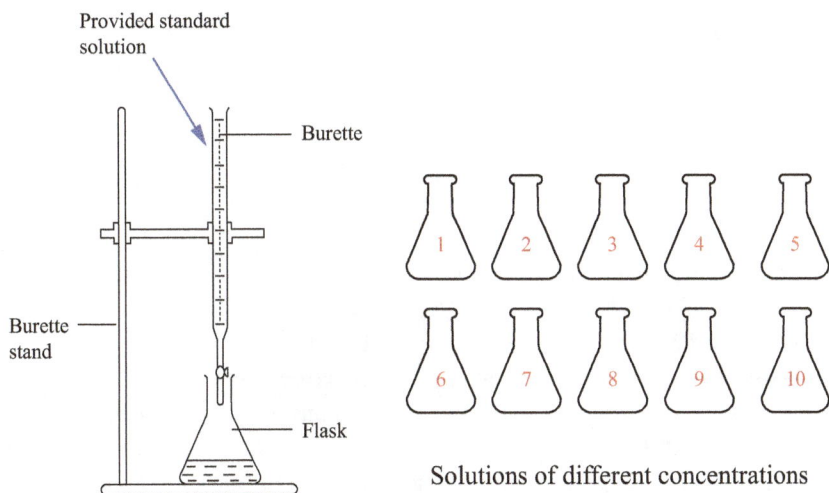

Figure 4.2: Experimental setup for the dilution of given standard solution to prepare different concentrations.

5. Fill another cuvette with reference solution (solvent only; blank) and place it in the blank/reference compartment.
6. Record the absorbance values at different wavelengths (Table 4.3) and plot a curve of the obtained values of absorbance at different wavelengths. A curve as shown in Figure 4.3 will be obtained. The wavelength at which the solution shown the maximum absorbance is called λ_{max} and is selected for analysis.

Table 4.3: Data recording of absorbance (A) versus wavelength (λ).

S. no.	Wavelength (nm)	Absorbance
1.	λ_1	A_1
2.	λ_2	A_2
3.	λ_3	A_3
4.	λ_4	A_4
5.	λ_5	A_5
6.	λ_6	A_6
7.	λ_7	A_7
8.	λ_8	A_8
9.	λ_9	A_9
10	λ_{10}	A_{10}

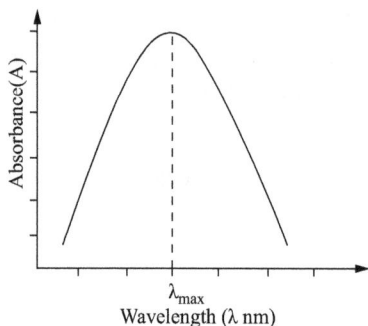

Figure 4.3: Absorbance versus wavelength plot.

4.6.1.2 Verification of Beer–Lambert's law

The following procedure can be followed to verify Beer–Lambert's law.
1. After getting the value of λ_{max}, set this value in spectrophotometer.
2. Prepare a range of solutions of different concentrations by taking a known amount of solution from stock and adding additional solvent to it, thus lowering the concentration. We can use the simple mathematical equation to calculate the volume of stock solution required to prepare a particular solution of desired concentration.

$$C_1 V_1 = C_2 V_2$$

where C_1 and V_1 are the concentrations and volumes of stock solution, and C_2 and V_2 are the concentrations and volumes of the required diluted solution.

3. First test tube with no sample (only solvent) will be used as blank (reference). Place this test tube in the reference compartment.
4. Place the test tube containing sample in the sample compartment and record the value of absorbance for all the samples, starting with the solution of lowest concentration. Make sure that the cuvette must be at least two-third filled for all the samples to get the accurate results.
5. Plot a standard curve by taking concentration of samples along X-axis and absorbance (A') along Y-axis. A standard curve is a plot of absorbance versus varying concentrations of known sample of a compound. It is used to determine the concentration of any unknown sample.
6. The molar extinction coefficient can be calculated using the relation:

$$A = \varepsilon c l$$

$$\varepsilon = \frac{A}{c l}$$

Suppose, 1 µM solution of sample gives A of x

$$\text{So, } \varepsilon = \frac{x}{1 \times 10^{-6} \text{ M} \times 1 \text{ cm}}$$

The value of molar extinction coefficient can be used to calculate the concentration of unknown solution.

Table 4.4: Protocol for verification of Beer–Lambert's law and determination of extinction coefficient.

Test tube no.	Stock solution (compound under study) (mL)	Solvent (mL)	Concentration of substock solution (M)	Absorbance (λ_{max})
1	0	10 (blank)	–	
2	1	9	C_1	A'_1
3	2	8	C_2	A'_2
4	3	7	C_3	A'_3
5	4	6	C_4	A'_4
6	5	5	C_5	A'_5
7	6	4	C_6	A'_6
8	7	3	C_7	A'_7
9	8	2	C_8	A'_8
10	9	1	C_9	A'_9
11	10	0	C_{10}	A'_{10}

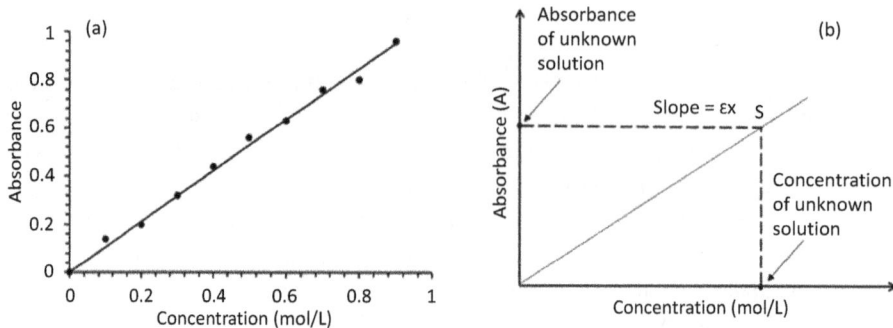

Figure 4.4: (a) Trial data points corresponding to data in Table 4.4; (b) graphical determination of concentration corresponding to absorbance of unknown solution.

4.6.1.3 Determination of concentration of unknown solution

1. After recording the absorbance of all sample solutions, take 10 mL of unknown solution in one cuvette and record the absorbance value at λ_{max}.
2. After recording the absorbance of unknown solution, the concentration of this solution can either be calculated mathematically or it can be determined graphically.
   Mathematically: Substitute the value of absorbance and slope (εx in the equation $A = \varepsilon Cx$) to get the value of concentration.
   Graphically: A standard curve is used to determine the concentration of any unknown sample. After obtaining the absorbance reading for known samples, each value is plotted (concentration on X-axis, and absorbance on Y-axis), and a linear line is obtained. The concentration of the unknown sample can then be determined by locating their absorbance on the line, dropping an imaginary line down from that point to the concentration axis as shown in Figure 4.4 (b).

Example 4.7: Determine the concentration of given unknown solution of $KMnO_4$ from the following data.

S. no.	Concentration (ppm)	Absorbance
1.	20	0.1
2.	40	0.2
3.	60	0.3
4.	80	0.4
5.	100	0.5
6.	120	0.6
7.	C (unknown solution)	0.45

Solution: The provided data when plotted gives the following graph:

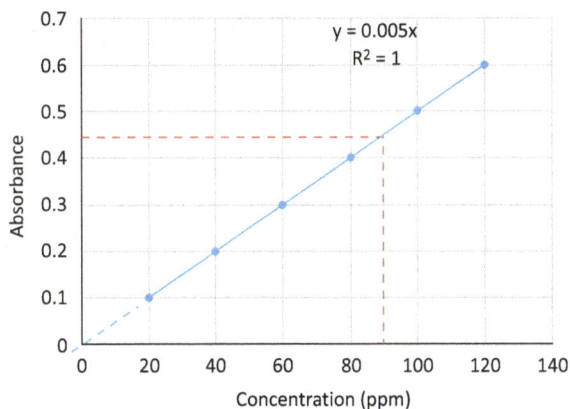

Graphically, the absorbance of unknown solution was marked on calibration curve and the concentration was determined. Concentration of unknown solution comes out to be 90 ppm.

Mathematically, molar absorptivity coefficient, $\varepsilon = 0.005$ (ppm cm)$^{-1}$.

Absorbance $(A) = \varepsilon Cx$ ($x = 1$ cm, width of cuvette)

Therefore, $C = A/\varepsilon = 0.45/0.005 = 90$ ppm

Both graphical and mathematical methods give same result.

4.7 Electronic transitions in organic compounds

The absorption of UV– or visible radiation results in the excitation of outer electrons in organic compounds, and three types of electrons are involved in these electronic transitions.

i. σ (sigma) – electrons – These electrons are involved in formation of sigma bonds.

ii. π (pi) – electrons – These electrons are responsible for double and triple bonds.

iii. n (nonbonding) - electrons – These are unshared or nonbonded electrons present on atoms such as nitrogen and oxygen.

Possible electronic transitions involving σ, π, and n electrons in organic compounds are:

i. $\sigma \rightarrow \sigma^*$ transitions – These types of transitions involve the excitation of an electron from a σ-bonding orbital to σ^* antibonding orbital. This type of transitions is found in saturated organic molecules without hetero atoms, for example, methane, which has only C–H bonds can only undergo $\sigma \rightarrow \sigma^*$ transitions. The energy required for such transitions is very high, and usually, they occur below 150 nm.

ii. $n \rightarrow \sigma^*$ transitions – These transitions are observed in saturated organic compounds containing atoms with lone pairs of electrons (nonbonding electrons) such as N, O, halogens, and are related to the promotion of an electron from a nonbonding orbital (n) to σ^* (antibonding) orbital. The saturated organic molecules with hetero atoms like N, O, S, and Cl show this type of transitions, which falls below 200 nm. Examples in this category include alcohols, amines, and ethers.

iii. $\pi \rightarrow \pi^*$ transitions – In this transition, an electron moves from a π-bonding orbital to an π^*-antibonding orbital. Organic compounds such as alkenes, alkynes, carbonyl compounds, nitriles, and aromatic compounds show this transition.

iv. $n \rightarrow \pi^*$ transitions – These types of transitions occur when an electron from nonbonding orbital to π^*-antibonding orbital. These transitions require lowest amount of energy and are observed in organic compounds containing double bonds involving hetero atom bearing lone pair of electrons such as carbonyl compounds. In aliphatic carbonyl compounds, $n \rightarrow \pi^*$ transition usually occurs in the wavelength range of 270–300 nm.

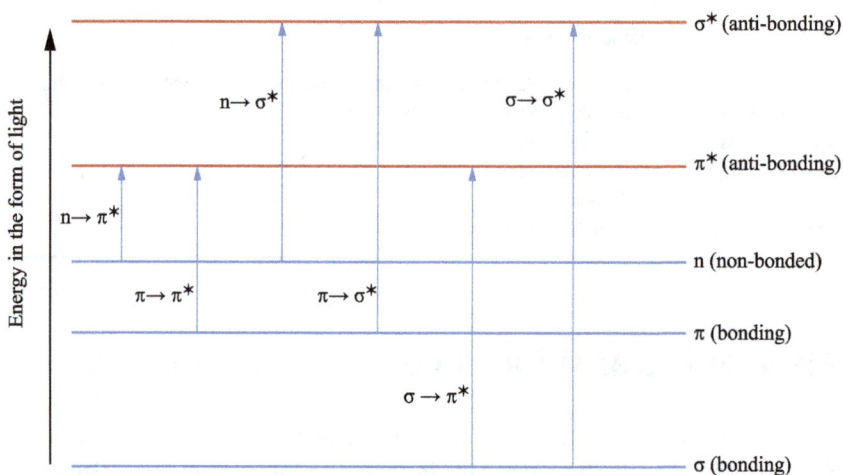

Figure 4.5: Various possible electronic transitions in organic compounds.

A simplified energy diagram depicting various electronic transitions is shown in Figure 4.5.

4.7.1 Chromophore

A chromophore is defined as a covalently unsaturated functional group which is responsible for absorption of radiations in the UV or visible region. For example, C=C, C=O, –N=N–, and NO_2. As stated earlier, compounds which absorb light of wavelength in the region 400–800 nm appeared colored to the human eye. The nature of chromophore is quite significant as it is responsible for the wavelength absorbed; thus a chromophore may or may not impart color to a compound. Chromophores like C=C have π electrons; thus they undergo $\pi \rightarrow \pi^*$ transitions whereas compounds such as C=O and

N=N contain both π and nonbonding (n)electrons; thus, they undergo $\pi \rightarrow \pi^*$ as well as $n \rightarrow \pi^*$ transitions along with $n \rightarrow \sigma^*$ transition.

4.7.2 Auxochromes

The presence of some functional groups with chromophores as substituent can change both the wavelength as well as intensity of the absorption maximum. Such substituents are known as auxochromes, and some important auxochromes are hydroxyl, amines, halogens, etc. We can thus say that a chromophore is responsible for absorption of radiations, and auxochrome enhances the absorbed radiation. For example, benzene shows λ_{max} at 256 nm, whereas aniline shows λ_{max} at 280 nm. Hence, $-NH_2$ acts as an auxochrome which resulted in the increased value of λ_{max}.

The effect of these substituents can be further categorized into four categories as shown in Figure 4.6:

Bathochromic shift (Redshift)

The presence of an auxochrome may shift the absorption maximum to a longer wavelength which is termed as bathochromic shift or redshift. For example, benzene shows λ_{max} at 256 nm and aniline shows λ_{max} at 280 nm, and hence, presence of $-NH_2$ group induced a bathochromic shift of 24 nm. Redshift can also be observed due to change in solvent polarity, and this effect is called solvatochromism. For example, λ_{max} of acetone is at 264.5 nm in water as compared to 279 nm in hexane.

Hypsochromic shift (Blueshift)

The shift of an absorption maximum to a shorter wavelength is termed as hypsochromic or blueshift. This is caused by the removal of conjugation or change in the solvent polarity. For example, λ_{max} of aniline is at 280 nm whereas anilinium ion (forms under acidic conditions) shows λ_{max} at 254 nm.

Hyperchromic shift

An increase in the intensity of the absorption maximum ε_{max} is called hyperchromic effect. The attachment of an auxochrome in an organic compound usually causes hyperchromic shift.

Hypochromic shift

A decrease in the intensity of the absorption maximum ε_{max} is called hypochromic shift. This type of effect is observed in compounds when introduction of a group distorts the geometry of the molecule.

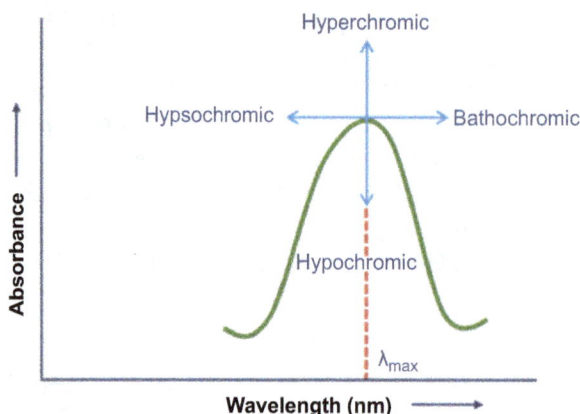

Figure 4.6: Shifts in absorbance and intensity.

Some of the examples of functional groups resulting in hyperchromic and hypochromic shifts are mentioned in Table 4.5.

Table 4.5: Examples of functional groups causing hyperchromic and hypochromic shifts.

Compound	Type of substituents	λ_{max}	Shift
Benzene	None	256 nm	–
Phenol	–OH	275 nm	Redshift
Phenoxide ion	–O$^-$	287 nm	Redshift
Aniline	–NH$_2$	280 nm	Redshift
Anilinium ion	–NH$_3{}^+$	254 nm	Blueshift

4.7.3 Effect of conjugation on electronic transitions

The presence of conjugation in an unsaturated system always leads to a bathochromic shift, and higher the extent of conjugation, the more will be bathochromic shift. This happens because an increase in conjugation results in an increase in the number of π electrons, which decreases the energy difference between highest occupied molecular orbitals and lowest unoccupied molecular orbitals (Figure 4.7). Thus, wavelength of light absorbed shifts to higher wavelength. The presence of conjugation not only results in a bathochromic shift but also enhances the intensity of absorption as displayed in Table 4.6. For example, ethylene absorbs at 175 nm, while 1,3-butadiene absorbs at 217 nm with two double bonds, whereas 1,3,5-hexatriene with three conjugated double bonds absorb at 258 nm.

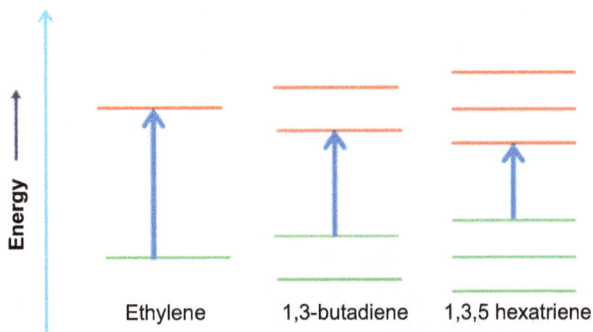

Figure 4.7: Comparison of the $\pi \rightarrow \pi^*$ energy gap on increasing conjugation.

Table 4.6: Effect of conjugation on electronic transitions.

Compounds	λ_{max} (nm)	ε (M^{-1} cm^{-1})
H_2C=CH_2	175	15,000
H_2C=====CH_2	217	21,000
H_2C=====CH_2	256	50,000
H_2C=====CH_2	290	85,000
H_2C=====CH_2	334	125,000
H_2C=====CH_2	364	138,000

4.8 Fluorescence and phosphorescence

In some compounds, electronic transitions in UV–visible region cause phenomena like fluorescence and phosphorescence, and in common language, these two processes are termed as luminescence. These are caused by emission of photons during the transition of excited electron from higher to lower energy states. Electrons have half spin (1/2). According to the Pauli's exclusion principle, electrons in ground state are paired (one with up spin and another with down spin), and thus, electronic ground states are singlet state. Excited states may be singlet or triplet, depending upon the number and spin of electrons in that energy level. S is spin quantum number. For two paired electrons with opposite spin, $S = (1/2) - (1/2) = 0$

Therefore,

Spin multiplicity $= 2S + 1 = 2 \times 0 + 1 = 1$, it is a singlet state.

For $S = 1/2 + 1/2 = 1$

Spin multiplicity $= 2 \times 1 + 1 = 3$, it is a triplet state.

In singlet state, electrons are present in same orbital with opposite spins, and in triplet state, electrons are present in different orbitals with same spin (either both up or both down). The repulsion between electrons in triplet state is less, and hence triplet state is more stable than singlet state. Knowing the stability of these two states is very important to understand the phenomena of fluorescence and phosphorescence.

These phenomena were explained by Jablonski through energy level diagram as shown in Figure 4.8. Electrons in the ground states (S_0) are excited to higher energy states after absorbing energy from incident UV–visible radiations. Depending upon the energy of incident radiations, electrons are promoted to different higher electronic states (S_1, S_2, S_3, . . .). Ground state is a singlet state because all electrons are paired in ground state. After going to a higher energy state, electron can come back to ground state through different ways. If it goes to some lower excited state from higher excited state and releases a photon of energy equivalent to the energy difference between the two states, this phenomenon is called *internal conversion*. If electron changes its spin in the same excited state, it is said to go to corresponding triplet state and the phenomenon is called *intersystem crossing*. When electrons release photon and go directly from singlet excited state to ground state, the phenomenon is called as *fluorescence*. Because excited singlet states are short-lived, fluorescence does not last for longer. As soon as the source of incident radiations stopped, fluorescence ceases. Triplet states are more stable than singlet ones hence, electron stay in triplet state for longer. When electron comes back to ground state from triplet state via radiative transition, this process is called *phosphorescence*. Phosphorescence lasts for longer and the substance continues to glow even after the removal of the source of incident radiations. The time period of various processes is given in Table 4.7, and the various electronic transitions are illustrated in Figure 4.8, through Jablonski's diagram.

Table 4.7: Various electronic transitions mentioned in Jablonski's diagram.

Phenomenon	Transition between energy states	Nature of transition	Time
Excitation	$S_0 \rightarrow S_n$ ($n = 2, 3, 4 \ldots$)	Radiative	10^{-15} s
Internal conversion (IC)	$S_n \rightarrow S_m$ ($m < n$)	Radiative/nonradiative	10^{-14}–10^{-11} s
Inter system crossing (ISC)	$S_n \rightarrow T_n$	Nonradiative	–
Fluorescence	$S_n \rightarrow S_0$ ($n = 2, 3, 4 \ldots$)	Radiative	10^{-9}–10^{-7} s
Phosphorescence	$T_n \rightarrow S_0$	Radiative	10^{-3}–10^{2} s
Quenching	$S_n \rightarrow S_0,\ T_n \rightarrow S_0$	Nonradiative	–

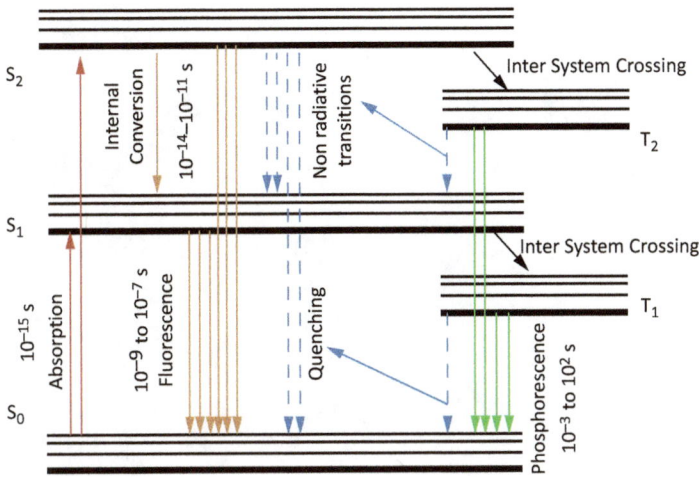

Figure 4.8: Jablonski's diagram.

References

[1] Aggarwal, A., and Gupta, A. Effect of electromagnetic radiations on humans: A study. In: IEEE technology students' symposium. IEEE, 2011, pp. 75–80.

[2] Itoh, T. Fluorescence and phosphorescence from higher excited states of organic molecules. Chemical Reviews. 2012; 112(8): 4541–4568.

[3] Lohman, F.H. The mathematical combination of Lambert's law and Beer's law. Journal of Chemical Education. 1955; 32(3): 155.

[4] McClure, D.S. Triplet-singlet transitions in organic molecules. lifetime measurements of the triplet state. The Journal of Chemical Physics. 1949; 17(10): 905–913.

[5] Mulliken, R.S. Intensities of electronic transitions in molecular spectra III. Organic molecules with double bonds. Conjugated dienes. The Journal of Chemical Physics. 1939; 7(2): 121–135.

[6]   Passos, M.L., and Saraiva, M.L.M. Detection in UV-visible spectrophotometry: Detectors, detection systems, and detection strategies. Measurement. 2019; 135: 896–904.

[7]   Swinehart, D.F. The Beer-Lambert law. Journal of Chemical Education. 1962; 39(7): 333.

[8]   Valeur, B., and Berberan-Santos, M.N. A brief history of fluorescence and phosphorescence before the emergence of quantum theory. Journal of Chemical Education. 2011; 88(6): 731–738.

Chanchal Singh
Chapter 5
Infrared spectroscopy

Abstract: Infrared (IR) spectroscopy is used for the identification of functional groups and hence plays a major role in elucidating the structure of an unknown compound. This branch of spectroscopy is widely used in research, industries, as well as in forensic and environmental analysis, as it is a simple, quick, and reliable method of analysis. All the functional groups give a band (strong, medium, weak) in IR spectra. IR spectroscopy deals with the IR region of electromagnetic spectrum. Although, IR spectra are often measured in terms of wave number from 12,500 to 400 cm^{-1}; however, organic chemists are interested in analyzing the region, referred to as middle IR which ranges from 4,000 to 400 cm^{-1}. The wave number is directly proportional to energy; the higher the energy of absorbed radiation, higher is the value of wave number.

Wave number can be defined as inverse of wavelength, and its unit is cm^{-1}:

$$\bar{v}\,(\text{cm}^{-1}) \;=\; \frac{1}{\lambda(\text{cm})}$$

In this chapter, we have discussed the different regions of infrared (IR) spectroscopy, basic principle governing IR spectroscopy, different modes of vibrations, and factors affecting IR frequency. The IR spectra have been displayed in the chapter at relevant place which makes it easier for the students to understand the topic, and they will learn the identification of different functional groups in a molecule.

5.1 Introduction

Spectroscopy is a versatile and significant method of understanding and elucidating the structure of molecules by measuring the interaction of light and matter. There are numerous benefits of spectroscopy, and it is employed in virtually every area of scientific research ranging from environmental analysis to biomedical sciences. Infrared (IR) spectroscopy is one of the most simple, reliable, and important analytical tools that is widely used in research and industry and provides qualitative information in a fast and cost-effective manner.

Any spectrum of molecule recorded by different analytical technique has its own unique pattern which corresponds to only that particular molecule. IR spectroscopy plays an important role in deducing the structure of an unknown compound, and it usually deals with the identification of functional groups present in a molecule. All the functional group gives characteristics bands in IR spectra, and the spectrum recorded is the

https://doi.org/10.1515/9783110794816-005

reflection of molecular structure under study which includes unique properties such as band position, band width, and band intensity.

IR spectroscopy deals with the IR region of electromagnetic spectrum. It is based on the study of absorption of IR radiations, and these radiations induce transition between vibrational energy levels (v_1, v_2, v_3, . . .) in the molecule. When the supplied energy is equal to difference between vibrational energy levels, absorption of IR radiation takes place, and a peak is observed. Almost any organic compound, having covalent bonds absorb electromagnetic radiations in the IR region of the electromagnetic spectrum, and it includes radiations that lies in the wavelength range from 0.8 to 1,000 μm (1 μm=10^{-6} m).

The energy of a photon is governed by the following equation:

$$E = hv \qquad (5.1)$$

and

$$v = \frac{c}{\lambda} \qquad (5.2)$$

where h is the Planck's constant, v is the frequency of the electromagnetic radiation, λ is the wavelength, and c is the speed of light. The radiations in the IR region are specified in terms of wave number $\bar{v}$ rather than the wavelength (μm). Wave number can be defined as inverse of wavelength and is expressed as reciprocal centimeters (cm^{-1}). It is directly proportional to energy, and higher the energy of absorbed radiation, the higher is the value of wave number. In IR spectrum, the transmitted frequencies are plotted versus intensity of the transmission. Frequency appears on the x-axis in units of inverse centimeter (wave number), and intensity is plotted on the y-axis in percentage units. Some of the relationships given further are worth remembering for simple conversions:

$$\bar{v}(cm^{-1}) = \frac{1}{\lambda(cm)} \qquad v(Hz) = \bar{v}c = \frac{c\left(\frac{cm}{s}\right)}{\lambda(cm)}$$

The IR region of the electromagnetic spectrum is usually divided into three regions as shown in Figure 5.1.

i. Near-IR: 12,500–4,000 cm^{-1} (0.8–2.5 μm)
ii. Mid-IR: 4,000–400 cm^{-1} (2.5–25 μm)
iii. Far-IR: 400–10 cm^{-1} (25–1,000 μm)

The mid-IR region deals with the fundamental vibrations of the molecule; hence IR is also known as vibrational spectroscopy. This region is further divided into functional group region (4,000–1,500 cm^{-1}) and fingerprint region (1,500–400 cm^{-1}).

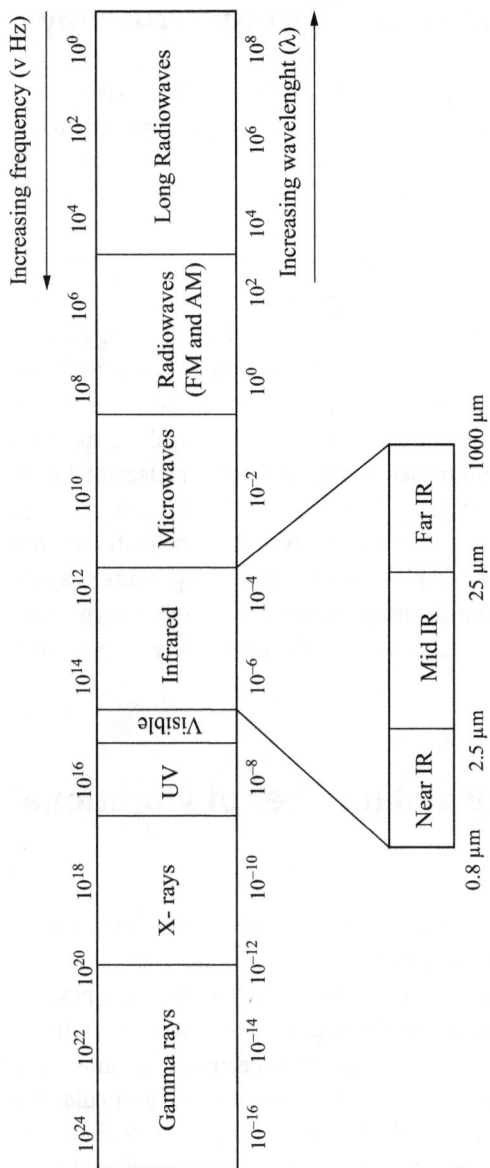

Figure 5.1: Electromagnetic spectrum showing IR regions. The most explored region is mid-IR (4,000–400 cm^{-1}).

5.2 Fourier-Transform infrared (FTIR) spectroscopy

IR spectroscopy deals with all types of samples such as liquids, solutions, pastes, powders, and gases, which make IR an important analytical technique. The instrument that records the absorption spectrum for a molecule is called an IR spectrometer or more precisely, a spectrophotometer. The IR spectrometers have been commercially available since the 1940s, and the two types of spectrometers which are commonly used are dispersive and Fourier transform (FT) instruments. The original IR instruments were of dispersive type. These instruments separated the individual frequencies of energy emitted from the IR source. The introduction of FT methodology greatly improved the way we study IR spectroscopy. This instrument employs an interferometer and exploits the well-established mathematical process of FT. Some of the IR radiation is absorbed by the sample and some of it is transmitted. The resulting spectrum represents the molecular absorption and transmission, creating a molecular fingerprint of the sample. It has dramatically improved the quality of IR spectra and minimized the time required to obtain data. It can identify unknown materials and determine the quality or consistency of a sample and amount of components in a mixture. FTIR methodology provides numerous advantages over dispersive-type spectrometer such as high data collection speed, great spectral quality, easy maintenance, and use.

5.3 Vibrational frequency and number of vibrational modes

The frequencies in the IR region of the electromagnetic spectrum correspond to energy required to cause vibrational motion in a molecule. A molecular vibration occurs when atoms in a molecule are in periodic motion while the molecule as a whole has constant translational and rotational motion. The frequency of the periodic motion is known as a vibrational frequency. For the excitation of molecules from one vibrational level to another, the molecule must absorb IR radiations of a particular frequency. The molecular vibrations lead to a change in the dipole moment of the molecule, resulting in the absorption bands in the IR region. Only those molecules which can undergo change in dipole moment (on vibration) are IR active, and the molecules having zero-dipole moment are IR inactive, that is, they do not show any band in IR spectra. Diatomic molecules such as hydrogen (H_2), oxygen (O_2), and chlorine (Cl_2) are IR inactive whereas carbon monoxide and iodine chloride absorb IR light and are IR active. A large change in dipole moment usually results in a strong band.

The degree of freedom (DOF) for a molecule with N atoms is defined as the number of coordinates required to identify the location of all atoms in a molecule; if a molecule

5.4 Vibrational frequency

The value of vibrational stretching frequency of a bond can be calculated by Hooke's law. According to Hooke's law, the vibrational frequency of a bond is directly proportional to the square root of the force constant of the bond and inversely proportional to the square root of the reduced mass (μ) of the system and can be written as

$$\bar{v} = \frac{1}{2\pi c}\sqrt{\frac{K}{\mu}}$$

where $\bar{v}$ is the vibrational frequency in cm^{-1} (wave number), c is velocity of light in cm/s, $\mu = \frac{m_1 m_2}{m_1 + m_2}$, m_1 is the mass of atom 1 in g, and mass of atom 2 in g, and K is the force constant in dyne/cm (g/s^2).

The force constant is related to the strength of the bond, and the stronger the bond, the greater will be the frequency of absorption. The force constants for triple bonds are three times more than those of single bonds whereas the force constants for double bonds are two times those of single bonds. The approximate value of the absorption band in the IR spectrum can be calculated by assuming K for single bond as 5×10^5 g/s^2 and 10×10^5 g/s^2 and 15×10^5 g/s^2 for double and triple bonds, respectively. Vibrational frequency or wave number also depends on the masses of the atoms and state of hybridization. Resonance also affects the length and strength of a bond and hence the frequency of absorption. Different groups are absorbed in different regions of IR spectra and not every group gives same kind of band. The intensity of band can be divided as strong (s), medium (m), and weak (w).

i. Bond strength: The vibrational frequency ($\bar{v}$) is directly proportional to the strength of the bond.

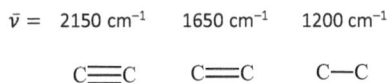

$$\bar{v} = \quad 2150 \text{ cm}^{-1} \quad 1650 \text{ cm}^{-1} \quad 1200 \text{ cm}^{-1}$$

$$\text{C}\equiv\text{C} \qquad \text{C}=\text{C} \qquad \text{C}-\text{C}$$

Triple bonds are stronger than double and single bonds between the same atoms and have higher vibrational frequencies.

ii. Mass of the atoms: The vibrational frequency is inversely proportional to the mass of an atom. As the mass of an atom attached to carbon increases, the reduced mass also increases, and hence vibrational frequency (wave number) decreases.

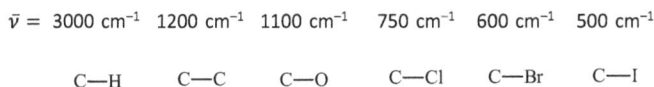

$$\bar{v} = 3000 \text{ cm}^{-1} \quad 1200 \text{ cm}^{-1} \quad 1100 \text{ cm}^{-1} \quad 750 \text{ cm}^{-1} \quad 600 \text{ cm}^{-1} \quad 500 \text{ cm}^{-1}$$

$$\text{C}-\text{H} \qquad \text{C}-\text{C} \qquad \text{C}-\text{O} \qquad \text{C}-\text{Cl} \qquad \text{C}-\text{Br} \qquad \text{C}-\text{I}$$

iii. State of hybridization: Bond strength is also affected by the hybridization, and bonds are stronger in the order sp > sp² > sp³. More the s character, more is the force constant thus higher the value of vibrational frequency.

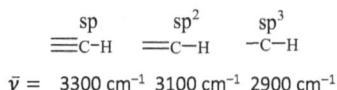

$$
\begin{array}{ccc}
sp & sp^2 & sp^3 \\
\equiv\!C\text{–}H & =\!C\text{–}H & -C\text{–}H
\end{array}
$$

$\bar{v}$ = 3300 cm⁻¹ 3100 cm⁻¹ 2900 cm⁻¹

iv. Resonance: Resonance affects the strength of a bond and hence the vibrational frequency. For example, a normal ketone (C=O) absorbs at 1,715 cm⁻¹, whereas the presence of C=C, in conjugation with C=O, lowers down the vibrational frequency to 1,680 cm⁻¹. The reason is resonance increases the C=O bond length and induces the more single-bond character, as depicted:

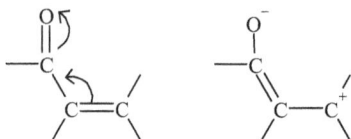

5.5 Instrumentation

IR spectrometer (or spectrophotometer) measures the vibrations of atoms and produces an IR spectrum. A schematic diagram of an IR spectrometer is shown in Figure 5.4. The main parts of IR spectrometer are:

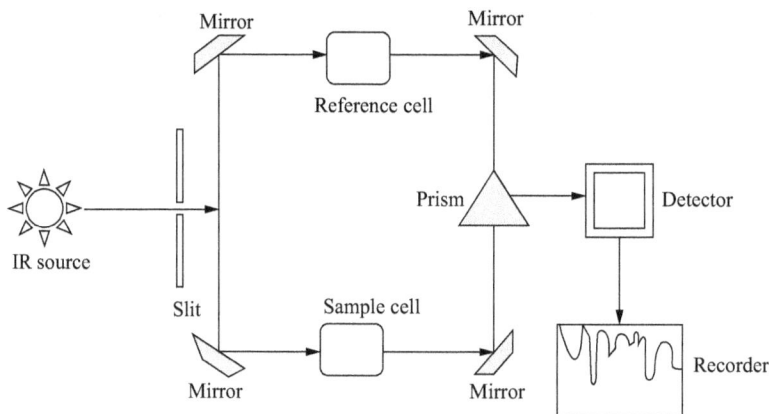

Figure 5.4: Schematic diagram of an IR spectrophotometer.

1. IR source
2. Sample cells and reference cells
3. Monochromators (prism)
4. Detectors
5. Recorder

i. IR source

For IR radiation, the wide wavelength range is used which is constant over long periods of time. Different types of IR sources are available like the Nernst glower, mercury arc, and incandescent wire source. The most commonly used is the Nernst glower. The heating temperature can reach up to 2,000 °C. It is composed of a mixture of rare earth oxides such as zirconium oxide (ZrO_2), yttrium oxide (Y_2O_3), and thorium. It consists of a cylindrical hollow rod or tube having a diameter of 1–2 mm and length of 30 mm.

ii. Sample cells and reference cells

IR spectra may be obtained for gases, liquids, or solids (neat or in solution). Salts like sodium chloride (NaCl), potassium chloride (KCl) and potassium bromide (KBr) are used, as sample must be transparent to the IR radiation.

iii. Monochromators

Prism is used as a dispersive element, and sodium chloride prism is most commonly used. Monochromators are subjected to mechanical and thermal instability or water solubility. Gratings give better resolutions than prisms. Gratings are nothing but rulings made on some materials like glass, quartz, or alkyl halides depending upon the instrument. The rays which are incident upon the gratings get reinforced with the reflected rays. Grating can be used over considerable wavelength range.

iv. Detectors

The detector used is thermocouple which can be divided into three categories:
a) Thermal detectors: Their responses depend upon the heating effect of radiation.
b) Pyroelectric detectors: Pyroelectric effect depends on the rate of change of the detector temperature rather than on the temperature itself.
c) Photoconducting detectors: Thermocouples consist of a hot junction and a cold junction. The difference in intensity of IR radiation falling on both junctions give rise to small electrical potential difference between the two ends. The developed alternating current is detected on recorder. In IR spectroscopy, the cold junction is carefully partitioned in a protective box and kept at a constant temperature.

v. Recorder

It records the IR spectrum as a plot of frequency of absorbed radiation and intensity of absorption in terms of transmittance. If I_s is intensity of sample radiation and I_r is intensity of reference radiation, then percent transmittance (%T) will be:

$$\% \ T = \frac{I_s}{I_r} \times 100$$

5.5.1 Preparation of samples

IR spectra of compounds can be obtained for gases, liquids, or solids (neat or in solution). Material containing sample must be transparent to the IR radiation. So, the salts alkali halides, such as sodium chloride (NaCl) and potassium bromide (KBr), are generally used.

5.5.1.1 Solid samples

There are different methods used for determining the IR spectra for solid samples. Some of them are as follows:

(a) *Solid run in solution*

In this technique, solid sample under study is dissolved in a nonaqueous solvent. Choice of solvent should be properly considered. There should not be any chemical interaction between sample and solvent, and the solvent should not absorbed in the range to be studied. The most commonly used solvents in IR spectroscopy are carbon disulfide (CS_2), carbon tetrachloride (CCl_4), and chloroform ($CHCl_3$). A drop of solution is deposited on the surface of alkali metal disc, and solvent is dried leaving a thin film of the solute.

(b) *Solid film technique*

If the solid is amorphous in nature, then the sample is deposited on the surface of a KBr or NaCl cell by evaporation of a solution of the solid. Ensured that the film is not too thick to pass the radiation.

a. Pressed pellet technique: In this technique, a small amount of finely ground solid sample is mixed with potassium bromide (in weight ratio 1:9) and compressed using a hydraulic press to form a thin transparent pellet (1–2 mm thick and 1 cm in diameter). These pellets are transparent to IR radiation and it is used for analysis. Precaution should be taken during sample preparation. The conditions must be anhydrous during whole process.

b. Mull technique: In this technique, the sample is grinded with Nujol (liquid paraffin) in a marble or agate mortar, with a pestle to make a thick paste. A thin film is applied onto the salt plates.

5.5.1.2 Liquid samples

For liquid samples, a drop of a liquid organic compound is placed between a pair of polished sodium chloride or potassium bromide plates. These plates are known as salt plates, and a pair of plates is inserted into a holder placed in the spectrometer. A thin liquid film forms between the plates, when squeezed gently. A spectrum obtained by this method is known as a neat spectrum as no solvent is used. As salt plates are water soluble and can break easily, organic compounds analyzed through this method should be free of water.

5.6 Origin of IR spectroscopy

When molecule absorbs IR radiation, transition between vibrational and rotational energy levels occurs. IR spectrum is the fingerprint of a molecule and used for its identification. Only IR active molecule gives rise to different mode of vibrations, these are mainly classified as (1) stretching vibrations and (2) bending vibrations (Figure 5.5).

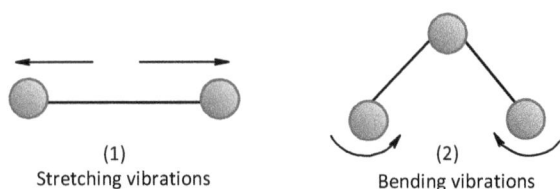

(1)
Stretching vibrations

(2)
Bending vibrations

Figure 5.5: Stretching and bending vibrations.

5.6.1 Stretching vibrations

The interatomic distance between two atoms changes continuously; it either decreases or increases on same bond axis, such that bond length changes without change in bond angle in regular interval. It requires more energy, so bands associated with stretching vibration appear at higher wave number (shorter wavelength). It is of two types: (a) symmetrical stretching; (b) asymmetrical stretching as shown in Figure 5.6.

(a) *Symmetrical stretching:* It occurs at low frequency than asymmetrical stretching; in this, both atoms move in the same direction with respect to central atom without a change in the bond angle or bond axis.

(b) *Asymmetrical stretching:* It occurs at high frequency than symmetrical stretching; in this, both atoms move in opposite direction with respect to central atom. One bond length is increased and other is decreased.

Figure 5.6: Two modes of stretching vibrations: (a) symmetrical and (b) asymmetrical stretching.

5.6.2 Bending vibrations

Bending vibrations involve movement of atoms which are attached to a common central atom. These are characterized by change in the angle between two bonds, also known as deformation vibration. It requires less energy, so absorption bands associated with these vibrations appear at lower wave number (longer wavelength). It occurs at considerable low frequency than stretching vibrations. It is of four types: (a) scissoring, (b) rocking, (c) wagging, and (d) twisting as shown in Figure 5.7.

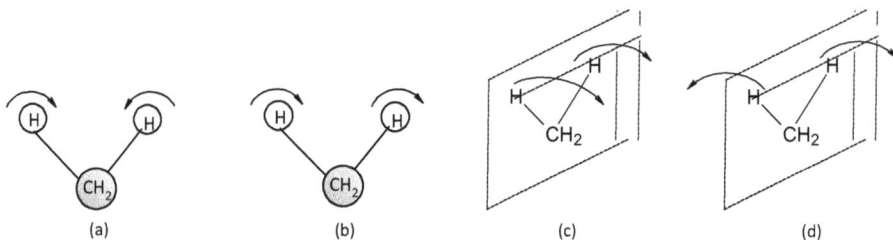

Figure 5.7: Different bending vibrations of C–H bond.

i.   *Scissoring* – It is in-plane bending, both atoms move in opposite direction with respect to central atom. The change in bond angle takes place with respect to bond axis.
ii. *Rocking* – It is in-plane bending, both atoms move in same direction with respect to central atom. The change in bond angle takes place with respect to bond axis.
iii. *Wagging* – It is out-of-plane bending, both atoms simultaneously move out, and move in the plane with respect to the central atom.
iv. *Twisting* – It is out-of-plane bending, one atom moves out of the plane and other atom moves below the plane with respect to the central atom.

5.7 Fingerprint region

The IR spectra ranges from 400 to 4,000 cm^{-1} which is divided into functional group region (1,500–4,000 cm^{-1}) and fingerprint region (400–1,500 cm^{-1}) as shown in Figure 5.8. Fingerprint region mainly includes bending and rocking vibrations unlike functional group region which mainly contains stretching vibrations. The importance of finger-print region is that every compound produces a different pattern of bands which is unique to molecular structure of that compound. This was confirmed by the IR spectra of 2-bromopentane and 3-bromopentane which are structural isomers with same molecular formula, same number of carbon, hydrogen, and halogens atoms. The only difference is in position of bromine atom. The C–H stretching absorption occurs at 2,950 cm^{-1} in both spectra but the fingerprint region has a clear difference in bending and rocking vibration frequencies (Figure 5.9(a) and (b)).

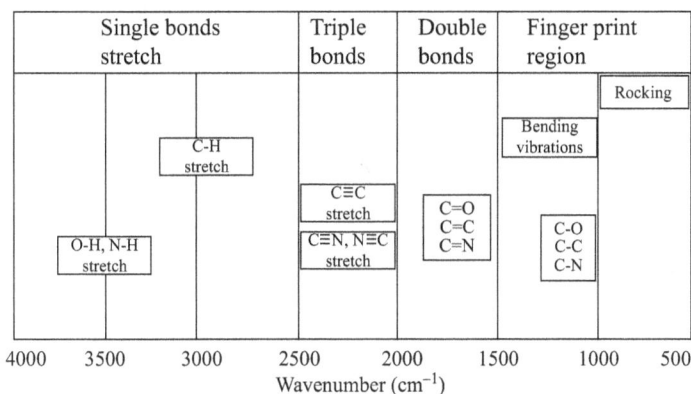

Figure 5.8: Vibrational frequency range of different functional groups in the IR spectrum.

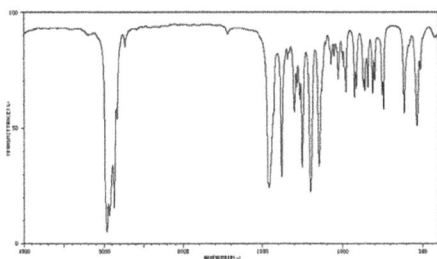

Figure 5.9a: IR spectra of 2-bromopentane.

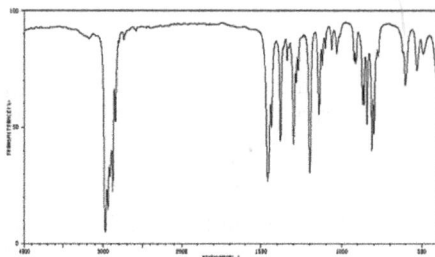

Figure 5.9b: IR spectra of 3-bromopentane.

5.8 Factors affecting IR frequency

There are various factors which affects the absorption peaks in the IR region such as
i.   Effect of bond order
ii.   Vibrational coupling/coupled vibrations
iii. Hydrogen bonding
iv.   Fermi resonance
v.   Electronic effects

5.8.1 Effect of bond order

Bond order is directly proportional to bond strength; the higher the bond order, higher is bond strength, thus higher is the band frequency. C≡C (triple bond) is stronger than C=C (double bond) which is stronger than C–C (single bond); thus, the stretching frequencies are in the order C≡C > C=C > C–C (Table 5.1). Similarly, a C=O bond is stronger than C–O, and C=N is stronger than C–N bond, and therefore, the vibrational stretching frequency of C=O is higher than C–O and C–N. Similarly, a C≡N has a higher stretching frequency than C=N bond which in turn has a higher stretching frequency than a C–N bond.

5.8.2 Vibrational coupling/coupled vibrations

Any isolated C–H bond exhibit has only one stretching frequency, but the stretching vibrations of C–H bond in –CH_2 (methylene) group combine to generate couple vibrations of different frequencies. Theoretically, in case of methyl, amines, and carboxylic anhydrides, a single value of stretching frequency for C–H, N–H, and C=O, respectively, should be obtained but experimentally, more than one value for each group is observed; one for symmetrical stretching and the other for asymmetrical stretching.

Asymmetric stretching occurs at higher frequency than the symmetric stretching. Stretching vibrations for each group (C–H, N–H, C=O) combine together to produce two stretching vibrations, and these are known as coupled vibrations. Both symmetric and asymmetric vibrations occur at different frequencies as illustrated in Figure 5.10. Strong coupling between stretching vibrations occurs only when the two vibrations have a common atom while for bending vibrations common bond should be present between the vibrating groups. If groups are separated by two or more bonds, no interaction occurs. For interaction to occur, the vibrations must be of same symmetry species.

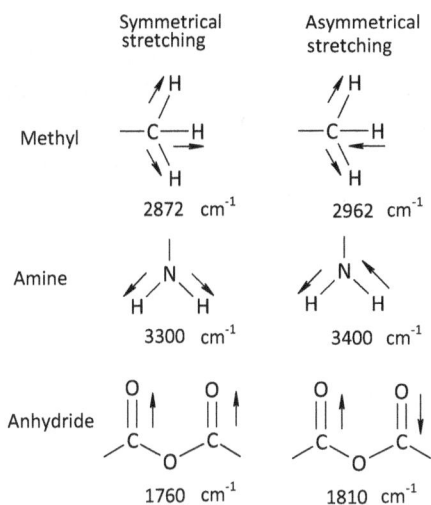

Symmetrical stretching | Asymmetrical stretching

Methyl

$2872 \ cm^{-1}$ | $2962 \ cm^{-1}$

Amine

$3300 \ cm^{-1}$ | $3400 \ cm^{-1}$

Anhydride

$1760 \ cm^{-1}$ | $1810 \ cm^{-1}$

Figure 5.10: Symmetrical and asymmetrical stretching of methyl, amines, and anhydrides.

5.8.3 Hydrogen bonding

Hydrogen bonding alters the vibrational stretching frequency of a bond. Stronger the hydrogen bonding, lower is the vibration frequency, and broader will be absorption band. The electronegative atoms F (fluorine), O (oxygen), and N (nitrogen) in the system form hydrogen bond with the hydrogen atom. The electronegative atom withdraws the electron density from hydrogen atom making it partially positive which lengthens the bond and decreases the bond strength, hence lowering the stretching frequency.

The O–H and N–H stretching frequencies in hydroxyl compounds and amines are affected by hydrogen bonding but frequency shift for amines is less than that for hydroxyl compounds as nitrogen is less electronegative than oxygen. So, hydrogen bonding in amines is weaker in comparison with that of hydroxyl compounds. When there is no hydrogen bonding, a "free" (nonhydrogen-bonded) O–H stretching band appears at about 3,600 cm^{-1}. This is obtained when the solution is very dilute or spectra are recorded in the vapor phase. Generally, a broad O–H stretching vibration in the range from 3,400 to 3,300 cm^{-1} is obtained for alcohols and phenols due to intermolecular hydrogen bonding. Free primary amines absorb near 3,400 cm^{-1} and 3,500 cm^{-1} due to symmetrical and asymmetrical N–H stretching modes, respectively. Hydrogen bonding in amines lowers the stretching frequency and band appears in the range of 3,250–3,330 cm^{-1} (symmetrical stretch) and 3,330–3,400 cm^{-1} (asymmetrical stretch) due to intermolecular hydrogen bonding.

Dilution has no significant effect on intramolecular hydrogen bonding; thus no significant change in absorption band is observed, whereas on dilution, the broad intermolecular hydrogen-bonded band is reduced considerably, and a free O–H stretching absorption is observed. Intermolecular hydrogen bonding weakens the O–H bond, thus shifting the absorption band to lower frequency (Figure 5.11).

The strength of hydrogen bonding is also affected by ring strain, molecular geometry, relative acidity, and basicity of proton donor and acceptor groups.

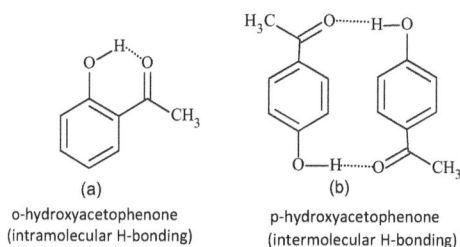

(a)
o-hydroxyacetophenone
(intramolecular H-bonding)

(b)
p-hydroxyacetophenone
(intermolecular H-bonding)

Figure 5.11: Example of intra- and intermolecular hydrogen bonding.

5.8.4 Fermi resonance

When a fundamental vibration couples with an overtone or a combination band, the coupled vibration is called a Fermi resonance, and this phenomenon was first explained by the Italian physicist Enrico Fermi. Coupling of two vibrational bands nearly of the same energy and symmetry occurs which result in a shift in frequency and change in intensity in the spectrum. As a result, two strong bands are observed in the spectrum instead of the expected strong and weak bands. Fermi resonance is shown by aldehydes in which C–H stretching absorption usually appears as a doublet near 2,850 cm^{-1} and 2,750 cm^{-1} due to the interaction between C–H stretching (fundamental) and the overtone of C–H deformation (bending).

5.9 Absorption frequency of different organic compounds

It has been discussed in the earlier sections that only IR active molecules display band in IR spectrum. The intensity of the bands depend on the magnitude of the dipole moment associated with the bond under study. Strong polar bonds such as carbonyl groups (C=O) produce strong bands, and medium polarity bonds and asymmetric bonds produce medium bands; while weakly polar bond and symmetric bonds produce weak or nonobservable bands. The IR absorption frequencies of some common functional groups are shown in Table 5.1.

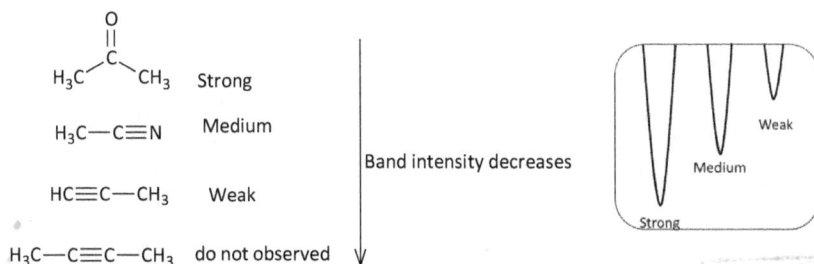

H_3C—C(=O)—CH_3 Strong

H_3C—C≡N   Medium

HC≡C—CH_3 Weak

H_3C—C≡C—CH_3 do not observed

Band intensity decreases

Weak

Medium

Strong

Table 5.1: IR absorption frequencies of common functional groups.

Group	Type	Frequency (cm^{-1})	Intensity
Alkanes	C–H stretch	2,850–3,000	m-s
	CH_3 bend	1,375 and 1450	m-s
	CH_2-bend	1,465–1,485	m
	C-C stretch	800–1,200	w
Alkenes	C=C stretch	1,620–1,680	m
	C=C–H stretch	3,000–3,100	m
	C=C–H bending	650–1,000	s
Alkynes	C≡C stretch	2,190–2,260	m-w
	C≡C–H stretch	3,300–3,310	s
	C≡C–H bending	615–650	s
Aromatic	C–C stretch	1,450–1,620	m-w
	C–H stretch	3,000–3,100	m
	C–H bend	1,000–1,300	w
C–O	Alcohol	1,000–1,260	m-s
	Ether	1,300–1,000	m
	Epoxide	1,280–1,230	m

Table 5.1 (continued)

Group	Type	Frequency (cm^{-1})	Intensity
C=O	Aldehyde (H–C=O) (aliphatic)	1,720–1,740	s
	Aldehyde (aromatic)	1,695–1,715	s
	Ketone (R–C=O) (aliphatic)	1,705–1,725	s
	Ketone (aromatic)	1,680–1,700	s
	Carboxylic acid (COOH) (aliphatic)	1,700–1,725	s
	Carboxylic acid (aromatic)	1,690–1,700	s
	Ester (COOR) (aliphatic)	1,735–1,750	s
	Ester (aromatic)	1,715–1,730	s
	Lactones	1,735–1,770	s
	Acid halide (COCl)	1,750–1,815	s
	Acid anhydride (CO–O–CO)	1,800–1,850	s
O–H	Alcohol (free) O–H stretch	3,590–3,650	s
	Alcohol (Intermolecular H-bond) stretch	3,200–3,400	s
	Alcohol (Intramolecular H-bond) stretch	2,500–3,200	s
	O–H bending	1,440–1,220	s
	Carboxylic acid O–H stretch	2,400–3,400	m
Amides	N–H stretch	3,100–3,500	m
	N–H bend	1,550–1,640	m
	C=O stretch	1,630–1,690	s
Amines (primary and secondary)	N–H stretch	3,300–3,500	m
	N–H bend	1,550–1,640	m
	C–N stretch	1,000–1,350	m-s
Nitrile and isonitrile	C≡N stretch	2,200–2,260	m
	C≡C stretch	2,070–2,220	m
Imine	C=NH	1,640–1,690	v
Thiocyanates and isothiocyanates	S–C≡N stretch	2,140–2,175	s
	N=C=S stretch	2,000–2,140	s
Azide	N≡N stretch	2,120–2,160	s
Nitro	NO$_2$ (aliphatic) stretch	1,370–1,570	s
	NO$_2$ (aromatic) stretch	1,290–1,550	s
Sulfur compounds	SH stretch	2,550–2,600	w
	S=O stretch	1,040–1,060	s
	S–O stretch	750–1,000	s
Phosphorus compounds	P–H stretch	2,440–2,350	s
	P=O stretch	1,250–1,300	s
	P–O stretch	725–845	m

Table 5.1 (continued)

Group	Type	Frequency (cm^{-1})	Intensity
Halogens	C–I	Around 500	s
	C–Br	500–750	s
	C–Cl	600–800	s
	C–F	1,000–1,400	s
	CH$_3$-X (X is any halogen)	Around 3,000	s
Amino acids	C=O stretch	1,730–1,755	s
	N–H stretch	3,030–3,130	s
	O–H stretch	2,380–3,333	s
	NH$_3$+ stretch	3,200–3,400	s
	NH$_3$+ bend	1,480–1,610	s

5.10 How to read IR spectra

IR spectra are mainly used for functional group analysis, and different functional groups give rise to medium (m) and strong (s) bands in IR spectroscopy which can be easily identified with the help of Table 5.1. The functional groups such as carbonyl (C=O), hydroxyl (O–H), and amines (N–H) give rise to strong absorption bands. The carbonyl (C=O) group is present in aldehydes, ketones, acids, amides, esters, acid chloride, and anhydrides and this group absorb strongly in the range from 1,850 to 1,650 cm^{-1}. The carbonyl group in these compounds can be differentiated on the basis of their band values. A broadband near 3,400 cm^{-1} is observed for hydrogen-bonded O–H group, whereas a sharp peak at about 3,600 cm^{-1} is observed for free nonhydrogen bonded O–H group. Amines also absorbed near 3,400 cm^{-1} but they can be differentiated easily from O–H as –NH$_2$ bands generally split into doubles. Aromatic region has distinguished band pattern in IR spectra near 1,450–1,650 cm^{-1}. NO$_2$ group gives two strong bands near 1,550–1,570 cm^{-1}.

5.10.1 Hydrocarbons

5.10.1.1 Alkanes

Alkanes have sp^3 carbons and show only few absorption bands in the IR spectrum. They show four or more C–H stretching peaks at around 3,000 cm^{-1}. The bending peaks for CH$_2$ and CH$_3$ can be observed in the range of 1,475–1,365 cm^{-1} (Figure 5.14). Generally, CH$_3$ and CH$_2$ bending vibrations occur as peak of medium intensity while C–H stretching vibrations are absorbed strongly giving sharp band. Since alkanes are inert molecules there is no effect on absorption frequency in case of change in chemical environment.

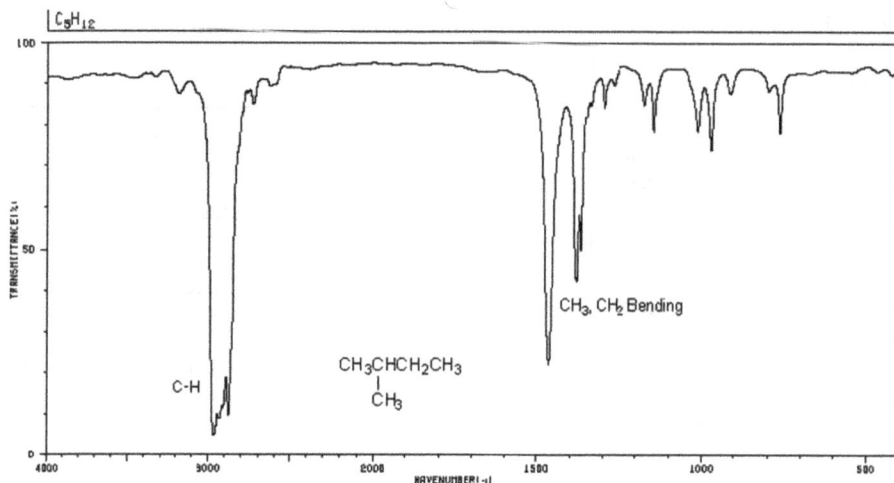

Figure 5.14: IR spectrum of isopentane.

5.10.1.2 Alkenes

Alkenes show the C=C stretch in the range of 1,620–1,680 cm^{-1} and C–H stretching peak for the sp^2 carbon at around 3,000–3,100 cm^{-1} (Figure 5.15). As mentioned, change in dipole moment is necessary for molecule to show IR spectrum; thus symmetrical trans alkenes do not absorb in IR region while asymmetrical trans alkenes absorb weakly than cis alkenes. Terminal alkenes absorb strongly than nonterminal alkenes. A very prominent feature of alkenes is the out-of-plane bending peaks that appears in the range of 1,000–650 cm^{-1}. Conjugated dienes absorb at lower frequency with increased intensity.

5.10.1.3 Alkynes

Terminal alkynes give a strong peak at 3,300 cm^{-1} for sp-hybridized "C–H" bond. A sharp band appearing near 2,150 cm^{-1} is a characteristic feature of terminal C≡C stretch (Figure 5.16 (a)). The sp^3-hybridized carbons C–H will give strong peak between 2,850 and 3,000 cm^{-1}. Bands appearing at 1,300–1,500 cm^{-1} correspond to –CH$_2$ bending while C≡C–H bending appears at 600–615 cm^{-1}. In case of nonterminal alkyne, strong peak for C–H band at 3,300 cm^{-1} will be absent (Figure 5.16(b)). The peak for C=C at 2,150 cm^{-1} is either very weak or absent from the spectrum.

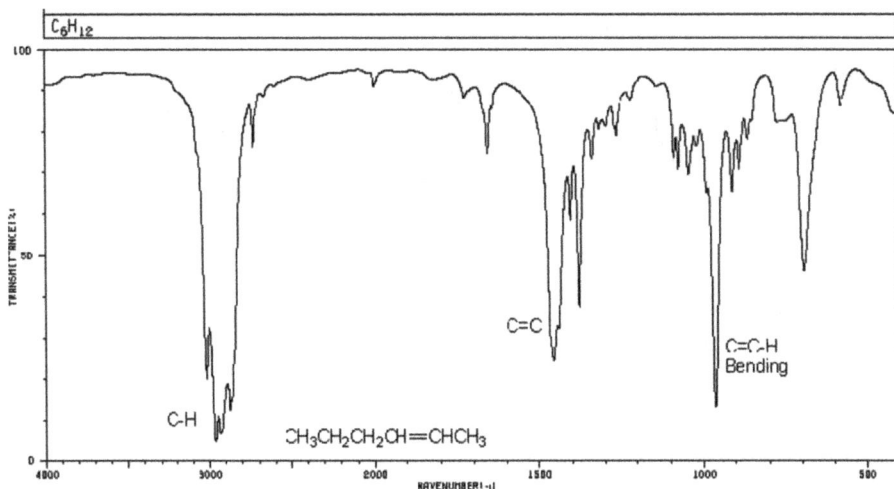

Figure 5.15: IR spectrum of 2-hexene.

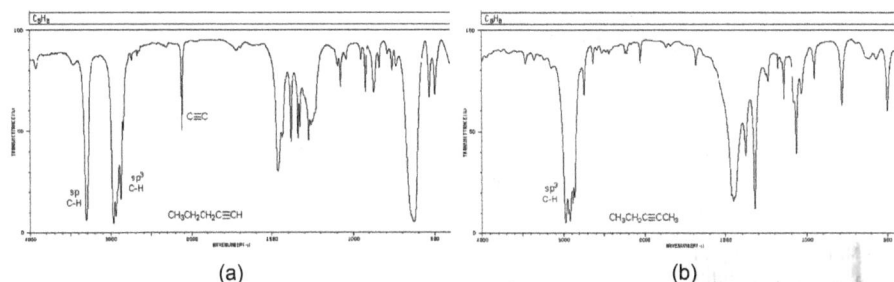

Figure 5.16: IR spectra of (a) pent-1-yne and (b) pent-2-yne.

5.10.2 Aromatic compounds

Aromatic compounds show many absorption peaks in the IR region, though only some of these are of diagnostic value. The C–H stretching frequency lies in the range of 3,100–3,000 cm^{-1}. Since C–H stretching bands for alkenes also lie in this region, it may be difficult to use C–H stretching frequencies to differentiate between alkenes and aromatic compounds. The characteristic pattern for C=C stretching bands for aromatic rings appears in the region 1,600–1,450 cm^{-1} (Figure 5.17); while C=C in alkenes appears at 1,650 cm^{-1}. The prominent out-of-plane bending peaks lie in the region 900–680 cm^{-1} along with some weak overtone/combination bands between 2,000 and 1,667 cm^{-1}. *Ortho-, meta-,* and *para-*disubstituted benzenes show their characteristic values which lie in the range of 770–735 cm^{-1}, 750–810 cm^{-1}, and 850–790 cm^{-1}, respectively. The bands correspond to aromatic pattern are strong and sharp in nature.

C_9H_{12}

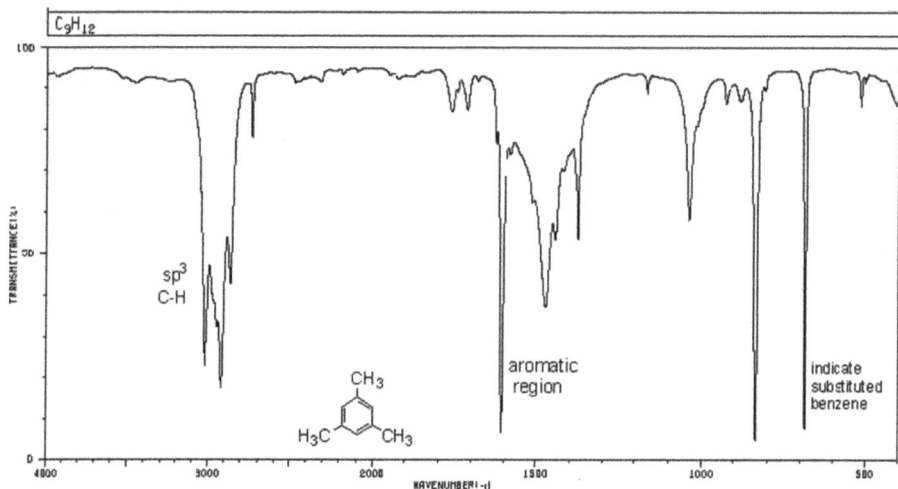

Figure 5.17: IR spectrum of mesitylene.

5.10.3 Carbonyl compounds

The carbonyl group (C=O) is present in aldehydes, ketones, acid chlorides, amides, esters, and acid anhydrides. This group shows a strong peak in the region 1,850–1,650 cm^{-1} because of its large change in dipole moment. The values for the C=O stretching vibrations for carbonyl group present in different compounds are shown in Figure 5.18, and this variation in values can be explained on the basis of inductive effect, hydrogen bonding, and resonance effect. The C=O frequency of an aldehyde which lies approximately in the middle of the range is generally considered as reference for comparison with other values.

Anhydride	Acid chloride	Ester	Aldehyde	Ketones	Acids	Amides

| Band I | 1810 cm^{-1} | 1800 cm^{-1} | 1735 cm^{-1} | 1725 cm^{-1} | 1715 cm^{-1} | 1710 cm^{-1} | 1690 cm^{-1} |
| Band II | 1760 cm^{-1} | | | | | | |

Figure 5.18: Base values for the C=O stretching vibrations for different carbonyl compounds.

The presence of an electronegative atom tends to withdraw electrons between carbon and oxygen atom, hence strengthening the bond, resulting in the higher frequency absorption, as observed in acid chloride. Esters exhibit higher vibrational frequency than aldehydes and ketones because of the presence of more electronegative oxygen,

–I effect raises the C=O frequency above that of aldehydes and ketones. In case of amides, resonance effect is observed, the unpaired electrons on nitrogen atom conjugate with the carbonyl group, resulting in increased single-bond character, and a lowering C=O absorption frequency.

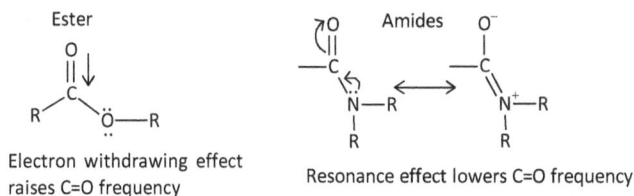

Ester

Electron withdrawing effect raises C=O frequency

Amides

Resonance effect lowers C=O frequency

Conjugation in molecules shifts the IR frequency to lower band value; thus aliphatic aldehydes and ketones appear between 1,705 and 1,740 cm^{-1}, while aromatic aldehyde and ketones show bands in lower region of 1,680–1,715 cm^{-1} (Figure 5.19).

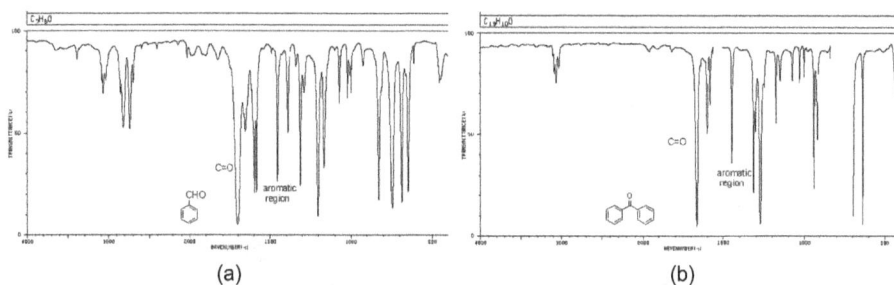

(a)

(b)

Figure 5.19: IR spectra of (a) benzaldehyde and (b) benzophenone.

Ring strain in cyclic compounds shifts the C=O stretching frequency to a higher value. As the ring strain increases, the C–CO–C bond angle decreases which increase the s character in the C=O bond, which is, thus, strengthened, and consequently, carbonyl stretching frequency increases (Figure 5.20).

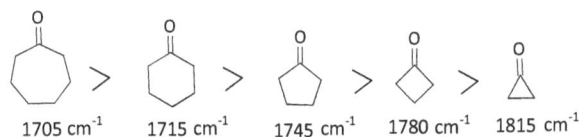

1705 cm^{-1} 1715 cm^{-1} 1745 cm^{-1} 1780 cm^{-1} 1815 cm^{-1}

Figure 5.20: Effect of ring size on C=O frequency.

The effect of electron releasing and electron withdrawing groups on C=O frequencies can also be understood from Figure 5.21. The electronegativity of nitrogen is lower than oxygen; thus lone pair of electrons in compound I participates more in conjugation (decreasing double bond character of C=O) as compared to compound II which result in lower absorption frequency for C=O in compound I than in compound II. In compound III, electron withdrawing nature of nitro (NO$_2$) group decreases the electron density on benzene ring which in turn withdraw electron from C=O group, thus increasing the double-bond character of C=O, resulting in higher stretching frequency. In compound IV, the lone pair of oxygen is delocalized in benzene ring and thus not available for conjugation with C=O group; thus, C=O absorption band appears at 1,770 cm^{-1} (higher frequency in comparison to compound I, II, and III).

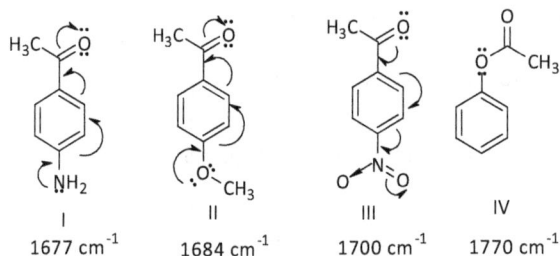

Figure 5.21: Effect on absorption frequency of C=O by electron withdrawing group.

5.10.4 Carboxylic acid

Carboxylic acid (RCOOH) shows three types of stretching frequencies associated with O–H, C=O, and C–O, and their corresponding absorption peaks appear in the regions 2,400–3,400 cm^{-1}, 1,700–1,725 cm^{-1}, and 1,050–1,250 cm^{-1}, respectively, as shown in Figure 5.22. Broadband at 2,400–3,400 for O–H along with a strong band in the region around 1,730–1,700 cm^{-1} shows the presence of aliphatic carboxylic acid in the dimeric form. This band shifts to lower values with conjugation to a C=C or phenyl group, and therefore absorption band for aromatic carboxylic acids appears in the range of 1,690–1,700 cm^{-1}. Carboxylic acid shows strong affinity to form hydrogen bond; thus they exist in dimer form in solution which is responsible for broadening of O–H band, and this broadband overlaps with C–H stretching bands at 2,900 cm^{-1}.

5.10.5 Esters and lactones

Esters and lactones show strong bands for the C=O and C–O groups (Figure 5.23((a) and (b)). The aliphatic esters show a strong band for C=O in the region 1,735–1,750 cm^{-1} while C=O band for aromatic esters appears at a lower value of 1,715–1,730 cm^{-1} due to conjugation with C=C bond of benzene. α,β unsaturation also shifts the C=O stretching

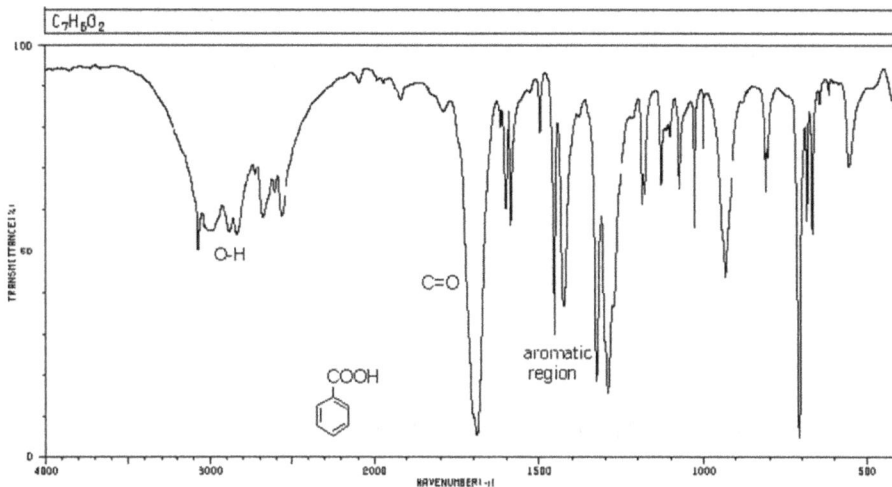

Figure 5.22: IR spectrum of benzoic acid.

frequency to lower value (1,700–1,715 cm^{-1}). Stretching frequency for C–O band appears at 1,050–1,250 cm^{-1}.

Lactones show C=O band in region of 1,735–1,770 cm^{-1} and stretching frequency for C=O shift to higher value with increase in ring strain. Conjugation due to C=C decreases the value of C=O stretching frequency.

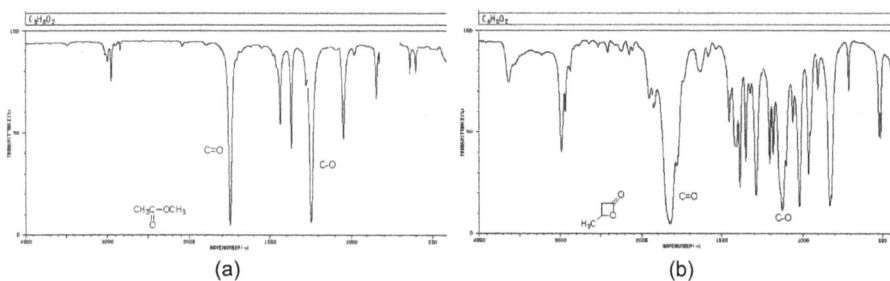

(a)   (b)

Figure 5.23: IR spectra of (a) an ester and (b) lactone.

5.10.6 Acid halides

Acid halides show stretching frequency due to C=O and C–Cl bonds. The strong band for C=O group appears in the region 1,750–1,815 cm^{-1}, while bands for C–Cl bond are observed in the lower frequency region 550–730 cm^{-1}. Aliphatic acid halides such as acetyl chloride show very strong band at 1,800 cm^{-1} for C=O, while aromatic acid chloride such as benzoyl chloride show strong band for C=O at slightly lower frequency

1,780 cm^{-1} due to conjugation with C=C bond. The IR spectra of acetyl chloride and benzoyl chloride are shown in Figure 5.24(a) and (b), respectively.

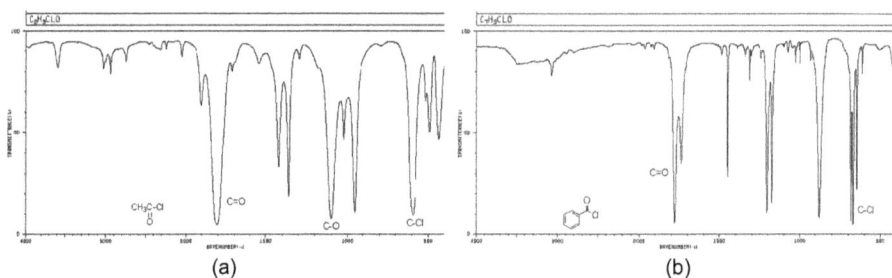

(a)   (b)

Figure 5.24: IR spectra of (a) acetyl chloride and (b) benzoyl chloride.

5.10.7 Acid anhydride

Acid anhydrides show the C=O and C–O stretching frequency. The strong band for C=O appears in the range of 1,800–1,850 cm^{-1} (band I) and 1,775–1,740 cm^{-1} (band II); whereas band in the region 1,050–1,250 cm^{-1} corresponds to C–O as displayed in Figure 5.25. The presence of two strong bands at 1,800–1,850 cm^{-1} and 1,775–1,740 cm^{-1} which looks like split band is due to asymmetrical and symmetrical C=O stretching. α–β unsaturation and aromaticity decreases the C=O stretching frequency. Also, ring strain in cyclic anhydride shifts the stretching frequency to higher region than the acyclic anhydride.

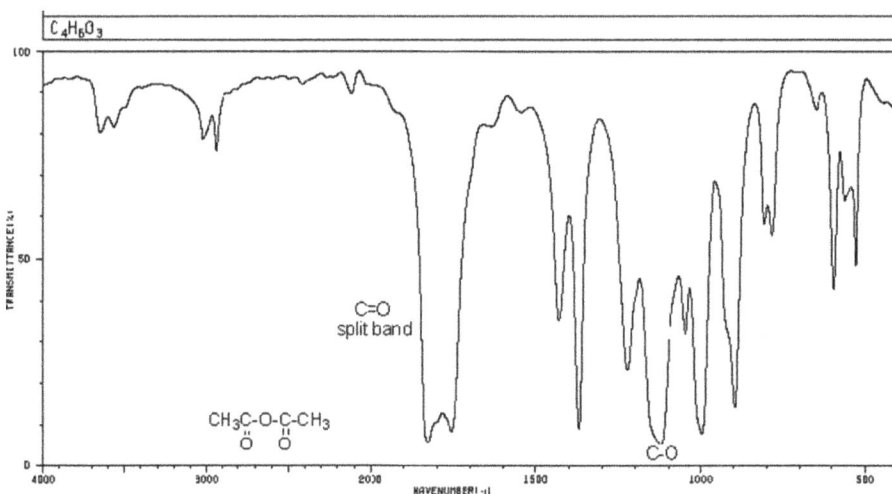

Figure 5.25: IR spectrum of an acid anhydride.

5.10.8 Alcohols and phenols

Both alcohols and phenols show C–O and O–H stretching frequency in the region 1,050–1,250 cm^{-1} and 3,200–3,400 cm^{-1}, respectively (Figure 5.26). The hydroxyl group O–H also shows bending at 1,230–1,420 cm^{-1}. The value of stretching frequency for O–H depends upon its nature of existence. When O–H is present in free form (nonhydrogen-bonded), a sharp peak that absorbs around 3,600 cm^{-1} is observed, while for hydrogen-bonded O–H group, the stretching frequency value shifts to lower region (~3,300 cm^{-1}). For intermolecular hydrogen bonding, O–H show broad and strong band at 3,200–3,400 cm^{-1} while for intramolecular hydrogen bonding, the value shifts to 2,500–3,200 cm^{-1}. In case of phenol, the aromatic region at 1,450–1,620 cm^{-1} can be used to distinguish it from alcohols.

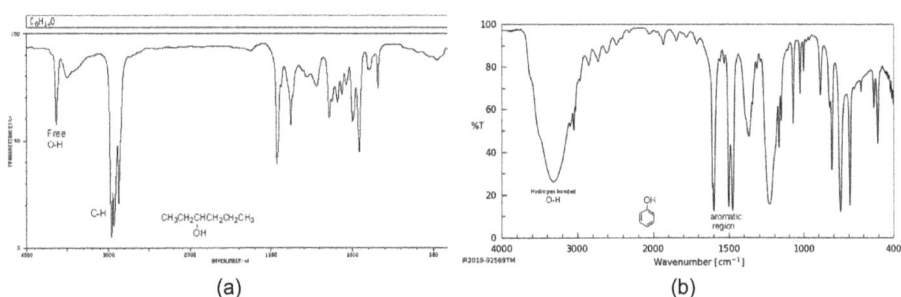

Figure 5.26: IR spectra of (a) 3-hexanol and (b) phenol.

5.10.9 Amides

Amides show the C=O and N–H stretching frequency at 1,630–1,690 cm^{-1} and 3,100–3,500 cm^{-1}, respectively. The N–H bending appears at 1,550–1,640 cm^{-1}. The primary amide, RCONH$_2$, shows two strong bands in the region 3,100–3,500 cm^{-1} for N–H bonds which sometimes observed as splitting of band into two whereas in secondary amide (RCONHR'), N–H appears as single strong band at same position. The strong C=O band for amide appears in the region 1,630–1,690 cm^{-1} which sometime overlap with N–H bending vibration that occurs at 1,550–1,640 cm^{-1}. Presence of α–β unsaturation, conjugation, and hydrogen bonding lowers the stretching frequency for C=O group. The IR spectrum of benzamide is shown in Figure 5.27(a).

5.10.9.1 Lactams (cyclic amide)

Lactams show the C=O stretching frequency at slightly higher values than amides (1,650–1,745 cm^{-1}) ((Figure 5.27(b)). As the ring size decreases, C=O frequency value increases.

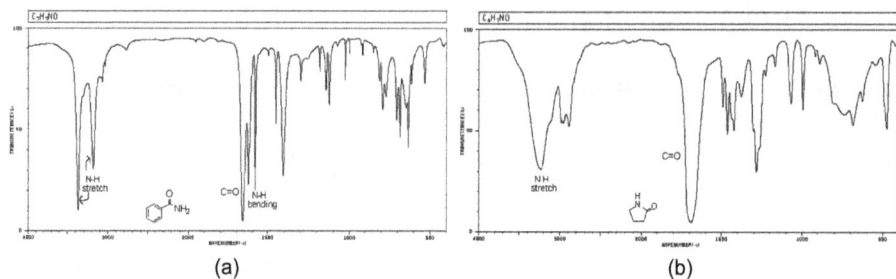

Figure 5.27: IR spectra of (a) benzamide and (b) lactam.

5.10.10 Amines

Primary amines (R-NH$_2$) show two N–H stretching bands in the region 3,300–3,500 cm^{-1} (Figure 5.28), one for symmetrical stretching and other for asymmetrical stretching. Secondary amines show only one weak band, whereas tertiary amines do not show any band due to absence of any N–H bond. The N–H bending band appears in the range of 1,500–1,640 cm^{-1} which is often not observed in primary aliphatic amines and for aromatic primary amines, this N–H band overlaps with aromatic region at 1,550–1,640 cm^{-1}. C–N stretching appears in the region 1,000–1,350 cm^{-1} for aliphatic amines while for aromatic amines it appears between 1,250 and 1,350 cm^{-1}.

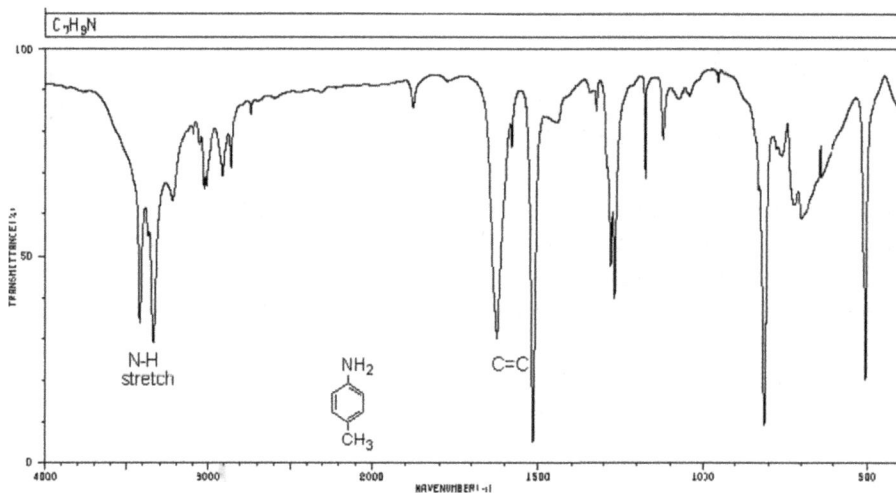

Figure 5.28: IR spectrum of 4-toluidine.

5.10.11 Nitriles, isonitriles, imines, azides, thiocyanates, and isothiocyanates

Nitriles show strong and sharp C≡N stretch in the range of 2,200–2,260 cm^{-1} while isonitriles N≡C stretching occur between 2,070 and 2,220 cm^{-1}. For imines (C=NH), N–H stretching occurs between 3,300–3,400 cm^{-1}. Thiocyanates show absorption frequency in the range of 2,140–2,175 cm^{-1} while isothiocyanates absorption frequency value lies between 2,000 and 2,140 cm^{-1}. It shows slightly broad peak than thiocyanates. The azides, N≡N stretching, occurs between 2,120 and 2,160 cm^{-1}. Conjugation due to α–β unsaturation or aromatic ring decreases the absorption frequency. The IR spectra of 2-methylbenzonitrile, ethylthiocyanate, and 1-isothiocyanatopropane are displayed in Figure 5.29.

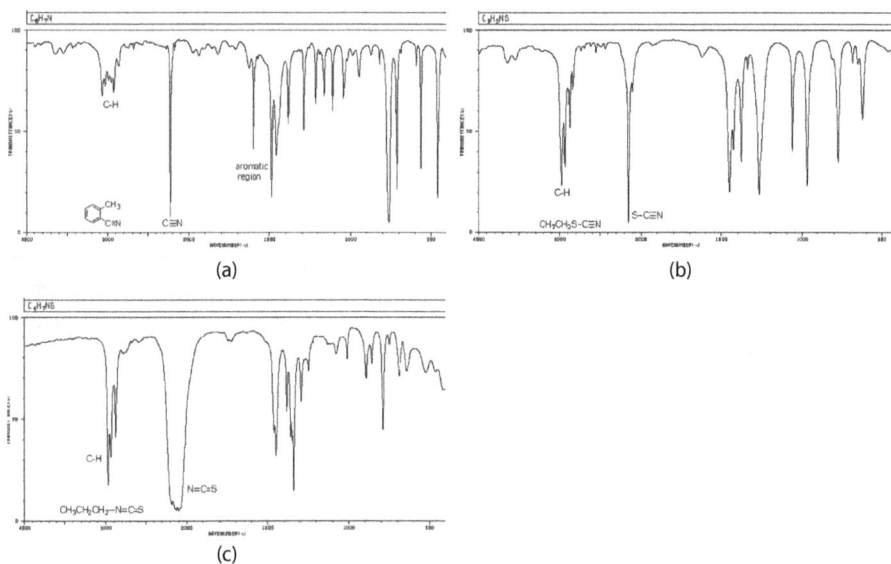

Figure 5.29: IR spectra of (a) 2-methylbenzonitrile, (b) ethylthiocyanate, and (c) 1-isothiocyanatopropane.

5.10.12 Nitro compounds

Nitro compounds show two strong bands in the region 1,350–1,550 cm^{-1} (Figure 5.30). Like amines, one band for symmetrical stretching and other for asymmetrical stretching. The strong band around 1,350 cm^{-1} appears for symmetrical stretching and other band around 1,550 cm^{-1} appears for asymmetrical stretching. The aliphatic nitro compounds absorb at slightly higher frequency than aromatic nitro compounds, as the presence of conjugation in the system decreases the absorption frequency to 1,290–1,570 cm^{-1}.

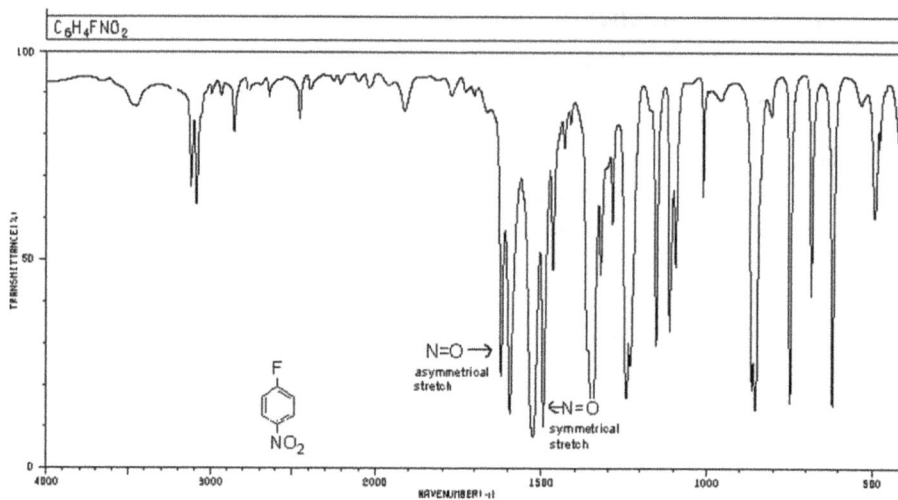

Figure 5.30: IR spectrum of 4-fluoronitrobenzene.

5.10.13 Sulfur compounds

The S–H stretch for mercaptans is weak and occurs near 2,550 cm^{-1} (Figure 5.31). The absorption band for S=O stretch in sulfoxides is very strong and occurs at 1,050 cm^{-1}. The strong S–O stretch for sulfonates and sulfonic acids appear in the range of 750–1,000 cm^{-1}.

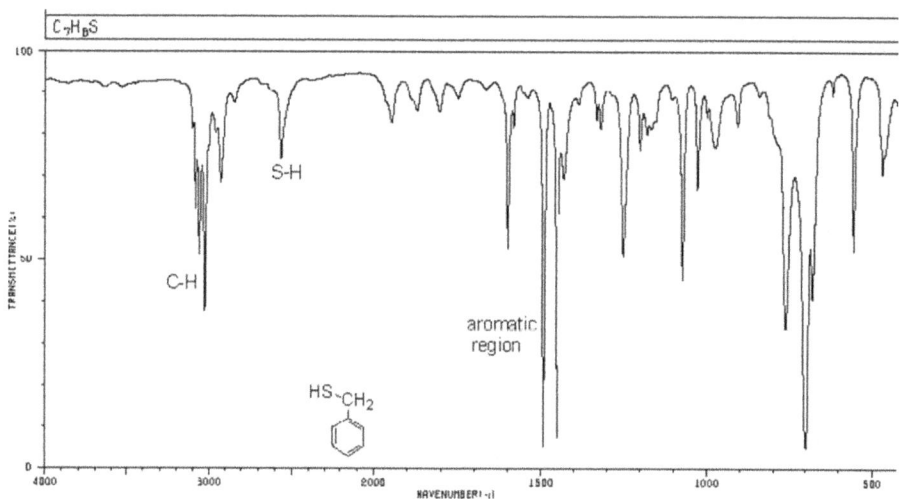

Figure 5.31: IR spectrum of phenylmethanethiol.

5.10.14 Halogens

As the electronegativity of halogens decreases down the group, the stretching vibration frequency decreases. The highest absorption frequency corresponds to fluorides which absorbs in the region 1,000–1,400 cm^{-1}, and chlorides absorb around 600–800 cm^{-1}, while bromide and iodides both show much lower absorption frequency around 500–650 cm^{-1}.

5.11 Applications of IR spectroscopy

5.11.1 Qualitative analysis (identification of a compound)

IR spectroscopy along with nuclear magnetic resonance and mass spectroscopy can be used for the identification of organic compounds. Development of various analytical techniques has reduced the time required to identify any compound. IR spectrum gives specific peak for each functional group such as carbonyl, hydroxyl, amines, nitro, and amides within the frequency region of 3,600 cm^{-1} to 1,200 cm^{-1} that helps in differentiating between different molecules. The presence of fingerprint region (4,000–400 cm^{-1}), which is unique for every molecule, thus, can be used to confirm the identity of compounds in vast database.

5.11.2 Structure elucidation

Different peaks observed in IR spectrum are sensitive to the chemical environment. Various factors such as conjugation, hybridization, cis–trans isomerization, and hydrogen bonding influence the absorption frequency and shift it to either higher or lower frequency region. IR spectroscopy, thus, has made the structure elucidation a much easier process. For example, functional group isomers can be easily identified by IR spectra. Both, dimethyl ether (CH_3–O–CH_3) and ethanol (CH_3–CH_2–OH) have same molecular formula but different functional groups. They can be easily differentiated with the help of IR spectroscopy; in spectrum of dimethyl ether, a peak for C–O group appears in the region 1,250–1,350 cm^{-1}, whereas in ethanol, the characteristic band for O–H group occurs in the range of 3,500–3,200 cm^{-1}.

Dichloroethane exists in two forms as shown in Figure 5.32. Absorption intensity of the bands at 1,289 cm^{-1} and 1,272 cm^{-1} assigned to the CH_2 wagging frequencies for the staggered and skew isomers, respectively. The staggered form predominates at low temperature while skew form exists mainly at higher temperature.

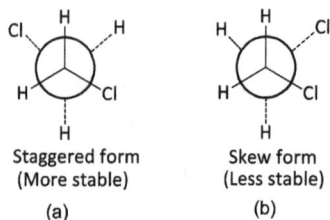

Staggered form
(More stable)
(a)

Skew form
(Less stable)
(b)

Figure 5.32: Staggered and skew conformations.

5.11.3 Differentiate between single (C–C), double (C=C), and triple (C≡C) bonds

IR spectroscopy gives different vibrational frequency value for single (C–C), double (C=C), and triple (C≡C) bonds. Triple bond absorbs at much higher frequency followed by double bond while single bond absorbs in the lowest frequency region.

5.11.4 Presence of hydrogen bonding in a molecule

The presence of hydrogen bonding in a molecule can be easily detected by IR spectroscopy, and it can also distinguish between intra- and intermolecular hydrogen bonding. In a hydrogen-bonded compound, the vibrational frequencies of the molecule (O–H and N–H) shift to lower absorption values on dilution. The O–H group in alcohol (in very dilute solution) gives sharp absorption peak in higher frequency region around 3,600 cm^{-1}, whereas in compounds where hydrogen bonding is extensive (in pure sample), one can observe a broadband at lower frequency, that is, around 3,300 cm^{-1}. Intermolecular hydrogen bonding decreases with dilution, whereas intramolecular hydrogen bonding shows no significant effect on the IR spectra of molecules.

5.11.5 Applications in medical field

Natural compounds such as lipids, proteins, carbohydrates, and nucleic acids have their unique vibrational fingerprints; therefore disease-induced changes in position, bandwidth, signal intensity, and area of spectral bands of these systems can also be detected by IR spectroscopy.

5.11.6 Study of the progress of a reaction

The progress of a reaction can be easily studied by IR spectroscopy. Since each functional group absorbs at different stretching frequency, reactions involving the interconversion of various functional groups such as oxidation of alcohols into ketone, reduction of acids into alkanes, or oxidation of alkanes into alcohols and then to ketones can be easily studied. If the sample shows more than one band in the functional group region in the IR spectrum, for example, in case of oxidation of alcohol into ketone, it indicates that reaction is not complete. The complete disappearance of O–H band and appearance of C=O band confirms the end of reaction.

5.11.7 Determination of group orientation in aromatic compounds

The substitution on benzene ring can be distinguished by IR spectroscopy. The substituents at ortho, meta, and para position give different bending vibration in fingerprint region in IR spectra. For example, in ortho xylene, a strong band is detected in the region 750 cm^{-1} and meta xylene substituted benzene shows strong band at 775 cm^{-1} while para xylene benzene exhibits band around 800 cm^{-1}.

5.11.8 Identification of drugs

Drugs can also be identified by IR spectroscopy; for example, steroids are difficult to distinguished from ketones by chemical analysis method or by UV spectroscopy as the chromophores are same in both, but they are easily distinguished by IR spectroscopy as they give completely different spectra in IR. Similarly, chloramphenicol exists in three polymorphic forms (same molecules but differ only in crystalline forms) which can be differentiated from each other by IR spectroscopy.

References

[1]   El-Azazy, M. Introductory chapter: Infrared spectroscopy-A synopsis of the fundamentals and applications. In: Infrared spectroscopy-principles, advances, and applications. 2019, p. 3.
[2]   Hayashi, M., Shiro, Y., Oshima, T., and Murata, H. The vibrational assignment, rotational isomerism and force constants of 1, 2-ethanedithiol. Bulletin of the Chemical Society of Japan. 1965; 38(10): 1734–1740.
[3]   (National Institute of Advanced Industrial Science and Technology) For all IR spectra thanks to SDBS, https://sdbs.db.aist.go.jp.
[4]   Mollaoglu, A.D., Ozyurt, I., and Severcan, F. Applications of infrared spectroscopy and microscopy in diagnosis of obesity. In: Infrared spectroscopy-principles, advances, and applications. IntechOpen; 2018.
[5]   Nicolet, T. Introduction to Fourier transform infrared spectrometry. Thermo Nicolet Corporation; 2001.

[6]   Pavia, D.L., Lampman, G.M., Kriz, G.S., and Vyvyan, J.A. Introduction to spectroscopy. Cengage learning; 2014. Stamford, CT 06902 USA.

[7]   Spectra Analysis Instrumental, inc by. www.spectra-analysis.com.

[8]   Willard, H.H., Merritt, L.L., Jr., Dean, J.A., and Settle, F.A., Jr. Instrumental methods of analysis. 7th edn, United States: Wadsworth Publishing Company; 1988.

[9]   Yadav, L.D.S. Organic spectroscopy. Springer-science business media, B. V; 2005.

Jaspreet Kaur and Mukesh

Chapter 6
Flame atomic absorption and emission spectroscopy

Abstract: The chapter offers an introduction to flame atomic absorption and emission spectroscopy, giving the basic principles of the instrumentation with focus on the techniques of atomization, sample introduction, methods of background correction, sources of chemical interferences, and method of their removal. The chapter also highlights the comparison between flame atomic absorption and emission spectroscopy as well as the quantitative estimation of trace level of metal ions from water samples using these techniques.

6.1 Introduction

In atomic spectroscopy [1–4], solution containing metallic salt (or metallic compound) is aspirated into flame which has temperature high enough to break the bond(s) present, resulting in dissociation of molecule to free metal atoms. Fraction of these free atoms may gain enough thermal energy that the outer electrons of atom are promoted to the higher energy levels (excited state). Since this excited energy level is unstable, atom immediately returns to the ground state by emitting radiation characteristic of metal, for example, sodium atoms (obtained from sodium compounds) impart yellow color to flame. This is the underlying principle of flame emission spectroscopy (FES), formerly known as flame photometry (Figure 6.1) [5]. For quantitative analysis of the element, the intensity of light emitted at the specific wavelength of the element to be determined is measured.

However, most of the atoms remain at the ground state in the flame. These atoms are capable of absorbing appropriate radiant energy of their own specific resonance wavelength. Radiant energy absorbed by atoms is equivalent to the energy which they would have emitted if they return to the ground state. If light of resonance wavelength passes through atoms in the flame, free ground state atoms may absorb part of light. Extent of absorption will be proportional to number of atoms in the ground state in the flame. This is the basis of flame atomic absorption spectroscopy (FAAS) (Figure 6.1) [6].

https://doi.org/10.1515/9783110794816-006

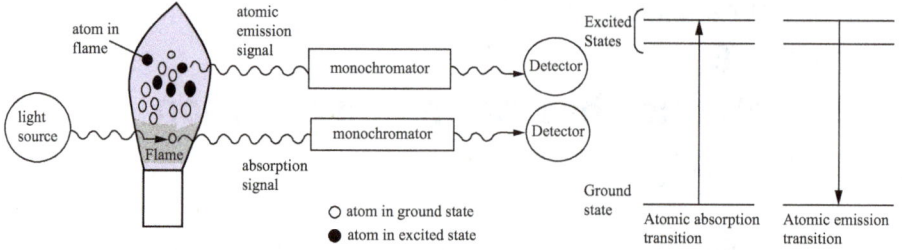

Figure 6.1: Absorption and emission of radiation by atoms in a flame. In case of atomic emission, radiation is emitted by excited atoms in the flame while in atomic absorption, atoms absorb radiation from the lamp and unabsorbed light reaches the detector [3].

6.2 Elementary theory

Population difference between the ground state and the excited state affects the signal strength. As the difference of population between two energy level increases, the signal strength increases, and as a result, sensitivity enhances. Boltzmann distribution given below describes the relationship between the relative populations of different energy levels at thermal equilibrium.

$$\frac{N_i}{N_0} = \left(\frac{g_i}{g_0}\right)e^{-\Delta E/kT}$$

where N_i and N_0 are number of atoms in the excited and ground state, respectively.

Number of states at each energy level is called degeneracy, g_i and g_0 are degeneracies of the excited and ground states, respectively;

$\Delta E = E_i - E_0$ is the energy of separation between two states, k is the Boltzmann constant, and T is temperature in Kelvin.

According to the Boltzmann equation, N_i/N_0 depend on ΔE and T. An increase in temperature and decrease in ΔE will cause an increase in the ratio N_i/N_0. Calculations show that even under most favorable conditions of low ΔE and high temperature, a small fraction of atoms are in the excited state (Table 6.1). Increasing temperature hardly affects the population of the ground state and would not affect the signal in atomic absorption. Flame emission methods measure the excited-state population while atomic absorption methods measure the ground-state population. It can be concluded that atomic absorption is not as sensitive as atomic emission to temperature variation (Table 6.1). However, wavelength of resonance line is also an important factor, elements whose resonance lines are with long wavelength, (sodium with emission line of 589.0 nm) and are associated with low energy value are more sensitive toward FES as compared to the elements with resonance lines having small wavelength, (zinc with emission line of 213.9 nm) and are associated with high energy value.

Table 6.1: Values of N_i/N_o for different resonance lines and temperature [4].

Element	Wavelength (nm)	N_i/N_o	
		2,000 K	4,000 K
Na	589.0	9.86×10^{-6}	4.44×10^{-3}
Ca	422.7	1.21×10^{-7}	6.03×10^{-4}
Zn	213.9	7.31×10^{-15}	1.48×10^{-7}

Temperature also determines the degree to which a sample breaks down to free atoms and the extent to which an atom is found in its ground, excited, or ionized state. Each effect influences the strength of the observed signal.

As in case of molecular spectroscopy, in atomic absorption spectroscopy, absorption of radiation (A) is given by the logarithmic ratio of the intensity of incident light (I_o) to the intensity of the transmitted light (I):

$$A = \log I_o/I = \varepsilon l N_o$$

where ε is absorption coefficient (M^{-1} cm^{-1}), l is path length through the flame (cm), and N_o is the concentration of atoms in flame (number of atoms per milliliter).

This is a linear function for small values of absorbance.

In flame photometry, the detector response E is given by the following expression:

$$E = kac$$

where c is the concentration of solution, a is efficiency of atomic excitation, and k is a constant containing factors such as efficiency of atomization and self-absorption.

Any electrical method that can increase the value of E (such as improved amplification) will make the technique more sensitive.

To summarize, the success of atomic absorption spectroscopy is governed by the factors that favor the production of gaseous atoms in the ground state while in FES there is a requirement of production of gaseous atoms in the excited state [4, 7].

6.3 Instrumentation

The components for the flame spectrophotometers are same as of the UV–vis (ultraviolet–visible) spectrophotometer. The basic components consist of a radiation source (for FAAS), a sample holder, a monochromator, a detector, a signal processor, and read-out device [9]. A sample holder in FAAS and FES is the flame (atomizer) that contains free gaseous atoms. FES does not require a line (radiation) source [4]. The components of the flame spectrophotometers (Figure 6.2a and b) are briefly discussed further:

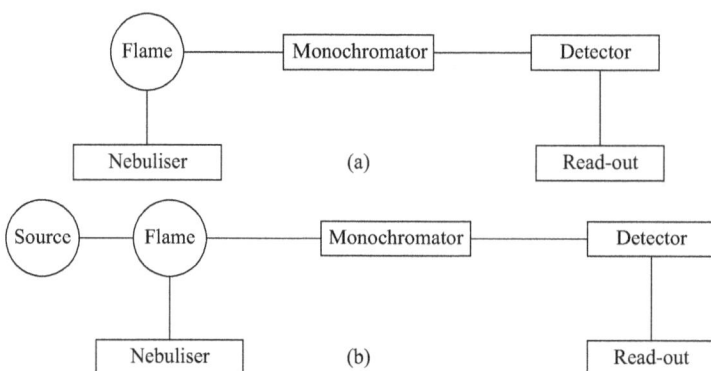

Figure 6.2: Block diagram of (a) FES and (b) AAS [4].

i. **Nebulizer-burner system** (sample introduction and atomizer) [8, 9]: Nebulizer is used to introduce homogeneous sample solution as mist into the flame. Sample solution in the presence of high temperature of flame undergoes evaporation and vaporization, and subsequently the gaseous molecules dissociate into free gaseous atoms. This is accomplished by choosing appropriate combustion flame involving fuel-oxidant gas mixture.

ii. **Monochromator** [10]: It helps to isolate the wavelength chosen for the measurement from other lines (occasionally from molecular band emissions that may interfere; interferences are discussed in Section 6.5). Instruments with diffraction grating (monochromator) have higher resolution over a long range of wavelength because of more uniform dispersion from grating than the dispersion from the prism.

iii. **Detector and display device** [11, 12]: Photosensitive detector such as photomultiplier tube is used to measure the intensity of the absorbed (FAAS) or emitted (FES) radiation. The output from the detector is fed into a suitable readout system.

iv. **Radiation source/resonance line source** [1]: It is required for each element to be determined by atomic absorption instrument and is usually modulated. Atomic absorption instruments are very specific because it involves the electronic transitions that are distinctive for each element which results in narrow absorption lines. These remarkably narrow absorption lines create a problem. For Beer's law to be obeyed, the line source should have narrow bandwidth as compared to the width of the atomic absorption line. So, if we use continuum source as in UV–vis spectrophotometer, it will give nonlinear calibration curve. Moreover, small calibration slopes are obtained as only fraction of radiation from the continuum source will be absorbed by the gaseous atoms that result in poor sensitivity. This problem due to limited width of the atomic absorption line can be overcome by using line sources with narrow bandwidth as compared to the atomic absorption line width. A separate source lamp is required for each element to be detected (or sometimes group of elements) and is placed in line with the detector. Examples of line sources that are used in AAS are hollow cathode lamp (HCL) and electrodeless discharge lamps (EDLs). These are discussed in Section 6.3.

In atomic absorption instrument, it is crucial to eliminate interferences caused by the emission of radiation within the flame. The effect of flame emission which is eliminated modulating the emission from the resonance line source using mechanical chopper is introduced between the source lamp and the flame. A chopper is a spinning wheel (Figure 6.3) that alternatively allows the radiation from line source to pass through the flame and then blocks the radiation from entering the flame [1]. When chopper allows radiation to pass through the flame, detector receives signal both from lamp and flame; but when it blocks radiations to reach the flame then detector detects signal from the flame. The difference between two signals gives the analyte signal.

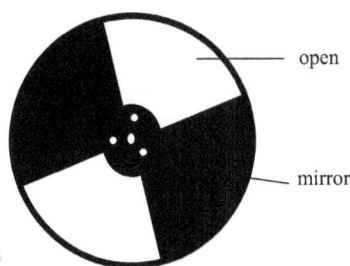

Figure 6.3: Illustration of chopper.

A typical version of single beam and double beam flame atomic absorption spectrophotometers are shown in Figure 6.4a and b [1–4]. In FES, lamp is not used.

i. **Single-beam instruments:** A typical single-beam instrument (Figure 6.4a) consists of several hollow-cathode sources (only one of which is shown in Figure 4a), an atomizer (such as flame), and a simple grating spectrophotometer with a photomultiplier transducer. In use, the dark current (refers to the constant response produced by a spectrochemical receptor, even in the absence of radiation) is first nulled. Subsequently, 100% transmittance adjustment is made while a blank is aspirated into the flame (or ignited in an electrothermal atomizer). Finally, the sample transmittance is measured for the atomized sample.

ii. **Double-beam instruments:** In double-beam-in-time instruments, the beam from the hollow-cathode source is split by a spinning wheel-mirrored chopper, one half passing through the flame and the other half around it as shown in Figure 6.4b. The two beams are then recombined by a half-silvered mirror and passed into a monochromator, a photomultiplier tube (detector). The output from the latter is the input to a lock-in amplifier that is synchronized with the chopper drive. The ratio between the reference and sample signal is then amplified and fed to the readout, which is usually a computer.

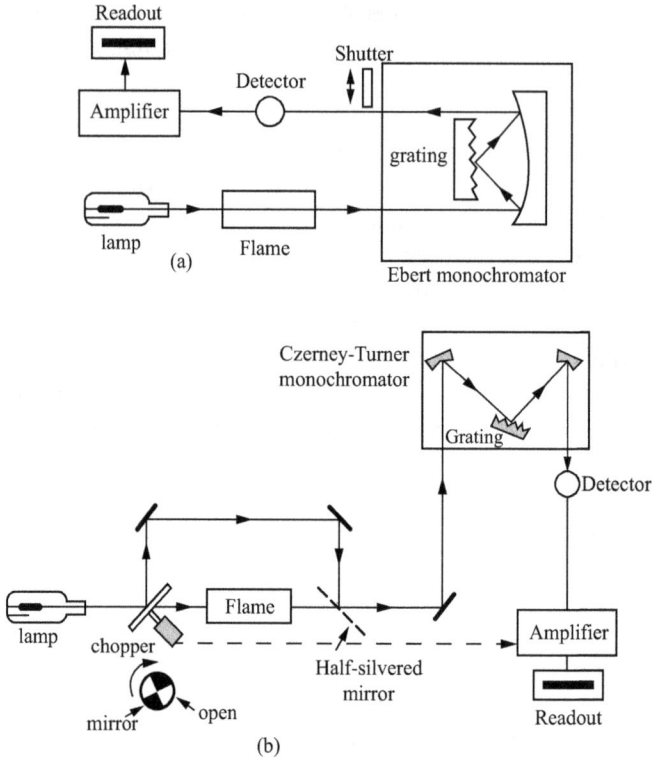

Figure 6.4: Single-beam flame atomic absorption spectrophotometer design and (b) double-beam flame atomic absorption spectrophotometer design [1].

6.4 Radiation sources for atomic absorption instruments

Atomic absorption instruments use two basic types of lamps (as a line source), HCLs and EDL, which are discussed below:

6.4.1 Hollow cathode lamps

HCLs consist of an anode made up of tungsten or zirconium, and a hollow cylindrical cathode sealed in a glass cylindrical glass tube filled with an inert gas such as neon or argon at a pressure of 1–5 torr having quartz window (made of borosilicate or UV transparent glass for wavelengths shorter than 400 nm) at one end (Figure 6.5) [1, 9]. The cathode is constructed from the metal (in pure form, 99.99%) whose spectrum is desired

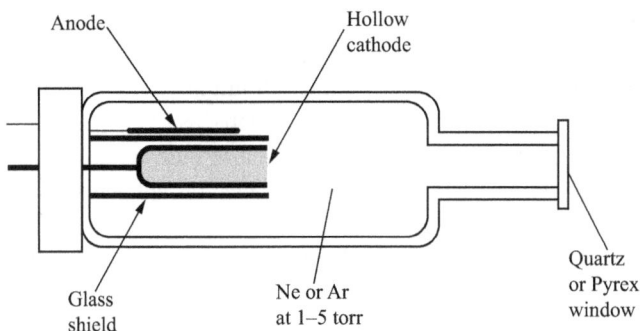

Figure 6.5: Schematic cross section of a hollow-cathode lamp [1].

(to measure an element such as lead, the cathode must contain lead). This means different lamp is required for each element to be measured. This assembly can neither be used for sodium as its melting point is too low nor for mercury due to its liquid state.

When a potential difference of the order of 300 V is applied across the electrodes, ionization of the inert gas within the lamp occurs, which results in the generation of current due to migration of ions and electrons to the electrodes. The gaseous cations (Ar^+ or Ne^+) acquire enough kinetic energy to strip some of the metal atoms from the cathode surface which produces an atomic cloud (*sputtering*). A fraction of the sputtered atoms which are in excited states emit their characteristic radiation as they return to the ground state. Emission corresponds to the following sequence:

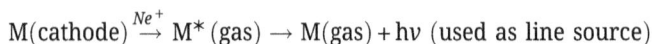

$$M(cathode) \xrightarrow{Ne^+} M^* (gas) \rightarrow M(gas) + h\nu \text{ (used as line source)}$$

where M(cathode) represents element in its metallic state (material from which cathode is made), M(gas) when element is in gaseous state, and M^* the excited gaseous atom. Ultimately, the metal atoms redeposit on the cathode surface or on the glass walls of the tube.

The cylindrical configuration of the cathode not only tends to concentrate the radiation in a limited region of the metal tube but also enhances the probability of redeposition at the cathode rather than on the glass walls. The efficiency of the HCL depends on its geometry and the operating voltage. Operating at high voltages and thus high currents lead to (a) greater intensities but this advantage is offset to some extent by an increase in Doppler broadening of the emission lines from the lamp and (b) the greater currents lead to produce an increased number of unexcited atoms, and these unexcited atoms have the potential to absorb the radiation emitted by the excited atoms which lead to lowering of intensity of the emission band. Various types of HCLs are commercially available. In some HCLs, the cathodes consist of a mixture of several metals (multielement lamps), and multielement lamps allow determination of more than a single element.

6.4.2 Electrodeless discharge lamp

An alternative to HCL is EDL. EDL provides higher radiation intensity (10–100 times greater than HCL) and narrow emission lines but EDL is not as stable as HCL [2, 3]. "EDLs consist of the element, or a salt of the element, sealed in a quartz bulb containing an inert gas atmosphere" (Figure 6.6). When a radiofrequency field of sufficient power is applied, the inert gas is ionized and the coupled energy vaporizes the element and excites the atoms inside the bulb, resulting in the emission of the characteristic spectrum [24]. These lamps are generally available for 15 or more elements such as As (arsenic), Se (selenium), Sb (antimony), Bi (bismuth), and P (phosphorous).

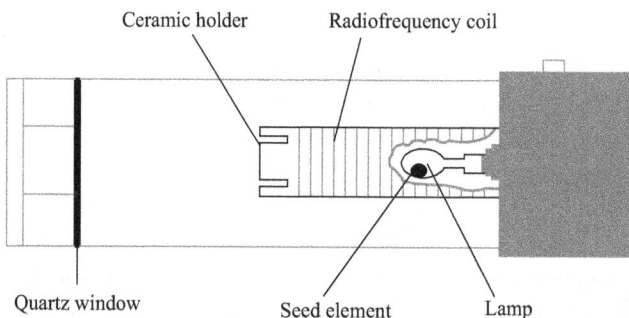

Figure 6.6: Design of electrodeless discharge lamp [1, 5].

6.5 Techniques of atomization and sample introduction

To obtain atomic spectra, the constituents of a sample must be converted into gaseous atoms (atomization). The method of atomization and introduction of the sample into the atomization region are very crucial as the accuracy, precision, and the detection limit of atomic spectrometric methods depend significantly on them.

6.5.1 Techniques of atomization: flames, furnaces, and plasmas

Atomization devices [1–4, 13] can be categorized as *continuous atomizers* and *discrete atomizers*. In continuous atomizers (such as plasmas and flames) samples are introduced in a steady manner while in case of discrete atomizers, samples are introduced in a discontinuous manner using a syringe or an autosampler. The electrothermal atomizer is most commonly used as discrete atomizer. Various types of atomizers and their corresponding atomization temperatures are tabulated in Table 6.2. In the following section, we have emphasized on some commonly used atomizers:

Table 6.2: Various types of atomizers used for atomic spectroscopy [1, 14].

Type of atomizer	Typical atomization temperature (°C)
Flame	1,700–3,150
Electrothermal vaporization (ETV)	1,200–3,000
Inductively coupled argon plasma (ICP)	4,000–6,000
Direct current argon plasma (DCP)	4,000–6,000
Microwave-induced argon plasma (MIP)	2,000–3,000
Laser-induced plasma	8,000–15,000
Glow-discharge (GD) plasma	Nonthermal
Electric arc	4,000–5,000
Electric spark	40,000 (?)

6.5.1.1 Flame atomizer

Flame operates both as an atomizer and excitation source [1–4]. The role of flame atomizer is to generate free gaseous atoms of analyte which get excited by the thermal energy of the flame and then emit characteristic radiation that is measured. Before entering the flame, the sample solution is drawn in and converted into a fine mist or an aerosol by nebulizer. An aerosol along oxidant and a fuel gas is then passed onto the next component, that is, the burner. Finally, in a flame, atomization occurs. In the next subsections, nebulizer, burner, flame, and its characteristics are discussed.

6.5.1.2 Nebulizer

It converts analyte solution into fine mist/aerosol (nebulization). Nebulization of liquid samples is the most popular sampling method to provide a steady flow of aerosol into a flame. Various types of techniques/methods [1, 15] are available for nebulization of a liquid sample, and a few are listed:
a) Pneumatic nebulization (concentric tube pneumatic nebulizer, cross-flow nebulizer, fritted-disk nebulizer, and Babington nebulizer)
b) Ultrasonic nebulization
c) Electrothermal vaporization
d) Hydride generation techniques

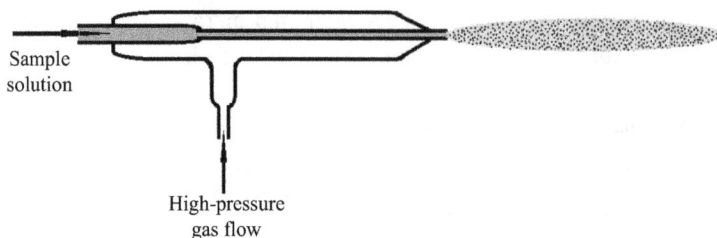

Figure 6.7: Illustration of concentric-type pneumatic nebulizer [1].

6.5.1.3 Concentric tube pneumatic type

This is the most common kind of nebulizer (shown in Figure 6.7). It consists of central capillary with sample solution surrounded by concentric tube with gas. The liquid sample is sucked through a capillary by high pressure gas stream flowing around the tip of the tube using the Bernoulli effect. This process of liquid transport is called aspiration. The high velocity of gas breaks up the liquid into droplets of different sizes, which are then carried into the atomizer. The other types of the pneumatic nebulizers also work on the same principle.

6.5.1.4 Burners

The sample is introduced in the form of a fine spray at a controlled rate into the flame of a burner with the help of a nebulizer. In flame photometry, different types of burners and fuel-oxidant combinations are used to produce analytical flame [4]. Two main types of burner systems: (a) the premix or laminar-flow burner and (b) the total consumption or turbulent-flow burner are discussed further.

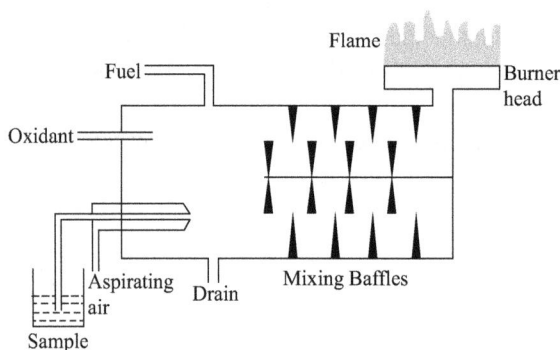

Figure 6.8: Premix or laminar flow burner [23].

6.5.1.4.1 Premix burner

Most flame spectrometers use a premix burner, in which sample is nebulized and simultaneously mixed with oxidant and fuel before reaching the burner opening and then entering the flame. In this type of burner, gases move in nonturbulent fashion, that is, laminar flow [4].

In premix burner, the solution of analyte is aspirated with the help of a nebulizer from the sample container into the mixing chamber in which fuel gas is also introduced (Figure 6.8). Mist flows past a series of baffles, which promote further mixing. The large droplets of liquid are blocked by baffles in the mixing chamber and are drained off, thereby resulting in loss of sensitivity. A fine mist containing about 5% of the initial sample reaches the flame and is easily decomposed (means an efficient atomization of sample in the flame takes place). The efficiency of this burner has been improved by putting an impact bead immediately after the nebulizer. The impact bead has the effect of breaking up large droplets as they come out from the nebulizer, thereby increasing the number of smaller droplets and the efficiency of nebulization.

In atomic absorption spectroscopy, the premix burner is a long horizontal tube with narrow slit along its length. This produces a thin flame of long path length.

Premix burner has the following advantages:

i.   Flame produced is nonturbulent, noiseless, and stable.
ii. As large droplets of solution are eliminated from the stream of gas reaching the burner, they possess little danger of incrustation around the burner head.
iii. Easy decomposition of the sample takes place which results in efficient atomization of the sample in the flame.

Premix burner has the following disadvantages:

i.   Solutions are made up of mixed solvents, and the more volatile solvents are evaporated preferentially, leaving the sample in the form of undissociated atoms in the less volatile component.
ii. Emission intensity will reduce as small number of atoms reaches the flame.
iii. Most of the sample solution goes down the drain which leads to loss of sensitivity in the determination of the given analyte.
iv. Potential explosive hazard exists since the burner uses relatively large volumes of gas (in modern instruments this type of hazard is minimized).

6.5.1.4.2 Total consumption burner

Total consumption burner consists of three concentric tubes (Figure 6.9). A fine capillary tube carries the sample solution directly into the flame. Fuel gas and oxidant gas reach top of burner from separate tubes so that they can turbulently mix, combust only at the tip of the burner, and produce flame. All the aspirated liquid sample reaches at the tip of the burner; here, it is broken into a fine spray (nebulization). This type of burner is called total consumption burner because the Venturi suction created by the fuel-oxidant flow aspirates entire sample directly into the flame. Desolvation, atomization, and excitation of the aspirated sample solution then follow in the flame. So, total consumption burner combines the function of both nebulizer and burner [4].

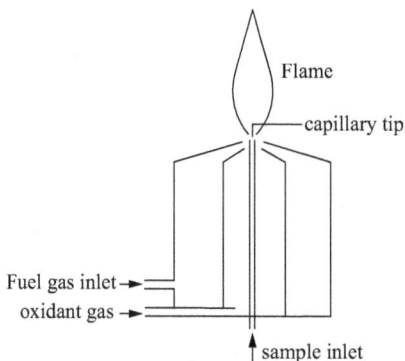

Figure 6.9: Representation of total consumption burner [4].

The total consumption burner has the following advantages.
i.   It is simple to manufacture.
ii. Total representative sample reaches the flame; for this reason, the emission (for a given concentration of analyte in solution) is higher than that from a premix flame.
iii. It is free from explosive hazards which may arise from unburnt gas mixtures.

Total consumption burner suffers from the following disadvantages.
i.   Aspiration rate varies with different solvents.
ii. Variations in the recorded signals may occur due to the tendency for incrustations to form at the tip of the burner.
iii. Droplets of various sizes are formed during aspiration. Many of the larger droplets may blow through the flame without totally evaporating or without pyrolysis of the solute.
iv. Flame produced is noisy and turbulent.

Total consumption burner gives a flame of relatively shorter path length hence these are predominantly used for flame emission studies.

6.5.1.5 Flame

When a sample solution is aspirated into the flame [1–4, 9], numerous complex set of interconnected processes occur in the flame, as illustrated in Figure 6.10.

i.   *Step one*: Desolvation

Solvent evaporates from the fine droplets by the heat of flame, leaving the dehydrated salt (MX(solution) $\rightleftharpoons$ MX(solid), solid molecular aerosol)

ii. *Step two*: Volatilization

The aerosol is volatilized to form gaseous molecules. $(MX(s) \rightleftharpoons MX(g))$

iii. *Step three*: Atomization

Most of the gaseous molecules formed in step two undergoes dissociation to produce an atomic gas $(MX(g) \rightleftharpoons M(g) + X(g))$. Some of the atoms in the gas ionize to form cations and electrons. Other molecules and atoms are produced in the flame as a result of interactions of the fuel with the oxidant and with the various species in the sample (molecules and atoms other than analyte results in interference during absorption/emission measurements, discussed in Section 6.5.1). In FAAS, free gaseous atoms absorb radiation from line source (HCL, EDL) and absorbed radiation is measured.

iv. *Step four*: Excitation

A fraction of the atoms and molecules ions are also excited by the absorption of energy from heat of the flame to yield atomic, molecular, and ionic emission spectra.

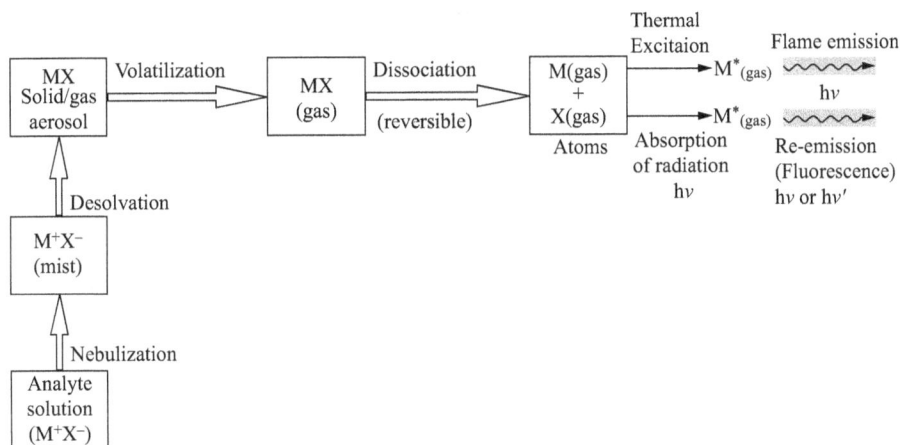

Figure 6.10: Various processes occurring during atomization [9].

It is essential to understand the characteristics of flames (atomizer) and the variables that affect these characteristics because atomization is the most critical step in flame spectroscopy due to many complex processes occurring in the flame as discussed above, limiting the precision of such methods.

A flame should have the following characteristics:

i.   It should provide sufficient energy for the sample to be atomized.

ii. It should be nonturbulent so that atom population shows the least spatiotemporal variation.

iii. The flame itself should have minimum emission and absorption at the wavelengths of interest.
iv. It should be able to operate at low gas velocities so the emitting atoms remain in the view volume as long as possible.
v. It should be safe and inexpensive to operate.

6.5.1.6 Types of flame

The essential requirement is that the flame shall have temperature greater than 2,000 K. In most cases, this requirement can be achieved by burning of fuel gas in an oxidant which is usually air, oxygen, or nitrous oxide (N_2O) diluted with either argon or nitrogen [1–4, 9]. The temperature of the flame depends on the fuel-oxidant ratio.

Table 6.3: Approximate flame temperatures of various fuel-oxidant mixtures [1, 4].

Fuel	Temperature (°C)		
	Air	Oxygen	Nitrous oxide
Natural gas	1,800	2,750	–
Hydrogen	2,300	3,080	2,900
Acetylene	2,250	3,150	2,800
Propane	2,200	–	3,000

Note (from Table 6.3) that temperatures near 2,000 °C occur with the various fuels when air is used as an oxidant. At these temperatures, only easily decomposed samples like Ca (calcium), Mg (magnesium), K (potassium), Na (sodium), Fe (iron), and other transition elements are atomized. Oxygen or nitrous oxide oxidants produce temperatures around 3,000 °C with the common fuels. These oxidants (oxygen or nitrous oxide) may be used for samples which are more difficult to atomize like Cr (chromium), Os (osmium), Rh (rhodium), and Al (aluminum), having high melting points and their oxides also dissociate at high temperatures (refractory samples).

6.5.1.7 Flame structure

Flame may be divided into three regions as shown in Figure 6.11; appearance and relative size of these regions may vary with the type of fuel and oxidant, the fuel–oxidant ratio, and the type of burner [1]:

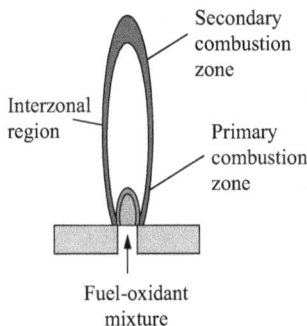

Figure 6.11: Regions in flame [1].

i. *The primary combustion zone*: This zone in a hydrocarbon flame is identifiable by its blue luminescence attributed to the band emission of C_2, CH, and other radicals. Thermal equilibrium is not achieved in this region. Therefore, it is rarely used for flame spectroscopy.

ii. *The interzonal zone region:* The interzonal area is the hottest part of the flame and lies immediately above the primary combustion zone. It may reach several centimeters in height in fuel-rich sources like acetylene-oxygen or acetylene-nitrous oxide. It is the most extensively used part of the flame for spectroscopy because higher temperature in this region favors both production of free atoms and maximum excitation of the free atoms.

iii. *The secondary combustion zone*: This is the outermost zone of the flame. Within this zone, the products of the inner core are converted to stable molecular oxides that are then dispersed into the surroundings.

6.5.1.8 Nonflame atomizers

Even though aspiration into a flame is the most well-suited and reproducible means of obtaining atomic vapor, it is not efficient in converting all the analytes into atomic vapor and to have atomic vapor present in the optical path for a long enough time to measure the absorption. Also in flame atomizer, volume of solution required is minimally of the order of a milliliter or more. A very small amount (as little as 0.1%) of the aspirated analyte may actually be atomized and measured.

An alternative to this is to use nonflame methods which involve the use of (i) electrically heated graphite tubes or rods (electrothermal atomizer), (ii) vapor techniques, or (iii) glow-discharge atomizer [1–4, 9, 14]. These methods are briefly discussed below:

6.5.1.8.1 Electrothermal techniques

In comparison to flame, electrothermal atomizer provides superior sensitivity as entire sample is atomized in a short period and the average residence time of the atoms in the optical path is of the order of few seconds or more. Electrothermal atomizers are generally used for atomic absorption and atomic fluorescence measurements but are also used for vaporizing analyte samples in inductively coupled plasma emission spectroscopy.

Graphite furnace consists of a graphite tube of about 50 mm in length and internal diameter of about 9 mm with a small cavity that can hold a precise quantity (few mg or μl) of sample (Figure 6.12). The central axis of graphite rod coincides with the optical axis of the spectrophotometer. It behaves as an ohmic resistor which can achieve operating temperatures to 3,000 K (Joule effect). The graphite tube is surrounded by metal jacket through which water is circulated (for cooling of device) and is separated from the graphite tube by gas space. Inert gas like argon is circulated in the gas space to protect the elements from oxidation.

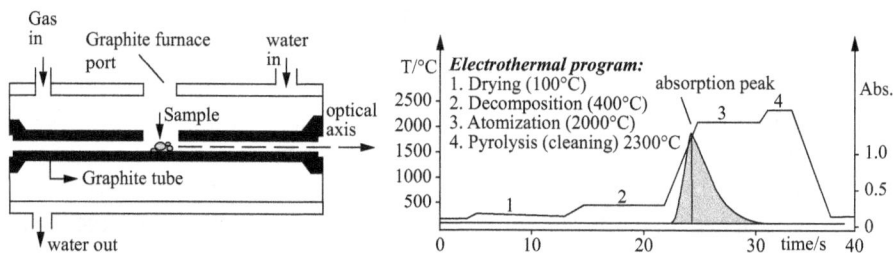

Figure 6.12: Illustration of graphite tube furnace and graph displaying temperature programming as a function of time showing the absorption signal. Two types of measurements can be chosen: the peak height absorbance or the integrated absorbance (in grey on the figure) [9].

The analyte solution is inserted using automatic syringe (or manually) through port and into the gas inlet at the center of the hollow graphite tube. Graphite tube is heated by passing electric current. The heating cycle in the tube generally consists of four steps as displayed in Figure 6.12. The temperature increased gradually first to accomplish desolvation of analyte solution, then ashing or charring (to decompose the sample and in order to volatilize some of the matrix), and finally to atomize the sample.

In graphite furnace atomization, sample matrix effects (caused by all other components of the sample except the analyte) and poor reproducibility can be minimized by reducing the natural porosity of the graphite tube. In graphite furnace, atomization process slows down as fraction of the analyte and matrix diffuse into the surface of the tube, giving smaller analyte signals. This effect can be overcome by coating graphite tube with a thin layer of pyrolytic carbon (layer-by-layer graphite is deposited by passing a mixture of an inert gas and a hydrocarbon such as methane through the tube held at a high temperature) thus sealing its pores. In addition to graphite furnaces, several other electrothermal devices have been used such as coils and loops made from platinum and tungsten. The tungsten coil atomizer has the advantages of low power as well as simplicity and is the most successful of these [25].

Advantages of electrothermal methods are as follows:

i.   It can be used to analyze very small amount of sample (as low as 0.5 µL). This is significant in the context of clinical samples.

ii. It generally requires very little or no sample preparation and certain solid samples require no prior dissolution.

iii. As compared to flame AAS, detection limits are improved between 100 and 1,000-fold in graphite furnace.

Disadvantages of electrothermal methods compared to flame atomic absorption are as follows:

i.   Background correction (discussed in Section 6.5.1) effects are more serious.

ii. Analyte may be lost during ashing especially when dealing with the volatile compounds such as chlorides of mercury and cadmium.

iii. The sample may not be completely volatilized during atomization and may create "memory effect" within the furnace.

iv. The precision is poorer in graphite especially when sample is injected manually. Autosamplers are used to enhance the precision.

6.5.1.8.2 Vaporization techniques

a) Cold-vapor atomization:

The cold-vapor atomization technique is confined to the determination of mercury which is a metallic element that has an appreciable vapor pressure at ambient temperature. Mercury compounds are determined by reducing Hg^{2+} to the elemental mercury with $SnCl_2$ (stannous chloride).

$$Hg^{2+} + Sn^{2+} \rightleftharpoons Hg + Sn^{4+}$$

The elemental mercury is flushed out of the reaction mixture by bubbling a stream of inert gas and then swept into absorption.

b) Chemical vaporization/hydride generation:

Some elements such as arsenic (As), bismuth (Bi), tin (Sn), or selenium (Se) in higher oxidation states are difficult to reduce to atoms by flame AAS. These can be analyzed by electrothermal method but an alternate preferred approach to electrothermal method is hydride generation. In this method, the analyte sample is reacted with a reducing agent such as sodium borohydride or tin chloride in an acidic medium (Figure 6.13). A volatile hydride of the element is formed which is swept up by a make-up gas into a quartz cell placed in the flame of the burner where hydride can be dissociated into atomic vapor (in relatively moderate temperatures such as of argon-hydrogen flame). This method can be used to determine other elements which can form volatile covalent hydrides that are easy to decompose thermally.

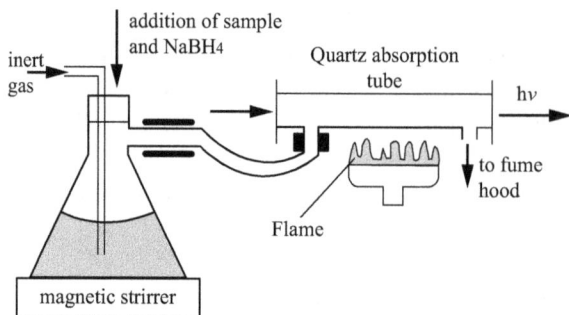

Figure 6.13: Illustration of hydride generation and atomization system for AAS [1].

6.5.1.8.3 Glow-discharge atomizer

A glow-discharge device, shown in Figure 6.14, is a source that performs both sample introduction and sample atomization. A glow discharge (plasma formed by the passage of electric current through a gas) of gas (at low pressure) takes place between a pair of electrodes maintained at dc voltage of 250–1,000 V. In the presence of applied voltage, argon gas breaks into argon ions and electrons. The electric field accelerates the positively charged argon ions to the cathode surface (contains sample) and neutral sample atoms are ejected (*sputtering*). The atomic vapor produced in a glow discharge consists of a mixture of atoms and ions (determined by atomic absorption or fluorescence or by mass spectrometry) as well as fraction of the atomized species is in an excited state (used for optical emission measurements). The most important applications of the glow-discharge atomizer is the analysis of metals, and other conducting samples, liquid samples, and nonconducting materials can also be analyzed by mixing them with a conductor such as graphite or pure copper powders. It also finds applications in surface imaging and elemental mapping [16].

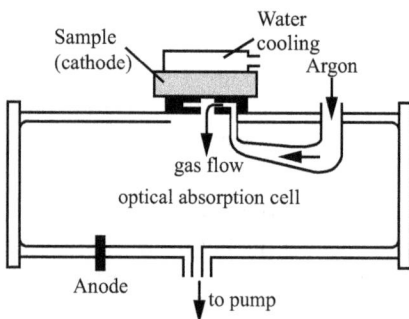

Figure 6.14: Illustration of glow-discharge lamp [1].

6.6 Sources of interferences, their removal, and method of background correction

6.6.1 Interferences

Interference is any effect that alters the signal when concentration of analyte remains unchanged, that is, radiation intensity/signal does not accurately represent sample concentration which may be because of the presence of some other material along with analyte in the sample. Interference must be controlled in order to get correct results. Interference can be corrected by either counteracting the source of interference or by preparing standards that exhibit the same interference [1–4, 17]. Interferences are broadly classified as (i) *chemical interference*, which arises as a result of various chemical processes that occur during atomization, such as chemical reactions and ionization of analyte atoms which decreases the concentration of neutral atoms. This changes the absorption characteristics of the analyte [18] and (ii) *spectral interference* arises when unwanted signals overlap or lies close to analyte signal and the monochromator cannot resolve them [19]. These are briefly discussed below:

6.6.1.1 Chemical interferences

Chemical interference is more common than spectral interference and is caused by any substance that decreases the extent of atomization of analyte. Different types of interferences caused by chemical reactions are discussed below [1–4, 17, 18]:

i. Interference due to formation of low volatile compounds

Certain *anions*, such as oxalate, sulphate, phosphate, and aluminate, form compounds of low volatility with analyte, thus reducing the fraction of the analyte that is atomized. This may lead to low results. For example, decrease in the absorbance of calcium in the presence of sulfate and phosphate anion. These anions form stable salts with calcium ($CaSO_4$ and $Ca_3(PO_4)_2$) which are difficult to volatilize and result in production of lesser free atoms.

Cation interference has also been observed, for instance, aluminum interferes in the determination of magnesium, apparently due to the formation of a thermally stable aluminum–magnesium compound (perhaps an oxide) which results in the decrease of absorption signal of magnesium.

Such interferences caused by formation of low volatile species can often be overcome by:

(a) Use of higher flame temperatures.

(b) Use of releasing agents: Releasing agents are cations that react preferentially with the interferant and prevent its interaction with the analyte. For example, lanthanum (or strontium) ion reacts with PO_4^{3-} (phosphate) and free the Ca^{2+}

(calcium) ions, thus minimizing the interference of phosphate in the determination of calcium.

(c) Use of protective agents: They prevent interference by forming stable but volatile species with the analyte. Examples of protective reagents are ethylenediaminetetraacetic acid (EDTA), 8-hydroxyquinoline, and ammonium salt of 1-pyrrolidinecarbodithioic acid. EDTA and 8-hydroxyquinoline protect calcium from the interfering effects of aluminum, silicon, phosphate, and sulfate.

(d) Extraction of analyte using separation methods such as solvent extraction.

ii. Interference due to oxide formation

This type of interference arises due to the formation of stable metal oxide (if oxygen is present in the flame). This reduces the intensity of signals as large amount of free atoms are removed from the flame as metal oxides. The alkaline earth metals form metal oxides and are subject to this type of interference. This interference can be reduced by either using very high temperature of the flame to dissociate the oxides or by using oxygen deficient environment (fuel-rich combustion mixtures) to produce excited atoms.

iii. Ionization interference

In analysis of alkali metals, high temperature of the flame may cause ionization of the ground state gaseous atoms which may be attributed to their lowest ionization potentials.

A gas-phase ionization reaction for any element (M) can be written as:

$$M(g) \rightleftharpoons M^+(g) + e^-(g)$$

$$K = \frac{[M^+][e^-]}{[M]}$$

Ionization of free atoms will reduce the extent of absorption in AAS or it will reduce the emission of the atomic spectra lines.

The ionization of an analyte may be reduced by adding excess of *ionization suppressant.* It is an element having lower ionization potential than analyte. For example, when a solution containing 1 mg/mL of CsCl (cesium chloride) is added to the sample containing potassium for the analysis, it creates an excess of ions because cesium is more easily ionized than potassium. A high concentration of electrons in the flame reverses the above reaction for potassium (Le Chatelier's principle), thus increasing the concentration of K.

The *method of standard addition* compensates for many types of interference by adding the known quantities of analyte to the unknown in its complex matrix (discussed in Section 6.6).

6.6.1.2 Spectral interferences

Spectral interferences may arise when the analyte and interfering *elements* (elements other than analyte) have line emissions of similar wavelength. The detector cannot differentiate between the sources of radiation and records the total signal, and this erroneous measured high absorbance can lead to a gross overestimation of the analyte concentration. For example, vanadium interferes in the determination of aluminum as vanadium (V) line at 308.211 nm overlaps with the aluminum (Al) line 308.215 nm. Such interferences can be overcome by choosing different resonance line which has no overlap, for example, in case of V and Al, the interference can avoided by using the aluminum line at 309.27 nm instead. It may also be avoided by removing the interfering elements by separation techniques such as ion exchange and solvent extraction. Alternatively, one may make a calibration curve, which is prepared from a solution having similar quantities of the interfering element.

Spectral interferences may also arise from the presence of *molecular species* such as combustion products or particulate products. Molecular species will result in broadband absorption while particulate matter will result in scattering of light from HCL and both will give erroneous result. If fuel-oxidant mixture alone is the source of these products, the error can be corrected by aspirating blank into the flame. (Note that the reference beam of a double-beam instrument does not pass through the flame.)

The sample matrix (molecular species) is also a source of interference. The power of the transmitted beam is reduced by the matrix components. An example for this type of interference can be observed in the determination of barium in alkaline earth mixtures. The wavelength used for atomic absorption analysis of barium appears in the center of a broad absorption band for CaOH (Figure 6.15). It can be easily anticipated from Figure 6.15 that calcium will interfere with barium determinations, but this effect can be easily eliminated by substituting nitrous oxide for air as an oxidant. The higher flame temperature of nitrous oxide decomposes CaOH and removes the absorption band. This is a case where an atomic absorption line gets superimposed over molecular absorption and detector is unable to identify broadband absorption (by molecular species) from line absorption (by atomic species).

Spectral interference may also be caused by organic species or organic solvents (sample matrix) that are used to dissolve the sample. Incomplete combustion of the organic matrix in the flame may lead to carbonaceous particles which are capable of scattering light.

Some metals such as Ti (titanium), Zr (zirconium), and W (tungsten) are prone to form solid refractory oxides in the flame that scatter radiation and detector cannot distinguish between scattered and absorbed light.

Several background correction methods such as the deuterium background correction, Zeeman correction system, and the Smith-Hieftie system have been developed to correct the spectral interferences (due to scattered radiation and molecular absorption from molecular and atomic species). These methods are discussed below:

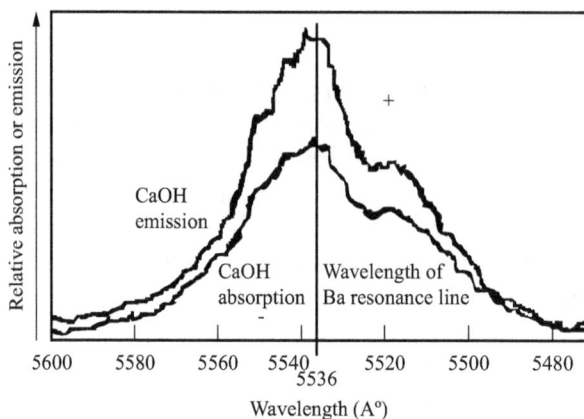

Figure 6.15: Molecular flame emission and absorption spectra for CaOH along with atomic emission wavelength of barium [1].

6.6.2 Background correction methods

The observed absorption may be higher than the true absorption signal of an analyte. As discussed above, this may occur by molecular absorption or by light scattering. If not corrected, this erroneously observed high absorbance can lead to a gross overestimation of the analyte concentration. The erroneously high signal of background absorption can be corrected by making sequential measurements of the absorbance due to the background and absorbance due to the analyte as well as background. The difference between two measured absorbance provides the absorption due to the analyte. Three techniques which can be used for background correction are (a) continuum source background correction method, (b) Smith–Hieftje correction method, and (c) Zeeman correction method [1–4, 17].

6.6.2.1 Continuum source background correction method

This method uses an additional light source, a continuum source such as a Deuterium lamp along with the HCL of element to be determined (Figure 6.16) [1–4, 9, 17], and deuterium lamp is a source of continuum radiation all over the ultraviolet region. Radiation from deuterium lamp follows the same double beam path as light from the resonance source (HCL). A chopper alternates the radiation passing between the deuterium continuum and the resonance source. The slit width is kept wide enough so that negligible amount of radiation from the continuum source is absorbed by the analyte atoms. As a result, the attenuation of radiation from the continuum source as it passes through the atomizer (containing atomized sample along with interfering molecules/matrix) and reflects only the broadband absorption or scattering by the sample matrix components. On the other hand, when the absorbance is measured with the HCL, it measures the absorbance due to the analyte atoms as well as the background absorbance at that wavelength. A background effect is eliminated by subtracting the absorbance of the deuterium lamp from that of the analyte lamp. Although it is inexpensive to implement and convenient to use it has some drawbacks: (i) it is difficult to achieve perfect alignment of D_2 lamp and HCL along an identical path through the sample cell, (ii) the low radiant output of the deuterium lamp in the visible region limits the use of this correction procedure for wavelengths less than 350 nm, and (iii) signal-to-noise ratio reduces due to addition of correction system.

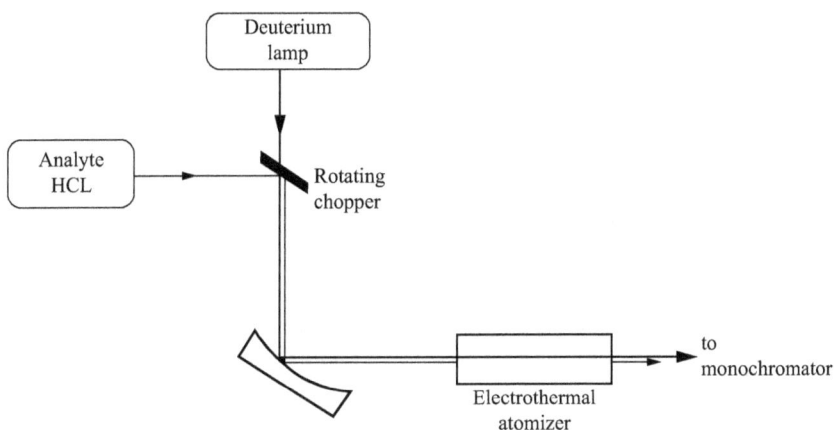

Figure 6.16: Schematic of a continuum source background correction system [4].

6.6.2.2 Zeeman background correction

The Zeeman correction method is likely the most sophisticated method [1–4, 9, 17]. It provides more accurate background correction than other methods described earlier. Its principle is based on the Zeeman effect, that is, energy levels of atoms may split in the presence of a strong magnetic field. The simplest form of splitting pattern observed with singlet transitions leads to a central π line (present at the original wavelength) and two equally spaced satellite σ lines, one shifts to a wavelength slightly greater while the other shifts to a wavelength slightly less relative to the original wavelength (Figure 6.17). The π-peak absorbs only radiation that is plane polarized parallel to the direction of the field while σ-peaks are polarized perpendicular to the direction of the magnetic field.

Unpolarized radiation from hollow cathode source is passed through a rotating polarizer, which separates the beam into two components: one component is polarized parallel to field and the other is polarized perpendicular to the field. These beams pass into atomizer (a tube-type graphite furnace). A permanent magnet surrounds the furnace and splits the energy levels into the three absorption peaks.

During that part of cycle when source radiation is polarized parallel to the field, the absorption of radiation by analyte and background takes place. During the other half cycle (radiation polarized perpendicular to the field), no analyte absorption can occur but broadband molecular absorption and scattering by the matrix products (background) occur. A background corrected value is obtained by subtracting measured absorbance during the perpendicular half cycle from that of the parallel half cycle.

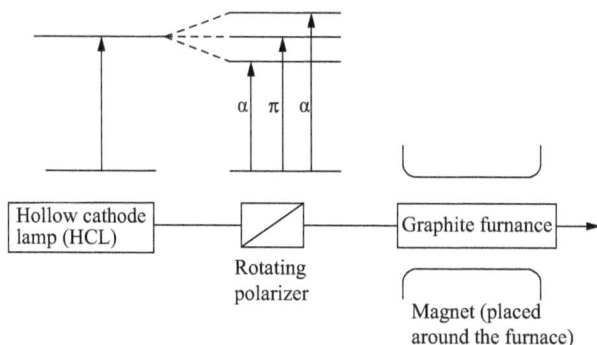

Figure 6.17: Illustration of Zeeman effect and schematic of an electrothermal atomic absorption instrument that provides a background correction based on the Zeeman effect [4].

6.6.2.3 Smith–Hieftje correction method

This method is based on the principle of the self-reversal or self-absorption. In this method of background correction, HCL is operated alternatively at low and high current [1–4, 9, 17]. The effect of high current is (a) it generates large concentrations of non-excited atoms, which can absorb radiation produced from the excited species and (b) it broadens the emission line of the excited species with minimum in its center, which corresponds exactly to the wavelength of the absorption line. As a result, at high current, background absorption is measured, and at low current, total absorption due to analyte and background is measured. The absorption of analyte is given by the difference between the absorbance measured at low current and the absorbance measured at high lamp current (Figure 6.18).

Zeeman and Smith–Hieftje background correction methods offer following advantages:

i. having only one light source unlike continuum source background correction method,

ii. not being limited to operating only in the UV region unlike continuum source, and

iii. the background is measured close to the sample absorption.

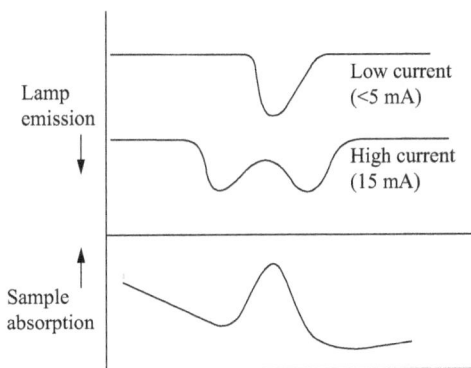

Figure 6.18: Emission profiles of HCL operated at low and high current along with the absorption profile by sample [4].

6.7 Techniques for the quantitative estimation of trace level of metal ions from water samples

FAAS is the most widely used technique employed for the estimation of transition metals, alkaline earth metals, Al and Pb, and FES is a preferred method for alkali and alkali earth metals [20, 21, 22]. Some of these elements can be qualitatively analyzed simply using flame, for example, sodium gives yellow and lithium imparts red color to the flame. But this method is not much reliable. To get authenticated results for the samples containing cations of alkali and alkaline earth metals, flame photometry is more reliable, fast, and easy to use technique. Flame AAS and FES follow Beer's law, that is, intensity of signal (for absorption/emission) is directly proportional to concentration. Thus, these techniques can be used for the qualitative as well as quantitative determination of these element(s) in the sample. Signal of specific wavelength represents the presence of metal and intensity of signal represents the amount of metal present in the sample.

In order to convert measured absorption or emission values into the concentration of analyte, the following methods may be used:
i.   Calibration curve
ii. The standard addition method
iii. The internal standard method

As in case of UV–vis spectroscopy, calibration curve here can also be used to determine the concentration of analyte in a sample by aspirating the test analyte solution and solutions of known analyte concentration (standard solutions) separately into the flame, and the absorption/emission intensity of each solution is measured. To get calibration curve, signal intensity is first plotted against concentration of standard solutions and then concentration of unknown analyte can be determined from the calibration plot. So, it is a simple matter to interpolate the concentration of analyte in a solution from the absorption/emission measurements of the solutions with the known concentration of analyte. Calibration plot may not always be linear due to uncontrollable variables in atomization and due to stray light. Modern instruments incorporate microcomputer which stores the calibration curve and allows the direct read-out of concentrations. This method can be used when we have single analyte in a deionized water but if other components are also present along with the analyte in a test solution, then they should also be incorporated in the standard solutions.

Matrix interference (e.g., seawater sample) can be reduced or overcome by using the method of standard addition. This method is widely used in FAAS and FES when the matrix of the sample solution is complex, its composition is not known, and the standard solutions having same composition to the sample are difficult to prepare. In this method, different known amounts of standard solution of analyte are added to the equal amount of test solution (except one test solution). This is called *spiking*. If the signal intensity is too high, the resulting solutions are diluted to same volume. Every solution (from low to high concentration) is aspirated into the flame and their emission/absorption signals are measured. The signal intensity (due to absorption or emission) is plotted on Y-axis and concentration on X-axis (Figure 6.19). A straight-line result on extrapolation to concentration axis gives the value of concentration of test solution.

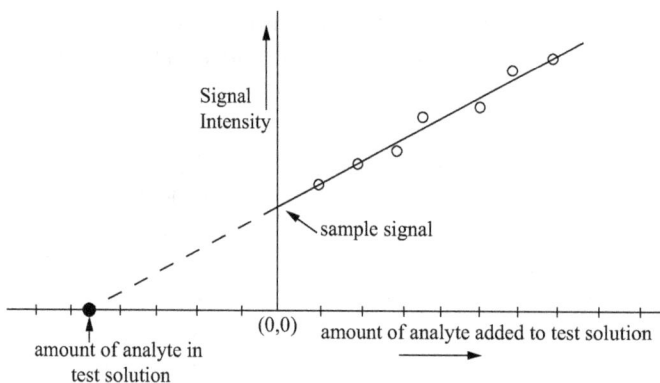

Figure 6.19: An illustration of standard addition method.

In the internal standard method, a known and fixed amount of the internal standard solution (solution of substance that is different from analyte and their wavelengths should not be too different) is added to the sample solution and calibration standards. These solutions (standard sample and analyte solution) are aspirated into flame and signal intensity is recorded for each of these solutions. Internal standard undergoes similar interferences as the analyte. Measurement of ratio of signal due to analyte and the internal standard cancels the interferences. If the aspiration rate, for example, fluctuates, each signal is affected to the same extent and the ratio, at a given analyte concentration, remains constant. Ratio of signal due to analyte and the internal standard signal are plotted against the concentration of the analyte in the standard solutions. Internal standards can be used with either the calibration curve or standard addition methods. This method is used to improve the precision in quantitative analysis and may also compensate for any variations in the nebulizer uptake or changes in the flow rates of both the fuel gas and the oxidant.

References

[1]   Skoog, D., West, D., Holler, F., and Crouch, S. Fundamentals of analytical chemistry, 9th edn, Cengage Learning; United States. 2013.

[2]   Christian, G., Dasgupta, P. and Schug, K. Analytical Chemistry, 7th edn, John Wiley & Sons; New Jersey. 2013.

[3]   Harris, D. Exploring Chemical Analysis, 4th edn, New York: W.H. Freeman; 2009.

[4]   Vogel. Textbook of Quantitative Chem 5e, 5th edn, Nashville, TN: John Wiley & Sons; 1989.

[5]   Fernández-Sánchez, M.L., Fernández-Arguelles, M.T. and Costa-Fernández, J.M. Atomic Emission Spectometry | Flame Photometry Encyclopedia of Analytical Science. 3rd edn, Elsevier Academic Press. Netherlands. 2019, pp. 160–168.

[6]   García, R. and Báez, A.P. Atomic absorption spectrometry (AAS). In: Farrukh, M.A. (ed) Atomic absorption spectroscopy [Internet]. London: IntechOpen; 2012, [cited 2022 Aug 05].

[7]   Bings, N.H., Bogaerts, A. and Broekaert, J.A.C. Atomic Spectroscopy: A Review. Anal Chem [Internet]. 2010; 82(12): 4653–4681.

[8]   Robinson, J.H., and Boothe, E.D. The development and evaluation of a new nebulization-burner system for flame atomic absorption spectroscopy. Spectrosc Lett [Internet]. 1984; 17(11): 653–671. Available from: http://dx.doi.org/10.1080/00387018408076116.

[9]   Rouessac, F., and Rouessac, A. Chemical analysis: Modern instrumentation methods and techniques, 2nd edn, Hoboken, NJ: Wiley-Blackwell; 2007.

[10] Palmer, C. Diffraction Grating Handbook, 6th edn, Rochester, NY: Newport Corp.; 2005.

[11] Grossman, W.E.L. A comparison of optical detectors for the visible and ultraviolet. Journal of Chemical Education. 1989; 66(8): 697–700.

[12] Polyakov, S.V. Chapter 3 – photomultiplier tubes. In: et al. (eds) Single-photon generation and detection physics and applications. Elsevier; 2013, pp. 69–82.

[13] Lagalante, A. Atomic Absorption Spectroscopy: A Tutorial Review. Applied Spectroscopy Reviews. 2007; 34(3): 173–189.

[14] Bings, N.H. and Bogaerts, A., and Broekaert, J.A.C. Atomic Spectroscopy. Analytical Chemistry. 2013; 85(2): 670–704.

[15] Browner, R.F. and Boorn, A.W. Sample introduction techniques for atomic spectroscopy. Analytical Chemistry. 1984; 56(7): 875A–888A.

[16] Gamez, G., Voronov, M., Ray, S.J., Hoffman, V. and Hieftje, G.M. Surface elemental mapping via glow discharge optical emission spectroscopy. Spectrochimica Acta Part B: Atomic Spectroscopy. 2012; 70: 1–9.

[17] Aller, A.J., Chapter 5 Interferences: types and correction.fundamentals of electrothermal atomic absorption spectrometry. World Scientific Publishing Co Pte Ltd; 2018, pp. 185–222.

[18] Hwang, J.Y. and Fuwa, K. Comparative study of chemical interferences of calcium in atomic absorption and flame emission spectrometry. In: Grove, E.L., Perkins, A.J. (eds) Developments in

applied spectroscopy developments in applied spectroscopy. Boston, MA: Springer; 1971, pp. 249–262.

[19] Koirtyohann, S.R. and Pickett, E.E. Spectral interferences in atomic absorption spectrometry. Analytical Chemistry. 1966; 38(4): 585–587.

[20] Narin, I., Soylak, M., Elçi, L. and Doğan, M. Determination of trace metal ions by AAS in natural water samples after preconcentration of pyrocatechol violet complexes on an activated carbon column. Talanta. 2000; 52(6): 1041–1046.

[21] Kiran, K. and Janardhanam, K. Determination of trace metals in water samples by flame atomic absorption spectrometry using column solid-phase extraction. Asian Journal of Chemistry. 2007; 19 (5): 3468–3474.

[22] Banerjee, P. and Prasad, B. Determination of concentration of total sodium and potassium in surface and ground water using a flame photometer. Applied Water Science. 2020; 113(5): 113.

[23] Chatwal, G.R., and Anand, S.K. (eds). Instrumental methods of chemical analysis. 4th edn, Himalayan Publising House; India. 2002.

[24] https://www.perkinelmer.com/category/electrodeless-discharge-lamps.

[25] Hanna, S.N. and Jones, B.T. A Review of Tungsten Coil Electrothermal Vaporization as a Sample Introduction Technique in Atomic Spectrometry. Applied Spectroscopy Reviews. 2011; 46: 624.

Section IV: **Thermoanalytical Methods**

Mukesh and Jaspreet Kaur

Chapter 7
Thermogravimetric (TG) and derivative thermogravimetric analysis (DTG)

Abstract: Thermogravimetric (TG) and derivative thermogravimetric (DTG) analyses are the most prominent and versatile techniques of thermal analysis that are developed and used to measure the variation in the mass (loss or gain) of a material sample arising through decomposition, degradation, vaporization, sublimation, oxidation, reduction, adsorption, desorption, etc., when a sample is subjected to the programmed temperature changes under controlled atmosphere. The chapter briefly describes the introduction of both the techniques, highlighting the differences between two. In practical aspects, the working principles of both the techniques and instrumentation have also been explained. The usefulness of the techniques in the quantitative estimation of calcium (Ca^{2+}) and magnesium (Mg^{2+}) in the sample mixtures have been covered in the chapter.

7.1 Introduction

After learning to control fire, inquisitive human beings heated various substances to observe the effect of heat on them, as it is not only one of the easiest tests to perform but also provides a lot of information about the substance. For example, when solid water (ice) is heated, it melts at 0 °C and if the heating is continued, then it boils at 100 °C and converts into gaseous water (steam). Heating of coal produces inflammable gases, tars, and coke. On heating sugar, it melts initially and then forms brown caramel [1]. Producing pottery, extracting metals, and making glass can be regarded as the earliest attempts of mankind in studying the effect of heat on different materials [2].

The list is endless and the take-home point is that such analysis of a change of a behavior of a material sample on heating can yield a great deal of useful information about the nature of the material. It is evident from the above paragraph that when a substance is heated, it may undergo certain physical changes like melting or boiling, and/or chemical change like decomposition reaction over a wide range of temperatures.

Thermal methods of analysis are the culmination of this scientific temperament of the mankind for the systematic study of effect of heating on different properties of a substance under controlled conditions.

https://doi.org/10.1515/9783110794816-007

7.2 Thermal analysis

As per the ICTAC (International Confederation for Thermal Analysis and Calorimetry), "Thermal Analysis (TA)" is the "study of the relationship between a sample property and its temperature as the sample is heated in controlled condition" [3].Thermal analysis, in general, includes a group of techniques that is used for analysis of a change in a property when it is heated or cooled under controlled environment. Each technique is defined according to the types of physical changes being analyzed, such as temperature, mass, temperature difference, heat flow difference, dimensions, and mechanical properties. While evaluating material characteristics, it is necessary to use different techniques or a combination of multiple techniques, depending on the purpose [4, 5].

Thermogravimetric analysis (TGA) or thermogravimetry (TG) is a type of thermal analysis technique in which "the mass change of a substance is measured as a function of temperature or time, while the substance is subjected to a controlled temperature program under a specified atmosphere" [6].

TG is an experimental technique in which the mass or weight of a sample is plotted as a function of sample temperature or time. In TGA, therefore, only those changes can be observed and studied that affect the mass of the sample; changes like condensed phase changes such as melting or crystalline transitions cause no mass change.

While studying the mass loss, the sample may be held at a constant temperature (isothermal measurement), or subjected to a constant heating rate (linear temperature program), or may be heated under nonlinear temperature program. It is the type of information that one seeks from the TG experiment that decides the choice of temperature program. The technique is termed as "isothermal mode" when the mass change is recorded as a function of time. The TG instrument, in general, is called a thermobalance. The type of atmosphere plays an important role in TG results. Sometimes, the TG instrument may include the analysis of evolved volatile or gaseous products called evolved gas analysis (EGA).

7.3 History of thermobalance

The credit for the development of earliest form of the thermobalance i.e., continuous measurement of mass change during heating may be given to Brill [7, 8]. But credit for the invention of first true thermobalance goes to Honda. [5, 9, 10] The earliest form of the commercial thermobalance, known as Talabot-Perso-Rogeat desiccator (Figure 7.1), was used in the middle of the nineteenth century for ascertaining the dry weight of silk, which is known to be hygroscopic [11].

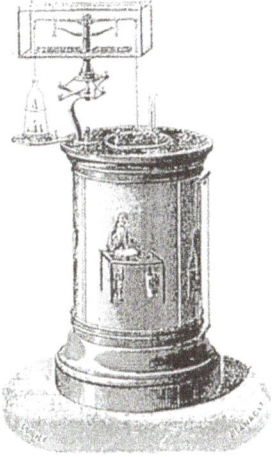

Figure 7.1: The Talabot-Persoz-Rogeat (TPR) desiccator [11] (reprinted with permission from SNCSC ©1995).

7.4 Understanding a TG curve / trace or a TG thermogram

A typical form of a result obtained from a TG instrument is given in Figure 7.2. You may note that the result obtained from a TG is in the form of a plot of mass or weight

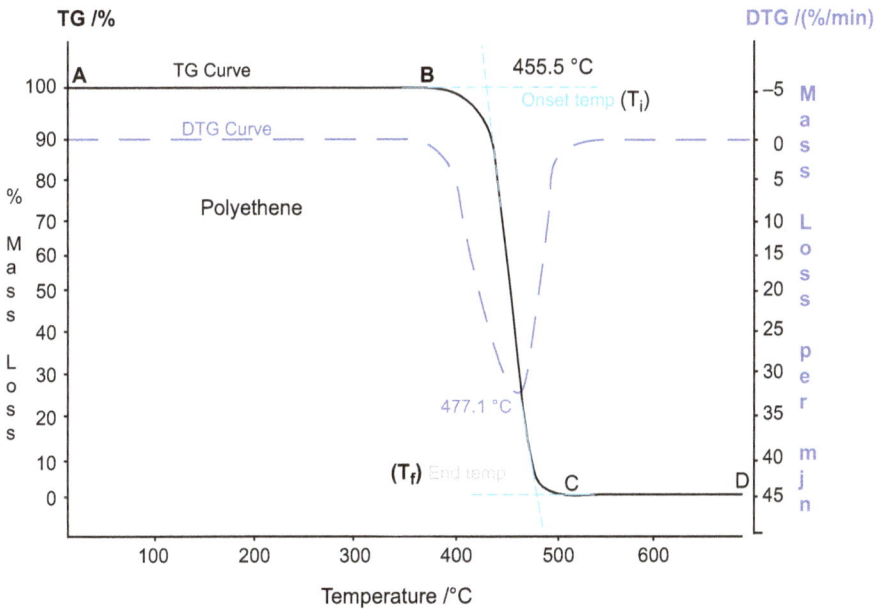

Figure 7.2: Typical TG and DTG curves obtained from a TG instrument.

of a sample, or % weight loss, or % mass loss against time or temperature. Such a plot is usually called a TG thermogram or TG curve or trace.

Derivative thermogravimetry (DTG) is different from the usual TG curve / trace as the results in DTG are presented in an alternative or complementary form. Usually, it is called a derivative plot of TG curve.

In a typical DTG curve, initially the rate of mass loss (or % mass loss) with respect to time or temperature (dm/dt or dm/dT) is obtained and then plotted against temperature or time. Production of such curves is called DTG. A typical DTG curve is also shown in Figure 7.2.

7.5 Interpreting TG and DTG curves

If one goes through the TG curve given in Figure 7.2, it may be noted that there are different regions on the curve:

i. **Horizontal portion or plateau region (A to B and C to D)** of the curve indicates no mass loss/weight loss of the sample in the corresponding temperature range specified on the ordinate. One can easily deduce that the material is thermally stable in this region.

ii. **Curved region of the curve (B to C),** where the *curve declines* indicate a weight loss/mass loss of the sample in the corresponding temperature range on the ordinate. The weight loss might have been a result of the physical change in the sample such as evaporation, sublimation, or desorption, or a result of chemical reaction in the sample such as dehydration, decomposition, degradation, etc., evolving volatile components.

You must have noticed that this process leading to weight loss completes in a wide range of temperature, which is different from what you might have expected, based on your previous experience of performing a reaction in solution state and completing them in a narrow range of temperature. It must be kept in mind that in TG, substances undergoing reaction are generally in solid state, and reactions are relatively slow in solid state compared to liquids/solutions and gases. The wide range of temperature signifies the slow nature of the reaction.

You may also have noticed the "S" shape of TG curve between B and C, which is shown in Figure 7.2. The shape is such, because at the start of the decomposition reaction (near the onset temperature, T_i), the value of slope of the declining TG curve, i.e., rate of change in mass with respect to temperature is small; but as the temperature progresses, the slope of the declining curve increases and reaches a maximum value, and thereafter, with further increase in temperature, the value of slope of the declining curve falls off rapidly and reaches zero value near the end temperature (T_f).

Occasionally, one may come across an inclining region with increase in temperature in a TG trace. Such an inclining curve indicates a weight gain in the sample. This weight gain may be attributed to processes such as adsorption and oxidation.

One can estimate the onset temperature (T_i) and the end temperature (T_f), i.e., the temperatures at which the decomposition reaction initiates and gets over, by extrapolating the horizontal and vertical portion of the curve to let them intersect at a point as shown in Figure 7.2. The term "reaction interval" of the thermal event corresponds to the temperature range between T_i and T_f.

There are seven different categories of TG curves that may be obtained, and these are given in Figure 7.3(a–g).

i.   Type "a" horizontal curve indicates no mass loss in the sample in the temperature range shown. This type of curve is obtained when either no decomposition reaction(s) is taking place in the sample leading to no loss of volatile products, or it may have undergone thermal events like phase transitions or polymerization, or other reactions may have taken place that do not generate volatile products. In such cases, the assistance of other thermal techniques may be taken to determine the type of thermal events taking place in the sample.

ii.   Type "b" curves exhibit initial mass loss, followed by a horizontal curve and this type of TG indicates desorption or drying of the sample. To confirm, a rerun of the sample should be done, and then it should give type "a" TG curve.

iii. Type "c" curve indicates decomposition reaction in a single stage in the corresponding temperature region shown. This type of the curve is useful to define the thermal stability limit of the sample.

iv.   Type "d" curves indicate that the sample is undergoing a multistage decomposition reaction in the displayed temperature range involving relatively stable intermediates. This type of curve is useful in defining the limit of thermal stability of the sample as well as of the intermediates of the decomposition reaction of the material.

v.   Type "e" curve indicates a multistage decomposition reaction of the sample in the exhibited temperature range, where unstable intermediates are formed. It becomes very important to check the heating rate if one gets this type of curve. Lowering the heating rate may convert the type "e" curves into type "d" curve. This type of curve is also useful in defining the limit of thermal stability of a substance of sample.

vi.   Type "f" and "g" curves represent gain in mass of the sample from the surroundings which may be a result of the oxidation of the sample in the exhibited temperature range. The difference between the two curves is that while in type "f" curve, the product of oxidation is stable, the oxidation product gets degraded in the type "g" curve, subsequently, resulting in mass loss. Type "g" curves are rarely seen.

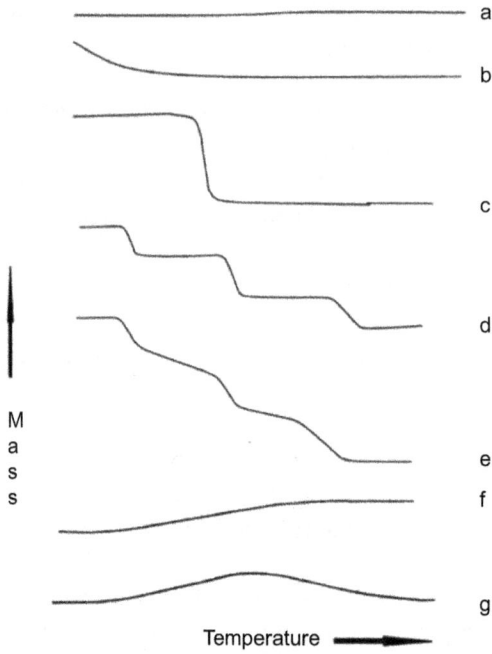

Figure 7.3: Comparison of main types of curves obtained from TG [4].

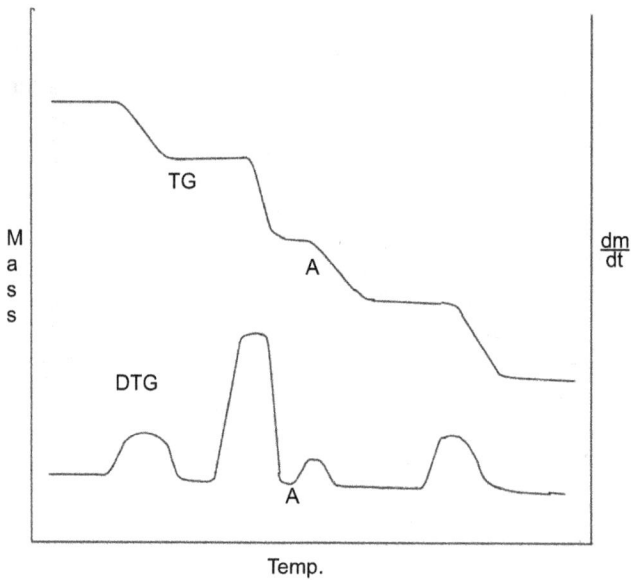

Figure 7.4: Resolution of TG curve with the help of DTG curve [4].

7.6 Importance of DTG curve

DTG is very helpful in the resolution of the individual stages of the TG curve (Figure 7.4). There are a few other points that are important to note:

i.   In the DTG curve, the value of the derivative of mass with respect to temperature, dm/dT is zero, when there is no weight loss in the corresponding TG.
ii. The peak of the DTG curve corresponds to a maximum slope in TG, i.e., maximum value of dm/dT.
iii. When dm/dT is minimum but is nonzero, there is an inflexion that means that there is change of slope on the corresponding TG curve (Figure 7.4, point A on the TG and DTG curves)
iv. When the spread of a TG thermal event (reaction) is over a wide range of temperature, a relatively broad peak is encountered in DTG.
v.   If there are overlapping reactions taking place in a sample, then the importance of the DTG curve becomes more apparent as it helps in simplifying the TG curve.
vi. DTG is also of great help when a double peak appears or a shoulder on a main peak appears in the TG trace.
vii. When there is a superimposition of the TG curves due to slow reactions and other simultaneous fast reactions, it appears as gradient changes in the DTG curve.

Further, it may be noted that the area under the DTG peak is proportional to the mass loss. The position of the DTG peak may not be indicative of any characteristic point in the mechanism of the reaction, but is only indicative of where mass loss is fastest. However, it may be used as a "fingerprint" of the presence of a substance in a mixture, for example, a particular mineral in a rock or soil sample.

TG and DTG analyses and their combinations together are the most prominent and versatile techniques of thermal analysis that have been developed. However, it is important to understand that not all types of thermal events lead to the mass change in the sample. These techniques are used to measure the variation in the mass (loss or gain) of a material (sample) arising through thermal events such as decomposition, degradation, vaporization, sublimation, oxidation, reduction, adsorption, desorption, etc., when a sample is subjected to the programmed temperature changes under controlled atmosphere.

7.7 Instrumentation

The block diagram of a typical TG instrument is displayed in Figure 7.5, and it consists of the following essential components:

i.   Balance
ii. Furnace or heating device
iii. Unit for temperature measurement and control

iv. Atmosphere control around the sample
v. Sample holder or crucible
vi. Unit for data recording and processing

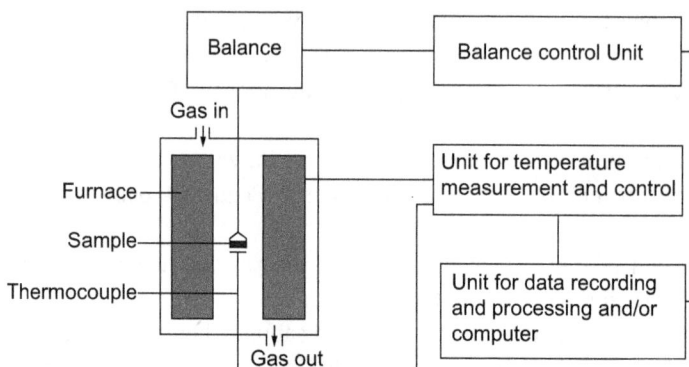

Figure 7.5: Block diagram of a typical TG instrument [12].

7.7.1 Balance

It is one of the most important indispensable components of the TG instrument or thermobalance. Most of the modern TG instruments typically use 10–100 mg of sample. Different types of balances are used in TG instruments. The measuring range of these balances is usually from 0.0001 mg to 1 g (gram). According to the resolution they provide, balances are generally classed as ultramicro- (0.1 μg), micro- (1 μg) or semimicro- (10 μg) balances. Balances having high accuracy, precision, sensitivity, stability, and reproducibility as well as those with fast response are the need of TG instruments. The sample of the substance that is to be studied thermogravimetrically is placed in a small inert crucible, which is either placed on top loading microbalance or is attached at the bottom to the double beam microbalance (bottom loading balances). Horizontal balances are rarely used and the atypical traditional spring type balance is not well suited for these instruments. Depending upon how a balance measures the changes in mass of sample in the thermobalance, these are categorized into two:

7.7.1.1 Null position type or compensation balance

These are the more widely used balances due to their ease of operation. The microbalance operates on a null balance principle. Deviation of the balance beam from the horizontal position, i.e., null position due to change of mass (weight) of a sample is detected by a sensor, which, in turn, triggers a restoring force such as current to bring the balance beam back to the null position. Thus, the amount of restoring force in terms of the current required to restore the balance to null position is directly proportional to the change in mass of the sample (Figure 7.6(a)).

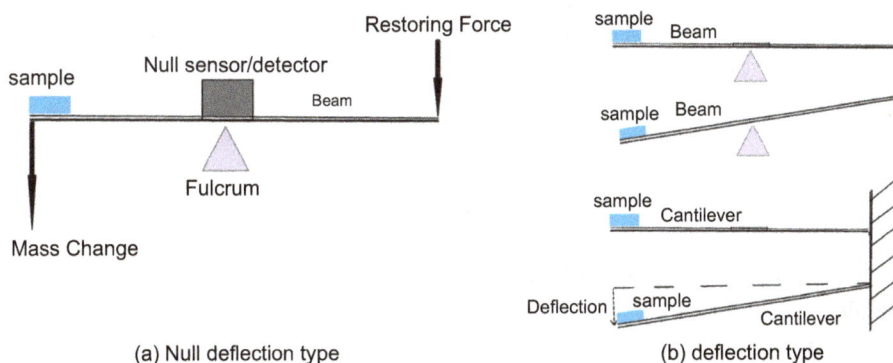

(a) Null deflection type   (b) deflection type

Figure 7.6: Different types of microbalances [6]: (a) null deflection type and (b) deflection type.

7.7.1.2 Deflection-type balance

Here, the mass change in the sample produces deflections in the beam of the microbalance (Figure 7.6(b)). The deflection about the fulcrum due to mass change of the sample is recorded exploiting photographic trace, and the corresponding signal is produced with the assistance of displacement-measuring transducers or electromechanically [6].

7.7.2 Furnace or heating device

The furnace is another indispensable component of TG instrument. Different furnaces with temperatures ranging from ambient temperature to 2,000 °C are used in modern TG instruments, depending on the model (Figure 7.7).

It is normally an electrical resistive heater. It must be designed in order to achieve a linear rate of heating over the whole working range. It should be kept in mind that the internal environment of the furnace can be affected by the temperature regime as well as the nature of the gaseous environment in the furnace. Hence, the furnace is chosen according to the requirements of the TG experiment. Different types of materials such as Nichrome or Kanthal wire or ribbon wound on ceramic or silica tube for temperature range up to 1,100 °C, platinum or platinum rhodium alloy for temperature up to 1,600 °C, and tungsten for even higher temperature are used for making heating coils.

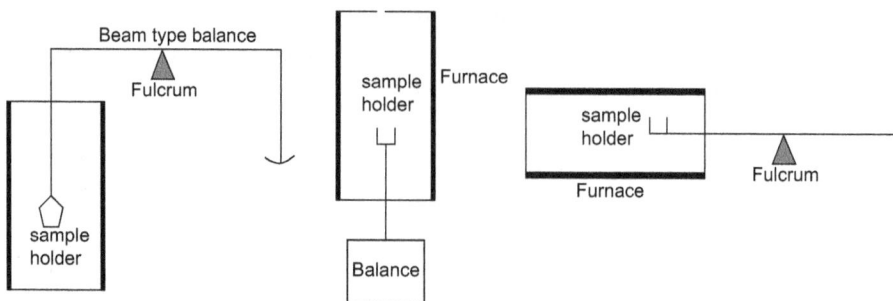

Figure 7.7: Relative position of the furnace with respect to different balances.

The relative position of the furnace varies in different TG instruments and is often decided by the type of the balance used in the instrument. In the case of beam balance, the furnace is generally at the bottom; it is at the top for spring type top leading balances, and in the case of horizontal balances, it is at the side of the instrument (Figure 7.7). Accordingly, the different thermobalances are classified as top-loading, bottom-loading, or side-loading. Inside the furnace, the crucible containing a sample is placed. A mechanism is provided in the TG instrument that allows the movement of the furnace away from the balance case to allow access to the sample. A thermocouple (used for measuring temperature) is placed near the sample. Therefore, it must also be noted here that the temperature of the furnace corresponds to temperature of the crucible containing the sample, and the heating rate of the furnace is regulated and governed by the temperature controller or furnace controller.

The size of the furnace also plays an important role in accurate TG measurements. A relatively heavy furnace not only withstands high temperature ranges but also provides a more uniform heating zone. However, it may require more time to get hotter and achieve targeted temperature, whereas a relatively lighter furnace acquires the targeted temperature more quickly but may lack effective control on the increase of rate of temperature. Furnaces mainly rely on conduction via solid or gas for heating of the sample and may have inevitably large temperature gradients. The magnetic field generated by the resistive heating coils must be kept in mind while working on magnetic samples, as this magnetic field may alter the accuracy of result of the TG experiment.

7.7.3 Unit for temperature measurement and control

7.7.3.1 Thermocouple

The temperature in the TG instrument is measured using a thermocouple temperature sensor. As the name suggests, a thermocouple consists of two different metals fused into a junction of a bead. The measurement of temperature is based on Peltier effect, that is, generation of standard EMF (electromotive force) across the junction when it is heated or cooled.

The material used in the thermocouple is chosen to produce very reproducible and accurate temperature measurement. The sensitivity and cost also determine the choice of the material. Chromel-Alumel, which are alloys of the aluminum are also used in TG instruments with temperature range below 1,100 °C. However, the most frequently used material in a thermocouple is a combination of tungsten and platinum alloys, due to their inherent advantages like their high melting point and their inertness to the most samples as well as purged and evolved gases. The placement of a thermocouple varies in different TG instruments. In some instruments, it is in direct contact with the sample holder and sample; however, in others, it is generally placed near the sample holder. The placement of the thermocouple should be as close to the sample as possible, but such placement should not affect the working of the balance or weight of the sample and should not damage the thermocouple.

7.7.3.2 Temperature controller or programmer

The aim of the temperature controller or programmer is to control the rate of heating during the analysis in such a manner as to accomplish a rate of heating (with respect to time) that is as linear as possible. To achieve the target, a thermocouple is placed very close to the resistance wire winding of the heating coils, almost touching it, by embedding it in the cement coating of the furnace, connecting it to the furnace temperature control circuit/unit.

7.7.4 Atmosphere control around the sample

One would prefer to have static air or dynamic air environment around the sample while heating, keeping the simplicity of the process in mind. But the sample may react with the various components of the air, such as oxygen present in its vicinity, while it is getting heated. Therefore, an inert dynamic environment of nitrogen or argon gas is provided. However, depending on the purpose of the TG study, specific environment such as inert, oxidizing, reducing, or corrosive may be provided to the sample via purging of different gases. To protect the balance from these different environments, it is important to house the balance in glass or metal. Such enclosure of balance in housing also enables the application of high vacuum and pressure up to 70 bars around the sample. The choice of the purge gas is determined by several factors such as reactivity or lack of reactivity, cost, availability, purity, density, thermal conductivity, etc. The denser inert gas, argon, is preferred over helium when the balance is above the furnace in a TG instrument, while the opposite is true for the reverse arrangement of the balance and furnace. Higher thermal conductivity of the purge gas can significantly improve heat transfer. To achieve the measurement and/or to control the flow of purge gas(es), various devices such as floats and tubes or soap film bubble towers are used. Valves are employed to switch flow of gas from one to another, for instance, from inert to oxidizing gas, etc. [13].

7.7.5 Sample holder or crucible

The sample which is to be studied thermogravimetrically is placed in a crucible, and the crucible is attached to the weighing arm of the weighing balance. These crucibles vary in shape, sizes, and material, and choice of material depends on the temperature range under consideration. The material chosen must be thermally stable under TG experiment conditions, inert to the sample and gases purged or evolved in TG experiment, and should have high thermal conductivity, so that heat can be transferred uniformly and efficiently to the sample. These crucibles are generally made of quartz, aluminum, or platinum. Also, thin crucibles are employed in order to keep their heat capacity low. Materials like stainless steel, glass, or graphite are also used occasionally. The shape of the crucible depends on the nature of the sample as well as the mass of the sample under consideration. Flat crucibles or shallow pan-shaped crucible are preferred for powder samples, as such crucibles allow effective escape of volatile substances generated on heating. Such a type of crucible is also used if diffusion of the volatile substances is a rate-limiting factor. For liquid sample, walled crucible or retort cup crucible is used.

7.7.6 Unit for data recording and processing

It varies in different TG instruments. In older instruments, it is a data recorder and chart plotter for recording and processing the mass and temperature data from the balance and sample thermocouple and to plot the temperature versus mass loss curve. Nowadays, computers do the entire job including the plotting of DTG curve. The control unit also has switches to control the mass range and to tare the mass to zero when the crucible is empty. It may also provide dampening control to dampen the swinging of the balance.

7.8 Factors that affect the results of TG

Although TG trace can be used to identify a substance, it should not be considered as a "fingerprint" for the substance as in the case of other methods like spectroscopy, because many experimental variables affect the results. TG traces of the same substance obtained from different instruments may also vary depending on the nature/value of different parameters used. Some of the prominent factors that affect the results of TG are discussed below:

7.8.1 Heating rate

The shape of the TG curve is considerably influenced by heating rate. Also, the temperature region in which the reaction occurs is very much dependent on the heating rate.

If a sample undergoes a single-stage decomposition reaction, following are the effects of the heating rate:
1) The observed values of T_i (observed temperature at which decomposition reaction begin) and T_f (observed temperature at which decomposition reaction ends) are higher at faster rates of heating than that at slower rates of heating.
2) The observed temperature range $(T_f - T_i)$, between which reaction takes place also varies with heating rate.
   a) (T_i) at faster heat rate of heating > (T_i) at slower heat rate of heating
   b) (T_f) at faster heat rate of heating > (T_f) at slower heat rate of heating
   c) $(T_f - T_i)$ at faster heat rate of heating > $(T_f - T_i)$ at slower heat rate of heating
3) It can also be said that at any temperature in the reaction interval, the extent of decomposition is more if the curve is recorded at slower heating rate as compared to that at faster heating rate.

The heating rate becomes more important if a sample also undergoes secondary decomposition reaction, with starting temperatures differing only by a small extent. By use of inappropriate heating rate, the reactions may overlap and remain undetected. However, with the exercise of appropriate rate of heating, one may resolve them.

7.8.2 Furnace atmosphere

The TG measurement of the sample cannot be done without allowing the exchange of material with its immediate surroundings, as mass of the sample in the closed system would remain the same even if it is undergoing decomposition. An important need is, therefore, the gaseous environment around the sample in the form of furnace atmosphere.

The furnace atmosphere may also have a significant effect on TG curve depending upon:
a) Type of thermal decomposition reaction taking place in a sample
b) The nature of the products in the decomposition reaction
c) Type of atmosphere employed in furnace

Example: If a sample of calcium oxalate monohydrate undergoes a thermal decomposition, then the effect of exercise of furnace environment of gases like N_2 (nitrogen) and O_2 (oxygen) would be different on the TG curve. The sample of monohydrate of calcium oxalate undergoes three thermal decomposition reactions, as shown in Figure 7.8, namely:

i. A reversible dehydration reaction:

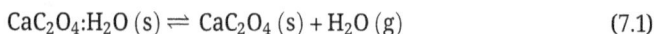

$$CaC_2O_4{:}H_2O\ (s) \rightleftharpoons CaC_2O_4\ (s) + H_2O\ (g) \tag{7.1}$$

ii. Decomposition of calcium oxalate to calcium carbonate and carbon monoxide:

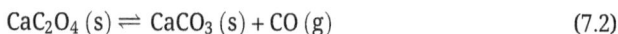

$$CaC_2O_4\ (s) \rightleftharpoons CaCO_3\ (s) + CO\ (g) \tag{7.2}$$

iii. Decarboxylation of calcium carbonate to calcium oxide and carbon dioxide:

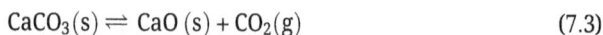

$$CaCO_3(s) \rightleftharpoons CaO\ (s) + CO_2(g) \tag{7.3}$$

TG traces due to decomposition reactions (7.1) and (7.3) largely remain unaffected under the dynamic furnace environments of N_2 and O_2 gases, as both the gases are equally effective in removing the evolved gases away from the sample surface and its neighborhood.

However, there is a slight difference in the TG decomposition traces for the reaction (7.2) under the two gases, which may be attributed to an exothermic reaction that takes place between the evolved CO (carbon monoxide) and O_2 gas present in the vicinity of the sample to produce CO_2 (carbon dioxide), resulting in an increase in temperature of the unreacted sample. This increase in temperature slightly accelerates the decomposition reaction, and it is completed at comparatively lower temperature. No such effect is seen in the case of inert N_2 gas environment.

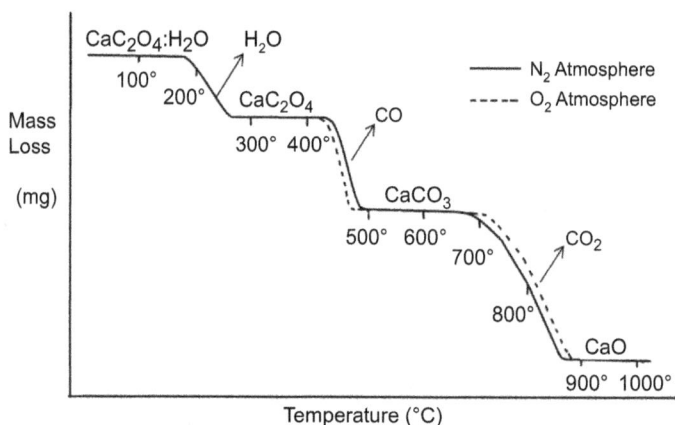

Figure 7.8: Schematic representation of TG curve of calcium oxalate hydrate ($CaC_2O_4{:}H_2O$) in N_2 and O_2.

7.8.3 Shape/geometry as well as the nature of the sample holder (i.e., Crucible)

As we have discussed earlier in this chapter, the shape of the sample holder or crucible varies from flat shallow pan to deep retort cup. The geometry of the crucible can alter the slope of the declining TG curve [14], i.e., the shape of the decomposition curve. Also, the weight of the crucible should be low to allow uniform increase in the temperature.

7.8.4 Sample characteristics such as weight, volume of the sample, and physical form of the sample

The shape of the TG curve is often affected by the weight, particle size, and mode of preparation (history) of the sample. For example, TG decomposition traces corresponding to decarboxylation of calcium carbonate, $CaCO_3$ (reaction (7.3)) given in Figure 7.8 differs slightly under different furnace environments of N_2 and O_2 gases. This difference may be attributed to the difference in the history of the $CaCO_3$ solid, i.e., different particle size, surface area, and other physical characteristics of the solid that may have formed during the previous step due to different environments.

The following are the effects of mass of a sample on a TG trace:
i.   With a large sample size, linearity of temperature rise is difficult to obtain.
ii. If the sample has low thermal conductivity, then the large sample tends to have considerable temperature gradient due to which outer regions of the sample may undergo decomposition earlier than the core of the sample.
iii. Moreover, a large sample may impede the diffusion of evolved gases.

7.8.5 Effects of particle size of a sample

Samples with bigger particle size can impede diffusion of evolved gases from the bulk of the solid crystals, which may be lead to decrepitation (spitting or spattering) or swelling or foaming of sample, which may further lead to spurious mass loss, if a few pieces of sample jump out of the crucible.

7.8.6 Effect of heat of reaction

Temperature gradient may further get enhanced when the sample is undergoing a reaction with large heat as in the case of the decomposition of calcium oxalate to produce CO, which is an exothermic reaction (Figure 7.8). Therefore, it is better to keep the weight/mass as well as particle size of the sample as small as possible. Also, it should be spread finely on the crucible.

7.9 Applications of thermogravimetry

The various applications of thermogravimetry are:
i. Determination of purity (e.g., residual solvents) and thermal stabilities of a material (e.g., oxidative and reductive stability)
ii. Investigation of appropriate drying temperature and conditions and suitability of various weighing forms in gravimetric experiments
iii. Determination of composition of alloys, polymers, rubbers, coal, and mixtures (compositional analysis), e.g., for quantitative estimation of calcium and magnesium in the given sample, determination of carbon black amount in polymers, determination of moisture, volatile components, and ash content of the given solid
iv. Ascertaining decomposition mechanism of inorganic salts
v. Direct application to analytical problems
vi. Study of kinetics of a thermal event or solid-state reactions
vii. Simulation of manufacturing (firing process) process in the case of ceramics as well as glass and study of evolved gas analysis
viii. Interaction with environment (e.g., reaction with environment gases) absorption and desorption studies of silica gel under humid environment or corrosion studies of steel (oxidation)
ix. Study of desolvation, sublimation, vaporizations, sorption, desorption, and chemisorption.

TG analysis is found to be important in materials like:
i. Polymers and plastics
ii. Building materials such as cement and concrete
iii. Ceramics and glass
iv. Inorganic materials like minerals, soils, and clays (for roasting and calcination)
v. Metals and alloys
vi. Pharmaceuticals
vii. Organics like fats, oils, soaps, and waxes
viii. Textiles and fabrics
ix. Explosives
x. Catalysts
xi. Flame retardants and
xii. Pyrolysis of coal, petroleum, and wood

7.10 Quantitative estimation of calcium and magnesium in the given sample

TG traces can be used to predict the relative amount of two substances in their binary mixture [15]. For doing so, it is required that the characteristic decomposition reaction(s) of the two substances lie in different temperature regions. This can be illustrated with the help of the following two examples.

7.10.1 Estimation of calcium and magnesium in a binary mixture of their carbonates

A typical TG trace of a mixture of calcium and magnesium carbonates, shown in Figure 7.9, depicts three declining regions signifying mass losses at three distinct non-overlapping temperature ranges.
i.   First, up to 210 °C, which is attributed to loss of moisture from the sample
ii. Second weight loss at about 470–480 °C is due to decarboxylation of magnesium carbonate ($MgCO_3$) to form magnesium oxide (MgO).
iii. Third weight loss takes place at around 850 °C, which is attributed to decarboxylation of calcium carbonate ($CaCO_3$) to calcium oxide (CaO).

The TG curve or trace given in Figure 7.9 also depicts three plateau regions, namely *ab*, *cd*, and *ef* representing $MgCO_3 + CaCO_3$, $MgO + CaCO_3$, and $MgO + CaO$, respectively. It must be recalled, here, that the plateau region of the TG curve represents a temperature region where no loss in mass of the sample takes place. If the mass (in mg) of the sample at the plateau regions, *cd* and *ef* is m_1 and m_2, respectively, then we may note that m_1 is due to the $MgO + CaCO_3$, and m_2 is due to $MgO + CaO$.

Therefore, change in mass in the declining region of the curve between plateau regions *cd* and *ef* is equal to ($m_1 - m_2$). This mass loss is due to the decarboxylation of $CaCO_3$.

If m_3 is the mass of the CaO at the plateau region, *ef*, then it can be written that

$$m_3 = (m_1 - m_2) \times \frac{M(CaO)}{M(CO_2)} \tag{7.4}$$

or

$$m_3 = 1.27 \times (m_1 - m_2) \tag{7.5}$$

where M(CaO) and M(CO_2) are the molar masses of the calcium oxide and carbon dioxide, respectively.

Now, if the mass (in mg) of the MgO in the plateau region *ef* of TGA trace (Figure 7.9) is m_4, then

$$m_4 = (m_2 - m_3) \tag{7.6}$$

Plugging the value of m_3 from eq. (7.5) in eq. (7.6), we get

$$m_4 = (m_2 - 1.27 \times (m_1 - m_2)) \tag{7.7}$$

Moreover, if m_{Ca} and m_{Mg} represents the masses of the calcium and magnesium in the binary sample, respectively, then,

$$m_{Ca} = m_3 \times \frac{M(Ca)}{M(CaO)} \tag{7.8}$$

Plugging the value of m_3 from eq. (7.6) in eq. (7.8), we get

$$m_{Ca} = 1.27 \times (m_1 - m_2) \times \frac{A(Ca)}{M(CaO)} \tag{7.9}$$

So,

$$m_{Ca} = 0.93 \times (m_1 - m_2) \tag{7.10}$$

and

$$m_{Mg} = m_4 \times \frac{M(Mg)}{M(MgO)} \tag{7.11}$$

Plugging the value m_4 from eq. (7.6) in eq. (7.11), we get

$$m_{Mg} = (m_2 - m_3) \times \frac{A(Mg)}{M(MgO)} \tag{7.12}$$

where $M(MgO)$ is the molar mass of the magnesium oxide, and $A(Mg)$ and $A(Ca)$ are the atomic masses of magnesium and calcium, respectively.

Thus,

$$m_{Mg} = 0.60 \times (m_2 - m_3) \tag{7.13}$$

7.10.2 Estimation of calcium and magnesium in binary mixture of their oxalates

A typical TG curve of the magnesium oxalate dihydrate ($MgC_2O_4{:}2H_2O$) is given in Figure 7.10. It may be noted that there are two declining regions on the TG curve between the temperature regions 160–240 °C and 500–510 °C, signifying mass loss to the sample of magnesium oxalate dihydrate.

i.   Removal of water from the magnesium oxalate dihydrate generating magnesium oxalate (MgC_2O_4)
ii. Conversion of magnesium oxalate (MgC_2O_4) to magnesium oxide (MgO)

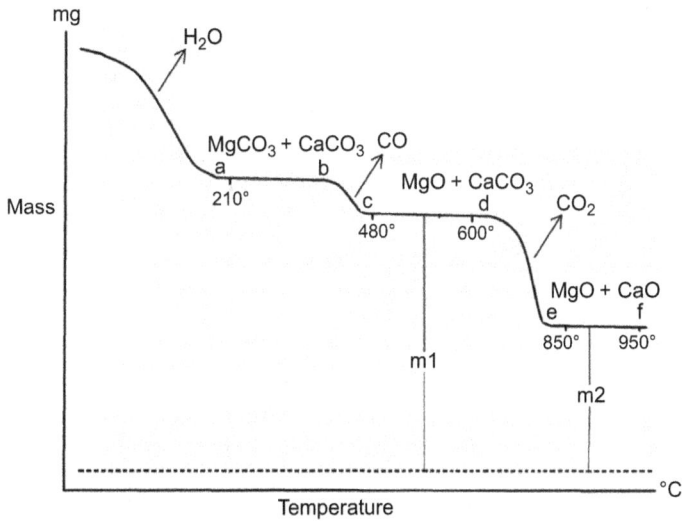

Figure 7.9: A typical TG trace of a calcium and magnesium carbonates (CaCO$_3$ and MgCO$_3$).

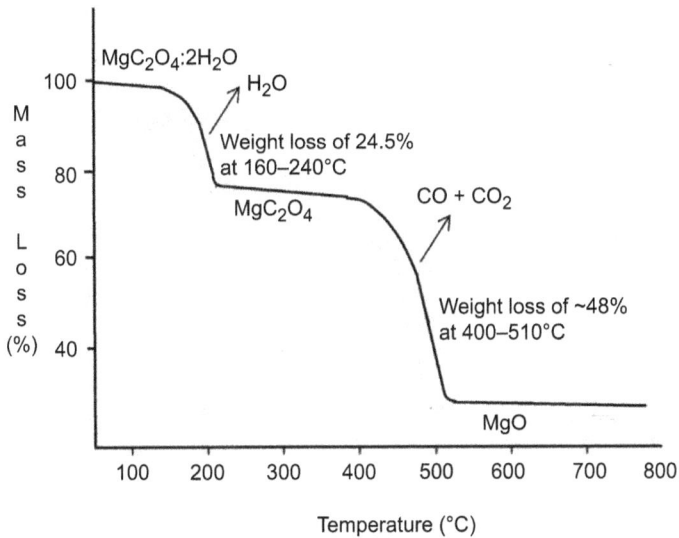

Figure 7.10: A typical thermogravimetric trace of magnesium oxalate dihydrate (MgC$_2$O$_4$:2H$_2$O).

It may be recalled, here, that a typical TG curve of the calcium oxalate hydrate (CaC$_2$O$_4$: H$_2$O) has already been discussed (see Figure 7.8). It has three different temperature regions where weight loss takes place, namely, dehydration of calcium oxalate hydrate at around 200 °C, removal of carbon monoxide from the resultant anhydrous form at around 500 °C, and decarboxylation of consequent calcium carbonate to calcium oxide at around 900 °C.

By comparing TG traces of two oxalates, one may easily deduce that although there is no weight loss in the TG curve of magnesium oxalate above 520 °C, the TG curve of calcium oxalate hydrate loses CO_2 at around 900 °C. Hence, the mixture of two oxalates will be in the form of $CaCO_3 + MgO$ above 500 °C. Therefore, if m_1 and m_2 are the masses of the binary mixture of the two oxalates at plateau above 500 °C and at plateau above 900 °C respectively, then the amounts of calcium and magnesium in the original binary mixture can be calculated as discussed in the example of mixture of calcium and magnesium carbonates (previous section).

Further readings

The topic "application of TG" is considerably large and we have only covered "quantitative estimation of calcium and magnesium in the given sample." Discussion of the rest of the subtopics is beyond the ambit of the chapter. However, there are many good books available on the topic [17–22], which may be consulted.

References

[1]   Haines, P.J. Introduction. In: Haines, P.J. (ed.) Principles of thermal analysis and calorimetry, Cambridge: The Royal Society of Chemistry; 2002, pp. 1–2.
[2]   Brown, M.E. Introduction to thermal analysis: Techniques and applications. Cambridge: Chapmann and Hall; 1988, pp. 1–2.
[3]   Koga, N. Introduction. In: Pielichowska, K., Pielichowski, K. (eds.) Thermal analysis of polymeric materials: methods and developments. Weinheim: Wiley-VCH; 2022, pp. 5.
[4]   Brown, M.E. Introduction to thermal analysis: Techniques and applications. 2nd edn, Dordrecht: Kluwer Academic Publishers; 2001, pp. 7–9.
[5]   Brown, M.E. Handbook of thermal analysis and calorimetry: Principles and practice. Oxford: Elsevier Science BV; 1998, pp. 25–30.
[6]   Loganathan, S., Valapa, R.B., Mishra, R.K., Pugazhenthi, G., and Thomas, S. Thermogravimetric analysis for characterization of nanomaterials. In: Thomas, S., Thomas, R., Zachariah, A.K., and Mishra, R.K. (eds.) Thermal and rheological measurement techniques for nanomaterials characterization: A volume in micro and nano technologies. Oxford: Elsevier Inc; 2017, pp. 67–105.
[7]   Brill, O. Zeitschrift für Anorganische und Allgemeine Chemie. 1905; 45: 275.
[8]   Szabadvary, F. and Buzagh-Gere, E. Historical development of thermoanalytical methods. Journal of Thermal Analysis. 1979; 15: 389–398.
[9]   Honda, K. The Science reports of the Tohoku University. 1915; 4: 97.
[10]   Saito, H. Progress of thermobalance and thermobalance-analysis in japan. In: Schwenker, F., Jr, Paul, D., and Robert, G. (eds.) Thermalanalysis: Instrumentation, organic materials, and polymers. New York: Academic Press Inc; 1969, pp. 10–11.
[11]   Keattch, C.J. Studies in the history and development of thermogravimetry: A search for the earliest commercial apparatus in the UK. Journal of Thermal Analysis. 1995; 44: 1211–1218.

[12] Kealey, D., and Haines, P.J. BIOS instant notes on analytical chemistry. BIOS Scientific Publishers; 2002, pp. 308.

[13] Gallagher, P.K. Thermogravimetry and thermomagnetometry. In: Brown M.E. (ed.) Handbook of thermal analysis and calorimetry: Principles and practice. Oxford: Elsevier Science BV; 1998, pp. 225–267.

[14] Mendham, J., Denney, R.C., Barnes, J.D., and Thomas, M., Text book of quantitative chemical analysis. 6th edn, Hongkong: Pearson education; 2000, pp. 475–486.

[15] Fozdar, B.I. Unit-8 Thermal Methods of Analysis[PDF]. Egyankosh: Indira Gandhi National Open University, New Delhi; 2021. [cited 2022 July 14]. Available from: http://egyankosh.ac.in//handle/123456789/82040.

[16] Mastuli, M.S., Kamarulzaman, N., Nawawi, A.Z., Mahat, A.M., Rusdi, R., and Kamarudin, N. Nanoscale Research Letters. 2014; 9: 134.

[17] Wagner, M. (ed.). Thermal analysis in practice: Fundamental aspects. Carl Hanser Verlag GmbH Co KG; 2017.

[18] Ebeid, E.M., and Zakaria, M.B. Thermal analysis: from introductory fundamentals to advanced applications. Elsevier; 2021.

[19] Russell, J., and Cohn, R. (eds). Thermogravimetric analysis, Book on demand, 2012.

[20] Surhone, L.M. and Timpledon, M.T., and Marseken S.F. (eds). Thermogravimetric analysis. VDM Publishing; 2010.

[21] Bottom, R. Thermogravimetric analysis. In: Gabbott, P. (ed.). Principles and applications of thermal analysis, Blackwell Publishing; 2008, pp. 88–118.

[22] Robinson, J.W., Frame, E.M.S., and Frame, G.M., Undergraduate instrumental analysis. 6th edn, New York: Mercel Decker; 2006, pp. 1003–1019.

Manju Saroj and Ritu Payal

Chapter 8
Differential thermal analysis

Abstract: Differential thermal analysis (DTA) is a method used in analytical chemistry to identify and quantitatively analyze the chemical composition of substances by monitoring the dependence of sample on associated heat. The principle behind the method is that when a substance is heated, reactions and phase transformation take place that entail the absorption or emission of heat (ΔH). DTA is typically thought of as a semiquantitative or qualitative method that has been used in research for a range of scientific investigations. We can foresee new applications of DTA and enhancing its functionality inside the "black box."

8.1 Introduction

In the early nineteenth century, conventional analytical techniques were used to analyze clay and related materials, but the results were not noteworthy. Even X-ray analysis of these samples typically identifies only the main components, while impurities remain undetected. Additionally, the petrographic analysis of clay and clay-like materials gave unsatisfactory results. The electron microscope method for estimating the size of these materials did not provide fruitful results. After attempting numerous well-known techniques without success, Le Chatelier (in 1884) found a different approach called thermal analysis. A description of thermometry from the sixteenth century is included in Mackenzie's articles on the history of the evolution of thermal technologies from their inception. In the seventeenth century, the introduction of practical calorimetry and the defining of thermometric scales, particularly by Lavoisier and Laplace, really brought about a change in how we think about and approach researching the effects of heat. It directly inspired Fourier's work on heat conduction and Joule's beautiful experiments on electrical heating and calorimetry. The main flaws of their equipment were that they frequently required big samples and took a long time to complete the experiment. The precursors to today's analytical procedures were chemical reactions and physical observations on gases.

Since then, thermoanalysis has been used in a very large range of scientific investigations like construction, electronics, engineering, geology, fine organic chemicals,

Acknowledgment: The author is thankful to the Library and Central Computer Laboratory of Rajdhani College, Raja Garden, West Delhi, for allowing access to their library and internet facilities.

https://doi.org/10.1515/9783110794816-008

polymers and pharmaceuticals, materials science, and quality control (Figure 8.1) [1–6]. Even a complex material, such as a polymer composite, will exhibit distinct instructive effects that are related to its origin, composition, and history which, in turn, provides information about its properties and shell life. The International Confederation for Thermal Analysis and Calorimetry (ICTAC) has created authoritative publications and articles, both on the nomenclature and on the calibration methods to be used, because it was necessary to agree on a common terminology as thermal methods have been developed by numerous researchers [7–11].

The term "thermoanalysis" refers to a series of techniques where a sample's property is measured against time or temperature by controlling the sample's temperature in a certain atmosphere. The reactions that occur during thermal treatment can easily be used to characterize and evaluate the weight loss as well as energy changes of a material or other chemical entity (heating or cooling). A thermobalance can be used to measure, record, and plot these changes over time or against temperature [12]. Differential techniques, on the other hand, involve the measurement of a difference in the property between the sample and a reference – for example, in DTA, the difference in temperature between the sample and a reference is measured. Derivative techniques imply the measurement or calculation of the mathematical first derivative, usually with respect to time. For instance, derivative thermogravimetry measures the rate of mass loss (dm/dt) as a function of temperature [13].

This article explores the history, applications, and a few facets of DTA and reports on how they are used in analytical chemistry.

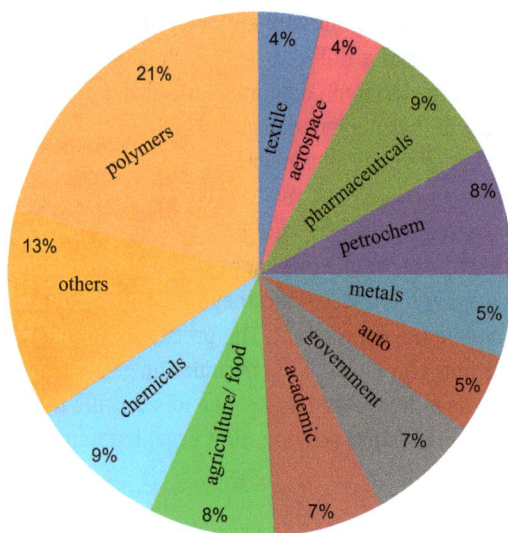

Figure 8.1: Range of scientific investigations of DTA.

8.2 Differential thermal analysis (DTA)

DTA is simply a development of the traditional method of observing phase changes while the system is uniformly heated or cooled, in order to examine phase transformations. DTA is gradually growing into a useful and widely used analytical chemistry tool for identifying and quantitatively assessing the chemical composition of substances by observing the thermal behavior of a sample, as it is heated. DTA measures the heat input required for increasing a sample's temperature. The principle behind the method is that when a substance is heated, it goes through reactions and phase changes that entail heat absorption and emission. When heating or cooling a sample, temperature differences (ΔT) compared to a thermally inert substance are measured in DTA. The term differential is used since changes in a sample are measured with respect to a standard reference material. In DTA, the differential temperature change ($\Delta T = T_{ref} - T_{sam}$) between the sample ($T_{sam}$) and reference ($T_{ref}$) for a fixed amount of heat input is measured [14, 15].

8.3 Instrumentation

The block diagram associated with DTA process is illustrated in Figure 8.2. The components of DTA process comprises a furnace, sample holder, thermocouple, sample containers and a ceramic or metallic block, temperature programmer, and recording system. The functions of these components are as follows:

i. **Furnace:** It supplies heat to both the sample and reference which are arranged symmetrically at the same rate. The furnace is controlled by a temperature program. A differential thermocouple is positioned throughout this procedure to find the temperature difference between the sample and the reference.

ii. **Sample holder (crucible):** The reference material and the sample are both held in the crucible. They are made from both metallic (nickel, platinum, stainless steel, and its alloys) and nonmetallic materials (glass, vitreous silica). The crucible does not interact with the sample during the course of the experiment.

iii. **Thermocouple:** It controls the temperature of the furnace and that of the reference and sample.

iv. **Programmer:** It steadily raises the temperature of the furnace.

v. **Furnace temperature control system:** This allows pure gas to maintain a suitable temperature in the furnace and around the sample holder.

    The sample is often placed in contact with a temperature sensor to record its temperature as the experiment progresses. The data are collected by the sensor system and, after processing, are shown as a thermal analysis curve on a screen or recorder.

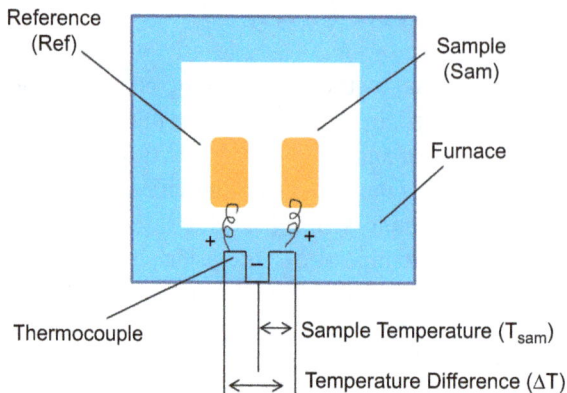

Figure 8.2: Block diagram representing the DTA process.

In Figure 8.2, a sample is placed in recessed heating blocks along with a reference material that is both chemically and thermally inert (usually α-alumina, precipitated silica, or sintered alumina), whereas helium or argon is used as inert gas. The thermocouples in the sample and reference are wired to opposing charges, and because of this, their temperature differences, T, are amplified and displayed against temperature as the "thermal analysis curve." The mass and thermal properties of the sample are revealed by the area under the curve, along with the change in enthalpy (ΔH). If small samples are completely indicative of the parent material, they can be used to produce sharper, narrower peaks. The specific heat capacity of the reference and sample should ideally be comparable [16, 17].

In DTA, the sample's temperature is raised linearly while ΔT is recorded as a function of that temperature. As the sample reacts, the DTA curve records these fluctuations and shows thermal impacts as deviations from the baseline (signal that is in a static state). The resulting DTA diagrams report the sample temperature (abscissa) and temperature (ordinate) in °C. As a result, the curve slopes toward one side of the base line or the other, depending on whether a reaction emits heat (exothermic process) or absorbs heat (endothermic process). When the sample goes through an exothermic process like crystallization or chemical reaction, the sample's temperature rises more than the reference's temperature, causing T to be more negatively skewed, as a result of a simple phase transition. The sample's temperature deviates from that of the reference materials during an endothermic reaction, such as the melting of a crystalline substance or the loss of water of hydration of a molecule, leading to a more positive ΔT value. The overall shape of the DTA curve is shown in Figure 8.3, with negative peaks denoting endothermic processes and positive peaks denoting exothermic processes [18, 19]. There is no physical or chemical change in the sample, though, if the reference and sample temperatures remain the same ($T = 0$). The enthalpy of the thermal process, ΔH, is determined by the region beneath the endotherm or exotherm.

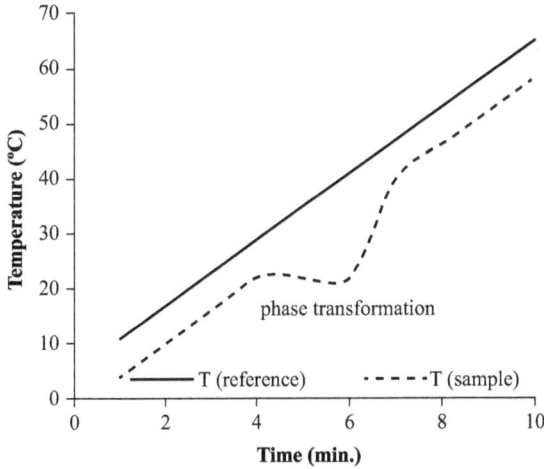

Figure 8.3: DTA curve showing variation of time and Δ*T* of specimen and reference for the same amount of heat associated. A negative peak of sample (dotted line) signals a phase transition.

The characteristics of the test sample can be inferred from the shape and size of the thermal analysis curves. A broad endothermic peak suggests dehydration process, whereas a sharp endothermic peak suggests a change in the fusion process or a change in the sample's crystal structure. Although endothermic curves show the physical changes, exothermic peaks mostly correspond to chemical reactions, showing the sample's oxidative nature.

8.4 Applications of DTA

The DTA process offers applications in various scientific explorations such as:

i. DTA curve provides data on transformations such as glass transitions, crystallization, melting, and sublimation. The heat capacity of the sample has no impact on the area under a DTA peak, which represents the enthalpy change. When a peak in a DTA curve is integrated, its area, *A*, is discovered to be proportional to *H* as

$$H = \kappa A$$

where the calibration constant, κ, is derived using a known reference material [20].

ii. DTA has applications in the research of medicinal substances, liquid crystals, and polymers [21].

iii. DTA method is frequently used for the identification of minerals and mineral combinations [22].

iv. Both food [23–25] and pharmaceutical sectors [26, 27] use DTA extensively.

v. DTA can be used in environmental investigations [28, 29], mineralogy research [30], and cement chemistry [31, 32].
vi. DTA curves can also be used to investigate archaeological materials or date bone remains [33–35].
vii. Only as a fingerprint for identification, a DTA curve can be employed to determine phase diagram determination, heat change measurements, and decomposition in various atmospheres, for example, in the study of clays, where the structural similarity of different forms renders diffraction experiments difficult to interpret [36, 37].

8.5 Advantages and disadvantages of DTA

DTA has several advantages and disadvantages. However, when computer-controlled, DTA has great advantages over traditional methods for proximate analysis.

8.5.1 Advantages of DTA

i. Highly sensitive instrument.
ii. Applicable at very high temperature (up to 2,400 °C) and relatively for large sample sizes.
iii. Accurate determination of transition temperature.
iv. DTA can also be used to measure two inert samples when their responses to the heat cycle are not identical.

8.5.2 Disadvantages of DTA

i. Reaction or transition estimations are only 20– 50% DTA.
ii. Uncertainty in heat of fusion (ΔH_{fus}).
iii. Highest sensitivity is attained at high heating rates.
iv. Small change in temperature cannot be determined and quantified.
v. The ΔH variations between sample and reference make DTA less sensitive.

8.6 Conclusions

Analytical techniques have been used as a complementary technique to a thermal method for many years. Thermal analysis is used to complement various techniques like optical and electron microscopy, and X-ray diffraction analysis, during the development of new materials and in production control. Simultaneous measurements are made possible by the flexibility of DTA, which has been extensively used in several methodical investigations. It is crucial that students receive instruction on how to apply DTA, and other approaches from practitioners who are versed in their usage, as well as that the executives and management overseeing testing, quality control, and research and development are made aware of these techniques.

References

[1]   Jansson, P., Renström, P., Sarnet, J., and Sjöland, A. A new methodology for thermal analysis of geological disposal of spent nuclear fuel using integrated simulations of gamma heating and finite element modeling. Annals of Nuclear Energy. 2022; 172: 109082.

[2]   Sathees Kumar, S. Effect of natural fiber loading on mechanical properties and thermal characteristics of hybrid polyester composites for industrial and construction fields. Fiber Polymer. 2020; 21(7): 1508–1514.

[3]   Kumar, A., Singh, P., and Nanda, A. Hot stage microscopy and its applications in pharmaceutical characterization. Applied Microscopy. 2020; 50(1): 1–11.

[4]   Kaushik, N., Saravanakumar, P., Dhanasekhar, S., et al. Thermal analysis of a double-glazing window using a nano-disbanded phase changing material (NDPCM). Materials Today: Proceedings. 2022; 62: 1702–1707.

[5]   Zainal, N.F., Saiter, J.M., Halim, S.I., Lucas, R., and Chan, C.H. Thermal analysis: Basic concept of differential scanning calorimetry and thermogravimetry for beginners. Chemistry Teacher International. 2021; 3(2): 59–75.

[6]   Rahbar, M., Han, M., Xu, S., Zobeiri, H., and Wang, X. Development of differential thermal resistance method for thermal conductivity measurement down to microscale. International Journal of Heat and Mass Transfer. 2023; 202: 123712.

[7]   Mackenzie, R.C., Keattch, C.J., et al. Nomenclature in thermal analysis- Part III. Journal of Thermal Analysis. 1975; 8: 197–199.

[8]   Mackenzie, R.C. Nomenclature in thermal analysis, part IV. Thermochimica Acta. 1979; 28(1): 1–6.

[9]   Mackenzie, R.C. Nomenclature in thermal analysis. Part V. Symbols. Thermochimica Acta. 1981; 46: 333–335.

[10] Mackenzie, R.C. Treatise on analytical chemistry, New York: Wiley; 1983.

[11] Hill, J.O. For better thermal analysis and calorimetry III, ICTA, 1991.

[12] Differential Thermal Analysis and Differential Scanning Calorimetry. California, 2002. (Accessed December 28, 2022, at https://chem.libretexts.org/Bookshelves/Analytical_Chemistry/InstrumentalA nalysis_(LibreTexts)/31%3A_Thermal_Methods/31.02%3A_Differential_Thermal_Analysis.).

[13] Haines, P.J. Introduction to thermal analysis. In: Differential thermal analysis and differential scanning calorimetry Dordrecht: Springer; 1995.

[14] Bard, A.J. Fundamentals of analytical chemistry (Skoog, Douglas A.; West, Donald M.). Journal of chemical education. 1963; 40(11): 614.

[15] Richardson, M.J. Thermal analysis. In: Jenkins, A.D., Loening, K.L. (eds.) Comprehensive polymer science and supplements, Pergamon Press plc; 1989, pp. 867–901.

[16] Smallman, R.E. and Ngan, A.H.W. Characterization and analysis. In: Modern physical metallurgy, 8th edn, Elsevier; 2014, pp. 159–250.

[17] Bud, R. and Warner, D. Instruments of science: An historical encyclopedia. Routledge; 1997.

[18] Bhadeshia, H.K.D.H., Thermal analyses techniques. Differential thermal analysis. University of Cambridge, Material Science and Metallurgy, 2002. Accessed December 28, 2022, at https://www.phase-trans.msm.cam.ac.uk/2002/Thermal1.pdf.

[19] Principle of Differential Thermal Analysis (DTA). Hitachi High-Tech Corporation, 2001. (Accessed December 28, 2022, at https://www.hitachi-hightech.com/global/products/science/tech/ana/thermal/descriptions/dta.html#:~:text=Definitions%20of%20Differential%20Thermal%20Analysis,a%20specified%20atmosphere%2C%20is%20programmed.&text=This%20is%20definition%20of%20DTA%20by%20ICTAC.).

[20] Willard, H.H., Merritt, J.L.L., Dean, J.A., and Settle, J.F.A. Instrumental methods of analysis, Univ. of Washington: Seattle; 1989.

[21] Skoog, D.A., Holler, F.J., and Crouch, S.R. Principles of instrumental analysis. Cengage learning; 2017.

[22] Wendlandt, W.W. Thermal methods of analysis. Wiley-Blackwell; 1974.

[23] Ramos-Sanchez, M.C., et al. DTG and DTA studies on typical sugars. Thermochimica Acta. 1988; 134: 55–60.

[24] de Angelis Curtis, S., Curini, R., et al. Grana Padano cheese: Thermoanalytical techniques applied to the study of ripening. Food Chemistry. 1999; 66(3): 375–380.

[25] Berger, K.G. and Akehurst, E.E. Some applications of differential thermal analysis to oils and fats. International Journal of Food Science & Technology. 1966; 1: 237–247.

[26] Craig, D.Q. and Reading, M. Thermal analysis of pharmaceuticals. CRC press, 2006.

[27] Akash, M.S.H. and Rehman, K. Differential thermal analysis. In: Rehman, K. (ed.) Essentials of pharmaceutical analysis. 1st edn, Singapore: Springer; 2020, pp. 207–213.

[28] Londo, J.P. and Kovaleski, A.P. Characterization of wild North American grapevine cold hardiness using differential thermal analysis. American Journal of Enology and Viticulture. 2017; 68(2): 203–212.

[29] Doctor, R. and Feinberg, J.M. Differential thermal analysis using high temperature susceptibility instruments. Journal of Geophysical Research: Solid Earth. 2022; 127(7), e2 021JB023789.

[30] Smykatz-Kloss, W. Application of differential thermal analysis in mineralogy. Journal of Thermal Analysis and Calorimetry. 1982; 23(1–2): 15–44.

[31] Dominguez, O., Torres-Castillo, A., et al. Characterization using thermomechanical and differential thermal analysis of the sinterization of Portland clinker doped with CaF2. Materials Characterization. 2010; 61(4): 459–466.

[32] Handoo, S.K. and Raina, S.J. Cement raw mix characterization by differential thermal analysis. Thermochimica Acta. 1985; 93: 609–612.

[33] Franceschi, E., Cascone, I., and Nole, D. Study of artificially degraded woods simulating natural ageing of archaeological findings. Journal of Thermal Analysis and Calorimetry. 2008; 92(1): 319–322.

[34] Votyakov, S., Kiseleva, D., Shchapova, Y., Smirnov, N., and Sadykova, N. Thermal properties of fossilized mammal bone remnants of the Urals. Journal of Thermal Analysis and Calorimetry. 2010; 101(1): 63–70.

[35] Boaks, A., Siwek, D., and Mortazavi, F. The temporal degradation of bone collagen: A histochemical approach. Forensic Science International. 2014; 240: 104–110.

[36] Li, X., Zang, X., Xing, X., Li, J., Ma, Y., and Li, T. Effect of Fe2O3 on the crystallization behavior of glass-ceramics produced from secondary nickel slag. Metals. 2022; 12(1): 164.

[37] Ndong, A., Cutard, T., Nzihou, A., Piotte, M., Cadot, J.C., and Minh, D.P. Dynamic characterization approach for the investigation of microstructural changes in clay-based materials during firing. Open Ceramics. 2022; 100326.

Section V: **Electroanalytical Methods**

Shramila Yadav

Chapter 9
pH Metry

Abstract: pH is an important chemical parameter that reflects the nature of a solution, and pH scale measures the concentration of hydrogen ions. It plays a significant role in daily life, and exhibits wide applicability in various biochemical processes. This chapter emphasizes the concept of pH, its important applications, and implications of varying the pH in our body. Different acid–base theories, dissociation of strong and weak acids, dissociation constant and its significance, and extent of dissociation have also been explained with the help of numerous examples. This chapter also includes buffer solutions, which regulate the various processes of our body, along with their buffer capacity, derivation, and application of the Henderson–Hasselbalch equation. Buffer solution is made using weak acid and its conjugate salt or weak base and its conjugate salt; preparation of different buffer solutions of the required pH, and their examples has also been discussed so that students can have a better understanding of the concept. Detailed discussion on colorimetric and electrochemical methods using pH meter, method used for the determination of concentration of a given acid, pH metric titrations, the pattern of curves for combination of different acids and bases, and the method to determine the equivalence point using derivative curves have also been elaborated.

9.1 Introduction

The substances that break up into ions (cations and anions), when dissolved in water, are known as electrolytes, and these can be categorized into strong or weak, depending upon their extent of dissociation into ions. A strong electrolyte dissociates completely when dissolved in water, while a weak electrolyte dissociates partially into its constituent ions. For example, hydrochloric acid (HCl) dissociates completely into H^+ and Cl^- ions, and sodium hydroxide (NaOH) dissociates into Na^+ and OH^- ions, and can be represented as

$$HCl \rightarrow H^+ + Cl^-$$

$$NaOH \rightarrow Na^+ + OH^-$$

So, they are categorized as strong electrolytes, while acetic acid (CH_3COOH) dissociates partially into CH_3COO^- and H^+ ions, as follows:

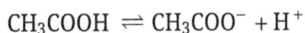

$$CH_3COOH \rightleftharpoons CH_3COO^- + H^+$$

Acids and bases form an important class of electrolytes. Proper functioning of enzymes, maintenance of osmotic pressure, excretion of waste from our body, and

https://doi.org/10.1515/9783110794816-009

almost all biological processes occurring in the cell are controlled by the precise balance of acid and base concentration (or pH) in our bodies. So, cell and organisms must maintain a specific and constant pH for the proper functioning of the body. Even a small change in pH can have a detrimental effect on the biological molecules, i.e., they lose their native (biological active) form, and will not function. This change can hamper the normal functioning of the body and can lead to disease or even death of a person.

9.2 Acid–base concepts

Different investigators put forward different concepts to characterize a substance as an acid or a base. Few of them are discussed in this section.

9.2.1 Arrhenius concept

According to the Arrhenius theory, an acid is a substance that dissociates in water to give hydrogen ions (H^+), and a base is a substance that dissociates in water to form hydroxyl ions (OH^-):

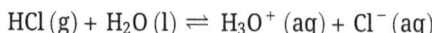

$$HCl\,(g) + H_2O\,(l) \rightleftharpoons H_3O^+\,(aq) + Cl^-\,(aq)$$

[Note: A proton is largely hydrated in aqueous solution and is present as hydronium (H_3O^+) ion, but here, it is represented as H^+ (aq), for the sake of convenience].

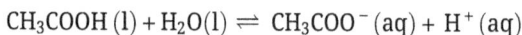

$$CH_3COOH\,(l) + H_2O(l) \rightleftharpoons CH_3COO^-\,(aq) + H^+\,(aq)$$

HCl and CH_3COOH are categorized as acids, as they give H^+ ions when dissolved in water.

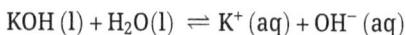

$$KOH\,(l) + H_2O(l) \rightleftharpoons K^+\,(aq) + OH^-\,(aq)$$

$$Ca(OH)_2(s) + H_2O\,(l) \rightleftharpoons Ca^{2+}\,(aq) + 2OH^-\,(aq)$$

KOH (potassium hydroxide) and $Ca(OH)_2$ (calcium hydroxide) are categorized as bases, as they give OH^- ions when dissolved in water.

In 1903, Svante A. Arrhenius (1859–1927), a Swedish Chemist was awarded Nobel Prize in Chemistry for his electrolytic theory of dissociation.

9.2.2 Bronsted–Lowry concept

According to the Bronsted–Lowry theory, an acid is a proton donor, while a base is a proton acceptor. To understand this better, let us take a simple example of hydrochloric acid:

$$HCl\,(aq) \rightarrow H^+\,(aq) + Cl^-\,(aq)$$

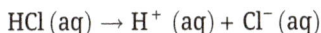

HCl has a tendency to donate a proton (H^+); so it is an acid, while Cl^- (aq) ion has a tendency to accept a proton; so it is a base. Consider another example:

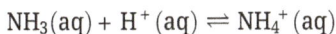

$$NH_3(aq) + H^+\,(aq) \rightleftharpoons NH_4^+\,(aq)$$

In the above equation, NH_3 has a tendency to accept a proton H^+; so it is a base, whereas NH_4^+ has a tendency to donate a proton (H^+); so it is an acid. In general,

$$Acid \rightleftharpoons H^+ + Base$$

The pair of substances that can be formed from one another by the gain or loss of a proton are known as conjugate acid–base pairs. Consider the following reactions:

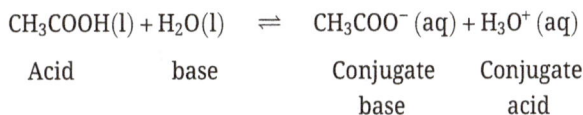

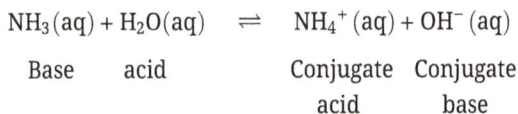

$$\begin{array}{ccccc} CH_3COOH(l) + H_2O(l) & \rightleftharpoons & CH_3COO^-\,(aq) & + & H_3O^+\,(aq) \\ \text{Acid} \qquad \text{base} & & \text{Conjugate} & & \text{Conjugate} \\ & & \text{base} & & \text{acid} \end{array}$$

$$\begin{array}{ccccc} NH_3\,(aq) + H_2O(aq) & \rightleftharpoons & NH_4^+\,(aq) & + & OH^-\,(aq) \\ \text{Base} \qquad \text{acid} & & \text{Conjugate} & & \text{Conjugate} \\ & & \text{acid} & & \text{base} \end{array}$$

Here, CH_3COOH is an acid, which on donating a proton forms CH_3COO^- which is a conjugate base of acetic acid. Similarly, NH_3 is a base which on accepting a proton forms NH_4^+, conjugate acid of ammonia. In the above two examples, we can see that H_2O acts as a base with CH_3COOH, whereas with NH_3, it acts as an acid. So, H_2O possesses both acidic as well as basic properties, and such substances are known as amphoteric or amphiprotic.

9.2.3 Lewis concept

Gilbert Lewis proposed a broader definition of acid and base in 1923. He defined an acid as a substance that can accept an electron pair, and base as a substance that can donate an electron pair. The main advantage of the Lewis concept is that it includes acid–base that does not involve protons. The earlier proposed concepts considered only those substances as acid or base, which contain either H^+ or OH^-. For example, BF_3 has a tendency to accept an electron pair; so it is considered as an acid, while F^- can donate an electron pair; so it is a base:

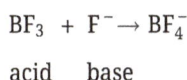

$$BF_3 + F^- \rightarrow BF_4^-$$
$$\text{acid} \qquad \text{base}$$

Another example is:

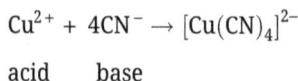

$$Cu^{2+} + 4CN^- \rightarrow [Cu(CN)_4]^{2-}$$

acid   base

In this chapter, we will consider acids/bases in aqueous solutions according to the Bronsted–Lowry concept.

9.3 Dissociation of acids and bases

Strong acids or bases are those that dissociate completely when dissolved in water. On the other hand, weak acids or bases dissociate partially or feebly when dissolved in water; higher the tendency of dissociation, greater will be the strength, i.e., the extent of dissociation of an acid/base determines how strong it is. So, relative strength of acids/bases can also be determined by studying the extent of dissociation.

9.3.1 Dissociation of a weak acid

Consider HA is a weak monobasic or monoprotic acid, which when dissolved in water, dissociates according to the equation:

$$HA + H_2O \rightleftharpoons H_3O^+ + A^-$$

HA is ionized into its constituent ions and after some time, the reaction attains equilibrium. The equilibrium constant for the reaction can be written by applying the law of chemical equilibrium:

$$k_c = \frac{[H_3O^+] \ [A^-]}{[HA] \ [H_2O]} \tag{9.1}$$

(Square brackets [] represent the concentration of the species enclosed within in mol/L (moles/liter).)

For simplification, in eq. (9.1), H_3O^+ (hydrated H^+ ion) may be replaced by $[H^+]$, and the concentration of water may be taken as constant (k), as it is present in excess; hence, eq. (9.1) can be simplified and rewritten as

$$k_c = \frac{[H^+] \ [A^-]}{[HA] \times k}$$

$$k_c \times k = \frac{[H^+] \ [A^-]}{[HA]}$$

$$k_a = \frac{[H^+]\,[A^-]}{[HA]} \tag{9.2}$$

The product of two constants, k_c and k, is also a constant, k_a, known as the dissociation constant of the acid, and is defined as the product of the concentrations of $[H^+]$ ions and anion $[A^-]$, divided by the concentration of the undissociated acid $[HA]$.

Significance of k_a: k_a is characteristic of an acid, and it varies with temperature. It is also the measure of the strength of an acid. Greater the magnitude of k_a, stronger is the acid.

The strength of an acid can also be measured by calculating the percent dissociation of an acid, which may be defined as

$$\text{Percent dissociation} = \frac{[H^+]}{[HA]_0} \times 100 \tag{9.3}$$

where $[H^+]$ is the hydrogen ion concentration present at equilibrium, and $[HA]_0$ is the initial concentration of the acid. Greater the percent dissociation of the acid, higher will be is its strength. The percent dissociation for strong acids, such as HCl (hydrochloric acid) and HNO_3 (nitric acid), is 100%; hence assumed to be completely dissociated. Therefore, we shall discuss only weak acids.

Another way to express k_a is in terms of degree of dissociation (a) and molar concentration of the acid. Degree of dissociation (a) may be defined as the extent of ionization of an electrolyte. It shows variation with temperature. a increases with increase in temperature, and decreases with decrease in temperature. The value of a ranges from 0 to 1. To understand it in a better way, consider the dissociation of the weak acid, HA:

$$HA \ \rightleftharpoons \ H^+ \ + \ A^-$$

At time, $t = 0$	c	0	0
(Initial concentration)			
At equilibrium,	$c(1-a)$	ca	ca

As,

$$k_a = \frac{[H^+]\,[A^-]}{[HA]} = \frac{ca \times ca}{c(1-a)}$$

$$k_a = \frac{c^2 a^2}{c(1-a)} = \frac{ca^2}{(1-a)}$$

For the sake of simplicity, $(1 - a) \approx 1$ in the denominator, as a is very- very small for weak acids:

$$\therefore \qquad\qquad k_a = ca^2 \ \text{ or } \ a = \sqrt{\frac{k_a}{c}}$$

The degree of dissociation, α, is the measure of the strength of an acid, as it is the capacity to furnish H^+ ions:

As,

$$H^+ = c\alpha = c\sqrt{\frac{k_a}{c}} = \sqrt{k_a c} \tag{9.4}$$

If HA_1 and HA_2 are two acids of the same concentration, their strengths can be compared using eq. (9.4):

$$\frac{\text{Strength of } HA_1}{\text{Strength of } HA_2} = \sqrt{\frac{k_{a1}}{k_{a2}}}$$

Some examples have been discussed below for better understanding of the concept.

Example 9.1: At room temperature, 0.09 M solution of acetic acid was found to be dissociated to an extent of 1.45%. Calculate the dissociation constant k_a of acetic acid at this temperature. What will happen if the temperature is raised?

Solution: Concentration of acetic acid = 0.09 M
Percentage dissociation of acid = 1.45
Degree of dissociation (α) of acetic acid = 1.45/100 = 0.0145
Dissociation constant of acetic acid, $k_a = \frac{c\alpha^2}{(1-\alpha)} \approx c\alpha^2$ (as α is very small)
$k_a = c\alpha^2 = 0.09 \times (0.0145)^2 = 1.89 \times 10^{-5}$
Effect of Temperature: the value of α increases as the temperature rises.

Example 9.2: The dissociation constant k_a of benzoic acid (C_6H_5COOH) and carbonic acid (H_2CO_3) are 6.26×10^{-5} and 4.52×10^{-7}, respectively. Predict which one is stronger and by what factor.

Solution: As we already studied in eq. (9.4), the concentration of H^+ ions is directly related to the dissociation constant, which can be written as

$$\frac{\text{Strength of } HA_1}{\text{Strength of } HA_2} = \sqrt{\frac{k_{a1}}{k_{a2}}}$$

$$\frac{\text{Strength of benzoic acid}}{\text{Strength of carbonic acid}} = \sqrt{\frac{k_{a1}}{k_{a2}}} = \sqrt{\frac{6.26 \times 10^{-5}}{4.52 \times 10^{-7}}} = 11.77$$

Thus, benzoic acid is 11.77 times stronger than carbonic acid.

9.3.2 Dissociation of polybasic acid

Polybasic acids are those that furnish two or more H^+ ions. Their dissociation always takes place in several stages. For example, the stepwise dissociation of phosphoric acid (H_3PO_4) is represented as

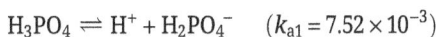

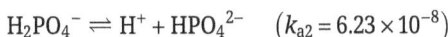

$$H_3PO_4 \rightleftharpoons H^+ + H_2PO_4^- \quad (k_{a1} = 7.52 \times 10^{-3})$$

$$H_2PO_4^- \rightleftharpoons H^+ + HPO_4^{2-} \quad (k_{a2} = 6.23 \times 10^{-8})$$

$$HPO_4^{2-} \rightleftharpoons H^+ + PO_4^{3-} \quad (k_{a3} = 4.80 \times 10^{-13})$$

$k_{a1} > k_{a2} > k_{a3}$; the first stage dissociation constant k_{a1} has the highest value, followed by the second stage dissociation constant k_{a2}, and the third one has least value. The reason for this decrease can be explained as H^+ ion in the first stage comes from a neutral molecule H_3PO_4, while in the second stage, it comes from a negatively charged $H_2PO_4^-$, and in third stage, it comes from the doubly charged HPO_4^{2-} ion.

9.3.3 Dissociation of a weak base

By analogy, with acid dissociation, we can study base dissociation. For example:

$$NH_3 + H_2O \rightleftharpoons NH_4^+ + OH^-$$

The dissociation constant k_b for a base can be written in the same way:

$$k_b = \frac{[NH_4^+]\,[OH^-]}{[NH_3]\,[H_2O]} \tag{9.5}$$

$[H_2O]$ is constant (as discussed earlier); therefore, eq. (9.5) can be rewritten as

$$k_b = \frac{[NH_4^+]\,[OH^-]}{[NH_3]}$$

The degree of dissociation for a weak base BOH can also be calculated in the similar way.

$$BOH \rightleftharpoons B^+ + OH^-$$

At time, $t = 0$ c 0 0
(Initial concentration)
At equilibrium, $c(1-a)$ ca ca

$$k_b = \frac{[B^+]\,[OH^-]}{[BOH]} = \frac{ca \times ca}{c(1-a)}$$

$$k_b = \frac{c^2 a^2}{c(1-a)} = \frac{ca^2}{(1-a)}$$

For the sake of simplicity, $(1 - \alpha) \approx 1$ in the denominator as α is very very small for weak bases:

$\therefore$

$$k_b = c\alpha^2 \ \text{ or } \ \alpha = \sqrt{\frac{k_b}{c}}$$

Degree of dissociation, α, is the measure of the strength of a base as it is the ability to furnish OH^- ions:

As,

$$[OH^-] = c\alpha = c\sqrt{\frac{k_b}{c}} = \sqrt{k_b c} \tag{9.6}$$

9.3.4 Dissociation of water

Water is amphoteric in nature. It behaves as an acid as well as a base, as it furnishes both H^+ and OH^- ions. Its dissociation can be represented as

$$H_2O \rightleftharpoons H^+ + OH^-$$

Water dissociates to a very less extent, so it behaves as a weak acid and a weak base as well. The dissociation constant of water is also known as ionic product of water, and is represented as k_w and is given by

$$K = \frac{[H^+]\ [OH^-]}{[H_2O]} \quad ([H_2O] = k, \text{ is constant})$$

$$K = \frac{[H^+]\ [OH^-]}{k}$$

$$K \times k = [H^+]\ [OH^-]$$

$$k_w = [H^+]\ [OH^-] \tag{9.7}$$

At 298 K, $k_w = 1.0 \times 10^{-14} \text{mol}^2/\text{dm}^6$. The concentration of H^+ and OH^- ions must be equal in pure water. So,

$$[H^+] = [OH^-] = \sqrt{k_w} = \sqrt{1.0 \times 10^{-14} \text{mol}^2/\text{dm}^6}$$

$$= 1.0 \times 10^{-7} \text{mol}/\text{dm}^3$$

9.4 pH scale

Water is a universal solvent, and furnishes both H^+ and OH^- ions. These ions play a central role in many chemical and biological processes. An aqueous solution can be acidic, basic, or neutral, depending on the relative concentration of H^+ and OH^- ions. If

$$[H^+] > [OH^-], \text{ the solution is acidic}$$
$$[H^+] < [OH^-], \text{ the solution is basic}$$
$$[H^+] = [OH^-], \text{ the solution is neutral}$$

Knowing the concentration of one, the concentration of other can be easily calculated.

The concept of pH was introduced by a Danish chemist, S. P. Sorensen in 1909. He introduced the pH scale, used to express the concentration of hydrogen ion. pH means the potential of hydrogen, and it is defined as the negative logarithm of the activity of hydrogen ion,

$$pH = -\log a_{H^+}$$

For very dilute solutions ($[H^+] \leq 0.1\,M$), the activity can be approximated to the concentration of $[H^+]$ ions in mol/L or M (molar)

$$pH = -\log([H^+]/M)$$

For simplicity,

$$pH = -\log[H^+]$$

Similarly, other logarithmic quantities can be expressed as follows:

$$pOH = -\log[OH^-]$$
$$pk_w = -\log k_w$$
$$pk_a = -\log k_a$$
$$pk_b = -\log k_b$$

According to eq. (9.7),

$$k_w = [H^+]\,[OH^-]$$

Taking negative log on both sides,

$$-\log k_w = -\log[H^+] + (-\log[OH^-])$$
$$pk_w = pH + pOH$$

As, at 298 K, $k_w = 1.0 \times 10^{-14}\,\text{mol}^2/\text{dm}^{-6}$

$$pk_w = pH + pOH = 14$$

This can be better understood from the following examples.

Example 9.3: Calculate the pH of (i) 2.0×10^{-5} M HCl solution and (ii) 1.0×10^{-4} M NaOH solution, assuming complete dissociation in both the cases.

Solution: (i) Assuming complete dissociation of HCl,

$$HCl \rightarrow H^+ + Cl^-$$

Concentration of HCl = $[H^+] = 2.0 \times 10^{-5}$ M

$$pH = -\log([H^+]/M) = -\log(2.0 \times 10^{-5}) = 4.699$$

(ii) Assuming complete dissociation of NaOH,

$$NaOH \rightarrow Na^+ + OH^-$$

Concentration of NaOH = $[OH^-] = 1.0 \times 10^{-4}$ M

$$pOH = -\log([OH^-]/M) = -\log(1.0 \times 10^{-4}) = 4.0$$

$$pk_w = pH + pOH$$

$$pH = pk_w - pOH = 14 - 4.0 = 10$$

So, the pH of 2.0×10^{-5} M HCl solution is 4.699, and that of 1.0×10^{-4} M NaOH solution is 10.

Example 9.4: A solution is obtained by mixing 50 mL of 0.1 M HCl and 50 mL of 0.3 M NaOH. Calculate the pH and pOH of the solution. (Given: $k_w = 10^{-14}$ mol^2/dm^6).

Solution: First, we will find out the concentration of the resultant solution.
Number of millimoles of HCl in the solution = $50 \times 0.1 = 5$
Number of millimoles of NaOH in the solution = $50 \times 0.3 = 15$
As number of millimoles of the base is greater than that of the acid, after neutralization, NaOH will be left in the solution.
Number of millimoles of base left in the solution after mixing the acid and the alkali = $15-5 = 10$
Total volume of the solution = $50 + 50 = 100$ mL
Thus, 10 millimoles of base in 100 mL of solution or 0.1 mole of base per liter of the solution.
∴ Concentration of OH$^-$ ions = 0.1 M

$$pOH = -\log([OH^-]/M) = -\log(0.1) = 1$$

$$pk_w = pH + pOH$$

$$pH = pk_w - pOH = 14 - 1 = 13$$

So, pH of the solution is 13, and pOH is 1.

Example 9.5: Calculate pH, pOH, and $[H^+]$ of a 4.3×10^{-4} M solution of Ca(OH)$_2$ in water at 25 °C. (Given: $k_w = 10^{-14}$ mol^2/dm^6)

Solution: Ca(OH)$_2$ is a strong base that furnishes two hydroxyl ions when dissolved in water:

$$Ca(OH)_2 \rightarrow Ca^{2+} + 2OH^-$$

$$[OH^-] = 2 \times 4.3 \times 10^{-4} M = 8.6 \times 10^{-4} M$$

$$pOH = -\log([OH^-]/M) = -\log(8.6 \times 10^{-4}) = 3.066$$

$$pH = pk_w - pOH = 14 - 3.066 = 10.934$$

$$k_w = [H^+][OH^-]$$

$$[H^+] = \frac{k_w}{[OH^-]} = \frac{1.0 \times 10^{-14} \, mol^2/dm^6}{8.6 \times 10^{-4} \, mol/dm^3} = 1.16 \times 10^{-11} \, mol/dm^3$$

So, pH of the solution is 10.934, pOH is 3.066, and $[H^+]$ is 1.16×10^{-11} mol/dm^3.

9.5 Buffer solutions

A solution that has the ability to resist change in its pH value upon the addition of small amount of either an acid or a base is called a buffer solution. These solutions have much significance in chemical and biological systems. The blood present in human body is a buffer, having pH between 7.35 and 7.45, and gastric juices present in the stomach also acts as a buffer, having pH 1–3.

Buffer solutions can be prepared in the laboratory by adding an equimolar solution of a weak acid and its conjugate salt, or a weak base and its conjugate salt to water. The buffer formed by mixing the weak acid and its salt is called an acidic buffer, while the one prepared by mixing a weak base and its salt is termed as basic buffer. Let us now understand how these buffers resist the change in pH, on adding an appreciable amount of an acid or a base.

Consider an acidic buffer that can be prepared by mixing acetic acid and sodium acetate. As we know, acetic acid is a weak acid and is feebly dissociated, while sodium acetate is almost completely dissociated in water:

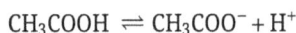

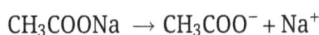

$$CH_3COOH \rightleftharpoons CH_3COO^- + H^+$$

$$CH_3COONa \rightarrow CH_3COO^- + Na^+$$

Thus, the mixture contains CH_3COOH, CH_3COO^-, H^+, and Na^+ ions. CH_3COO^- ion from acetic acid hydrolyses as

$$CH_3COO^- + H_2O \rightleftharpoons CH_3COOH + OH^- \tag{9.8}$$

The presence of CH_3COOH suppresses the hydrolysis of CH_3COO^-, and the equilibrium in eq. (9.8) shifts to the left hand side (Le Chatelier's principle), while the presence of CH_3COO^- ions (from salt) suppresses the dissociation of CH_3COOH. Therefore, the concentration of the acid CH_3COOH and its conjugate salt, CH_3COO^-, can be approximated to be the same as their initial concentrations.

$$[CH_3COOH] \approx [CH_3COOH]_0 \text{ and } [CH_3COO^-] \approx [CH_3COO^-]_0$$

If an acid is added to this buffer solution, H^+ ions of the acid combine with acetate ion and form acetic acid, which is feebly dissociated:

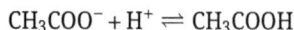

$$CH_3COO^- + H^+ \rightleftharpoons CH_3COOH$$

Most of the H^+ ions (from added acid) will be taken up by CH_3COO^- ions, and hence the pH of the buffer changes only slightly. Now, if a base is added, CH_3COOH will neutralize its effect, and more of CH_3COO^- ions and water molecule will be formed.

Similarly, a basic buffer can also be prepared by mixing a weak base such as ammonium hydroxide with its salt ammonium chloride. A buffer solution of a desired pH can be prepared with the help of the Henderson–Hasselbalch equation, or the pH of a buffer solution can also be determined using the same.

9.5.1 Henderson–Hasselbalch equation

The equation can be simply derived by considering the equilibrium equation of a weak acid, HA:

$$HA \rightleftharpoons H^+ + A^-$$

$$k_a = \frac{[H^+]\,[A^-]}{[HA]} \tag{9.9}$$

Taking the negative logarithm of eq. (9.9):

$$-\log k_a = -\log[H^+] - \log\frac{[A^-]}{[HA]}$$

$$pk_a = pH - \log\frac{[A^-]}{[HA]}$$

or

$$pH = pk_a + \log\frac{[\text{Conjugate base}]}{[\text{Acid}]}$$

or

$$pH = pk_a + \log\frac{[\text{Salt}]}{[\text{Acid}]} \tag{9.10}$$

Equation (9.10) is known as the *Henderson–Hasselbalch equation*.

Similarly, Henderson–Hasselbalch equation can also be derived for the basic buffer solution. For this, consider the dissociation of a weak base, BOH

$$BOH \rightleftharpoons B^+ + OH^-$$

$$k_b = \frac{[B^+]\,[OH^-]}{[BOH]} \tag{9.11}$$

Taking the negative logarithm of eq. (9.11),

$$-\log k_b = -\log[\text{OH}^-] - \log \frac{[\text{B}^+]}{[\text{BOH}]}$$

$$pk_b = p\text{OH} - \log \frac{[\text{B}^+]}{[\text{BOH}]}$$

or

$$p\text{OH} = pk_b + \log \frac{[\text{Conjugate acid}]}{[\text{Base}]}$$

or

$$14 - p\text{H} = pk_b + \log \frac{[\text{Salt}]}{[\text{Base}]} \tag{9.12}$$

Equation (9.12) is another form of the Henderson–Hasselbalch equation.

Now, the question arises why does a solution of a strong base like NaOH and its salt NaCl not work as a buffer? NaCl, when dissolved in water, dissociates completely into sodium (Na^+) and chloride (Cl^-) ions. If an acid is added to it, the concentration of H^+ ion increases. These hydrogen ions combine with Cl^- ions to form HCl; HCl is a strong acid that furnishes H^+ ions in solution, and hence, reduces the pH of the solution immediately. If a base is added, OH^- ions combine with Na^+ to form NaOH, which is again a strong base, and dissociates completely into Na^+ and OH^- ions, and increases the pH of the solution.

9.5.2 Buffer capacity

The effectiveness of a buffer solution can be measured in terms of buffer capacity, which is denoted as β. It is defined as the amount of acid or a base that must be added to the buffer to produce a unit change in its pH. It can be mathematically represented as

$$\beta = \frac{d[\text{B}]}{d\text{pH}}$$

$d[\text{B}]$ corresponds to the change in the concentration of the base, and $d\text{pH}$ is the corresponding change in the pH. $d[\text{B}]$ is given in moles per 1,000 g of solvent. If a base is added, both $d[\text{B}]$ and $d\text{pH}$ will be positive, and so will be the buffer. If an acid is added, there is a decrease in the base concentration as well as in pH, and is given as

$$\beta = \frac{-d[\text{B}]}{-d\text{pH}}$$

So, the resultant β is positive in this case too. As we have already discussed, a buffer resists change in pH, and how this resistance can be achieved is best explained with the help of the Henderson–Hasselbalch equations (9.10) and (9.12). A buffer will maintain its pH almost constant as long as the ratio of the concentration of the conjugate salt to that of the acid/base remains between 0.1 and 10. The reason for this is the slow variation of the logarithmic values between this range. Which combination of a weak acid and its conjugate salt, or a weak base and its conjugate salt, will form a buffer of what pH is governed by the pK_a value of the weak acid/base. We cannot prepare a buffer of any pH by using any combination of acid/base. There is a certain range of every buffer in which, it works as an effective buffer, and this is known as the buffer range. The buffer range is given as

$$pH = pk_a \pm 1$$

For example, the pK_a of acetic acid is 4.76; so, the solution of acetic acid and sodium acetate can act as a buffer in the pH range of 3.76–5.76.

9.5.3 Preparation of buffer solution

In this section, we will learn to calculate the relative amount of acid/base and its conjugate salt that are required to prepare the buffer solution of a desired pH. First of all, we have to choose a weak acid/base whose pk_a value is ± 1 of the desired pH. Then, obtain the ratio [conjugate salt]/[acid/base] by plugging in the values of pH and pk_a in the Henderson–Hasselbalch eqs. (9.10) or (9.12), whichever is applicable. Convert this ratio into molar quantities to calculate the amount of acid/base and its conjugate salt needed for preparing the buffer solution. Let us understand the procedure with the help of an example of an acidic buffer solution of pH = 4.8. From the data of pk_a values, the pk_a of acetic acid is 4.76; so the combination of acetic acid and its conjugate salt, sodium acetate, would be the best choice for the buffer. Plug in these values in eq. (9.10) to obtain the ratio of [conjugate salt]/[acid]:

$$pH = pk_a + \log \frac{[Salt]}{[Acid]}$$

$$4.8 = 4.76 + \log \frac{[Salt]}{[Acid]}$$

$$\log \frac{[Salt]}{[Acid]} = 4.8 - 4.76 = 0.04$$

$$\frac{[Salt]}{[acid]} = antilog\ (0.04) = 1.1$$

Therefore, the acidic buffer of pH 4.8 can be prepared by dissolving sodium acetate and acetic acid in a mole ratio of 1.1: 1.0 in water to make up a solution of the desired volume.

Let us discuss some of the examples based on the Henderson–Hasselbalch equation:

Example 9.6: Calculate the pH of a buffer system containing 0.2 M acetic acid (CH_3COOH) and 0.35 M sodium acetate (CH_3COONa). What would be the pH of the buffer after the addition of 0.005 mole of HCl to 500 mL of the solution? Assume no change in the volume of the solution. (Given: $pK_a = 4.76$).

Solution: As we know that the pH of a buffer is governed by the Henderson–Hasselbalch equation:

$$pH = pk_a + log\frac{[Salt]}{[Acid]}$$

$$pH = 4.76 + log\frac{0.35\ M}{0.2\ M} = 5.003$$

Addition of 0.005 M HCl will produce 0.005 M H^+ ions, which will neutralize 0.005 M CH_3COO^-, and produce 0.005 M CH_3COOH, according to the reaction:

$$CH_3COO^- + H^+ \rightarrow CH_3COOH$$

Before the addition of HCl	0.35 M	0.2 M
After the addition of 0.005 M HCl	(0.35–0.005) M	(0.2 + 0.005)
Therefore, pH of the buffer becomes		

$$pH = 4.76 + log\frac{(0.35 - (0.005)M}{(0.2 + 0.005\ )M} = 4.986$$

So, the pH of the buffer is 5.003 and becomes 4.986 after the addition of 0.005 moles of HCl to 500 mL of the solution.

Example 9.7: A buffer solution is obtained by mixing 4.5 gram of acetic acid and 6.5 gram of sodium acetate in water to make up a volume 500 mL. What would be the pH of the resultant solution? (Given: $k_a = 1.75 \times 10^{-5}$ at 25 °C).

Solution: Using the Henderson–Hasselbalch equation,

$$pH = pk_a + log\frac{[Salt]}{[Acid]}$$

$$[Salt] = \frac{wt \times 1,000}{MM \times vol} = \frac{6.5 \times 1,000}{82 \times 500} = 0.1585\ M$$

{MM: Molecular mass of salt (sodium acetate) = 82 g/mol}

$$[Acid] = \frac{wt \times 1,000}{MM \times vol} = \frac{4.5 \times 1,000}{60 \times 500} = 0.15\ M$$

{MM: Molecular mass of acid (acetic acid) = 60 g/mol}

$$pk_a = -log(1.75 \times 10^{-5}) = 4.76$$

$$pH = 4.76 + log\frac{0.1585}{0.15}$$

$$pH = 4.76 + 0.0239 = 4.784$$

The pH of the buffer is found to be 4.784.

9.6 Methods for the determination of pH

There are mainly two methods employed for the determination of pH of a solution:
i.   Colorimetric method and
ii. Electrochemical method using pH meter

Electrochemical methods are more precise and accurate, and have many advantages over colorimetric methods. If the exact pH value of the solution is not required, then colorimetric method can be used. It only gives an idea about the nature of the solution, whether it is acidic or basic. The use of litmus paper, which is a type of acid–base indicator, is the one way to detect the nature of the solution. They are available in three variants – red, blue, and neutral. They are made with a dye, derived from lichens. Red litmus paper turns blue in alkaline medium, blue litmus turns red when dipped in acidic medium, and neutral, which is originally purple, changes color either to red or blue, depending upon the nature of the solution. To find the pH of the solution, either dip a strip of the litmus paper in the test solution or put one or two drops of solution on the litmus paper with the help of a dropper, and then note the color change to know the nature of the solution. Usually, red litmus turns blue for pH greater than 8, and blue turns red for pH range of 1.0 to 5.0. Between pH 5.0 and 8.0, the color change is too small to detect. Although this method is very quick, portable, and inexpensive, it has certain limitations:
i.   It does not tell the exact pH of the solution.
ii. It can provide only a rough idea of the pH – whether it is less than 5 or greater than 8.
iii. It cannot be used for solutions that are nearly neutral.

9.6.1 pH metric titrations

The most important and common technique for the determination of pH in the area of analytical chemistry is the electrochemical method using pH meter. When a base is added to an acid, there occurs a change in the pH values. This change in the pH of the solution is plotted against the volume of alkali added, and the obtained curve is called a titration curve. It is the simplest experimental procedure to find out the concentration of an unknown acid solution – titrating it against the base solution of the known concentration or vice-versa. The same titration can be performed volumetrically by using a suitable indicator for a particular acid–base solution. Even color strips or pH paper can be used to perform such titrations, but pH metric titrations are preferred over volumetric or colorimetric titrations; it has the following advantages:
i.   pH metric titrations are more accurate, in comparison to volumetric titrations, as equivalence point can be determined instead of the end point.

[Note: Equivalence point corresponds to the amount of alkali exactly used to neu-
tralize the acid taken, while the end point is always higher than the equivalence
point, as it comes when an indicator shows the color change].

ii.   Once the pH meter is calibrated, it gives more accurate and precise readings.
iii. pH meter is easy to use and handle.
iv.   The readings on the pH meter are much less subjective, in comparison to reading
      a burette meniscus, or reading a color strip.
v.   Use of pH strips or pH indicators is messier than using the pH meter.
vi.   Use of pH meter decreases the extent of errors as well. When an indicator or a test
      strip is used, different people may report different readings, as different people see
      colors differently, and also different lighting conditions may alter the color.
vii. Indicators or color strips can be used only once; they also deteriorate with time.

9.6.1.1 Titration of a strong acid (HCl) against a strong base (NaOH)

The titration of a strong acid against a strong base involves the following steps:
i.   Calibrate the pH meter using the standard buffer solution.
ii.   Pipette out, say 50 mL of the acidic solution in a 100 mL beaker, and take the alkali
      solution in a burette.
iii. Start adding the alkali solution, dropwise; stir the solution, and note down the
      pH value.
iv.   After each addition of alkali, wait for a few seconds for the stabilization of the pH
      meter reading, and then repeat the same process till the pH values changes from
      the acidic to the basic range (around 12).
v.   The equivalence point will be at pH 7, when the solution becomes completely neu-
      tral. When the pH values of the solution are plotted against the amount of alkali
      added, the graph obtained will be "S" shaped as shown in Figure 9.1.

Now, let us understand the various stages of the neutralization process. Initially, the pH
of the solution is very low (between 1.0 and 2.0) due to the high concentration of H^+ ions,
as HCl is a strong acid and present in a completely dissociated state. Then, the pH starts
increasing very slowly (from 1.0 to 4.0) till about 99.9% of the alkali is added. Addition of
the last 0.01 mL of alkali raises the pH of the solution to 7.0, where the acid is completely
neutralized. Now, addition of a single drop of alkali makes the solution alkaline or basic,
which raises the pH to 9.0–10.0. The addition of alkali should be continued till a flat
curve is obtained. When all the points are joined, an almost vertical line is formed near
the equivalence point. The volume of the alkali corresponding to pH 7.0 is the actual
volume needed to neutralize the acid, and is the equivalence point.

To locate the equivalence point accurately and more precisely, it is better to plot the
derivative curves. The first derivative can give the equivalence point, but for more

pH vs Volume of NaOH graph

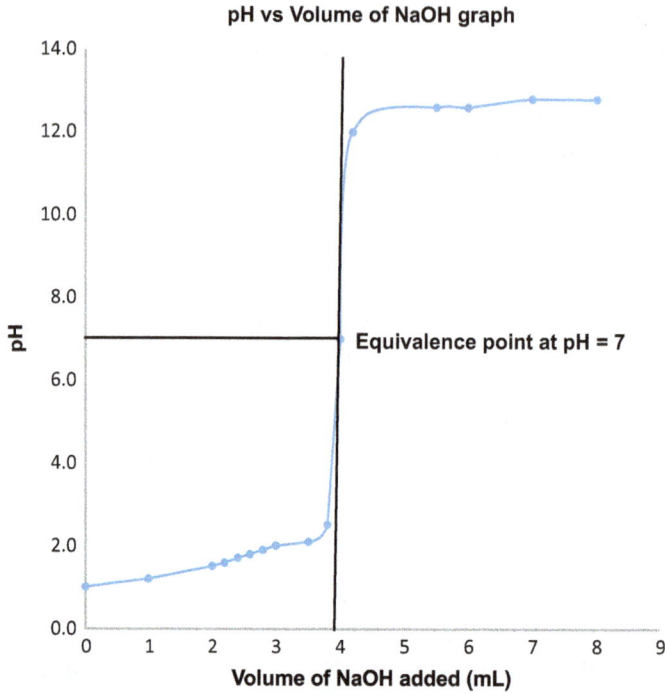

Figure 9.1: pH metric titration curve of strong acid versus strong base.

pH vs Volume of NaOH graph

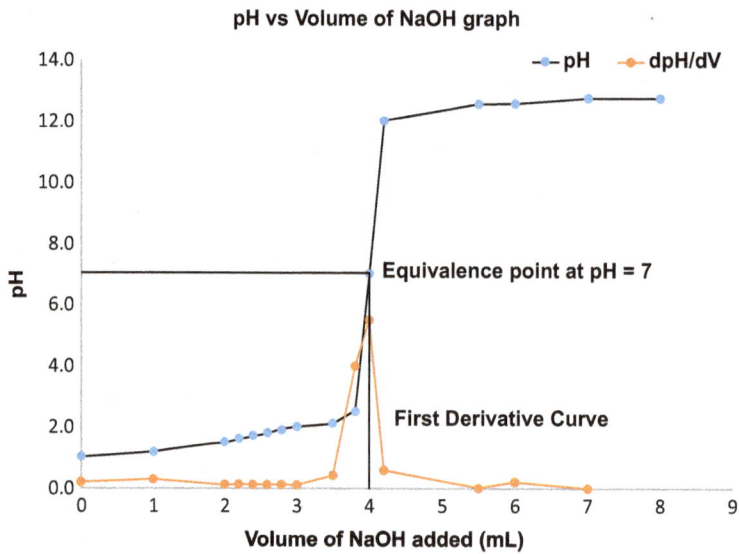

Figure 9.2: pH metric derivative curve of strong acid versus strong base.

accuracy and to minimize the chances of error, the second derivative curve is preferred. The first derivative curve for titration of strong acid vs strong base (HCl vs NaOH) is shown in Figure 9.2. There are two methods to plot the derivative curves. Suppose pH1, pH2, pH3, pH4, . . . are the values of pH obtained at the addition of V1, V2, V3, V4, ... volumes of alkali, respectively. In the first method, we can plot (pH2–pH3) against (V1 + V2)/2, (pH3–pH2) against (V2 + V3)/2, and so on. In the second method, we can plot (pH3– pH1) against V2, (pH4–pH2) against V3, and so on.

9.6.1.2 Titration of weak acid (CH₃COOH) against a strong base (NaOH)

The titration of a weak acid against a strong base can be best illustrated by the reaction of acetic acid with sodium hydroxide. The procedure of obtaining the curve is the same as strong acid vs. strong base but the curve obtained is different Figure 9.3. In this case, the initial pH value of the solution is around 3.0 as CH_3COOH is partially dissociated; so $[H^+]$ is quite less in the solution. The vertical portion begins around pH 7.0, and the equivalence point lies between 8.0 and 10.0. The first and second derivative curves for weak acid and strong base are shown in Figure 9.4.

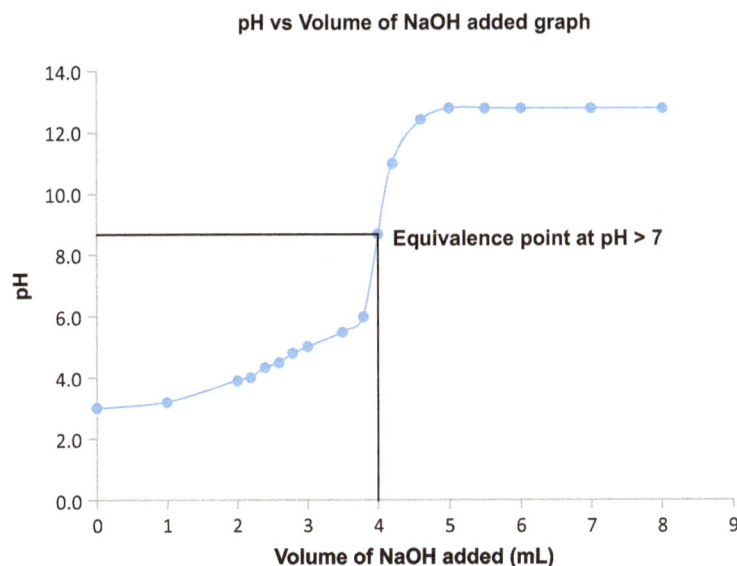

Figure 9.3: pH metric titration curve of weak acid vs strong base.

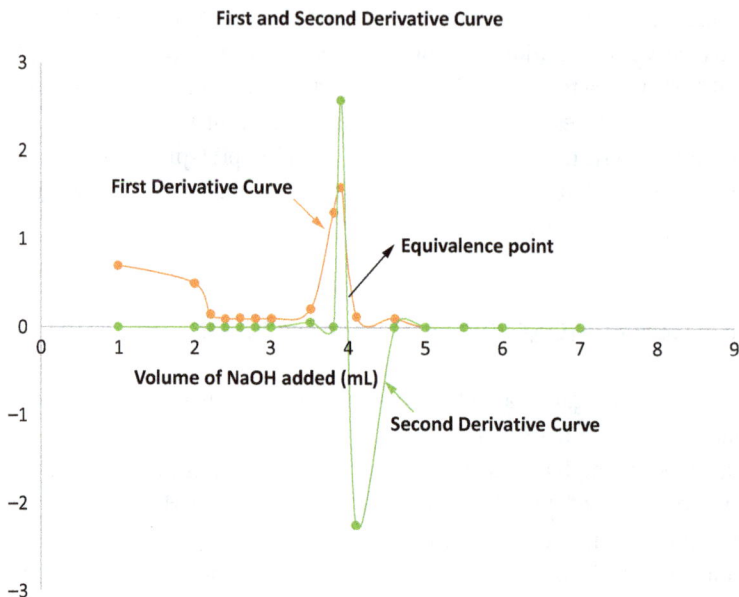

Figure 9.4: pH metric derivative curve of weak acid vs strong base.

To understand this behavior, consider the neutralization reaction,

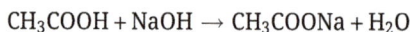

$$CH_3COOH + NaOH \rightarrow CH_3COONa + H_2O$$

CH_3COONa is the salt of a weak acid and a strong base; therefore, it undergoes hydrolysis according to the equation:

$$CH_3COO^- + Na^+ + H_2O \rightarrow CH_3COOH + Na^+ + OH^-$$

or

$$CH_3COO^- + H_2O \rightarrow CH_3COOH + OH^-$$

The formation of OH^- ions makes the solution basic at the equivalence point, instead of neutral.

9.6.1.3 Determination of pK_a of a weak acid

pK_a of the weak acid can be determined from the titration curve plotted from the data of pH metric titration. As we have discussed earlier, weak acids are partially ionized; therefore, there exists an equilibrium, which can be represented as

$$HA \rightleftharpoons H^+ + A^-$$

$$k_a = \frac{[H^+]\,[A^-]}{[HA]}$$

The relation between pH and pk_a is governed by the Henderson–Hasselbalch equation (9.10):

$$pH = pk_a + \log\frac{[A^-]}{[HA]}$$

The concentration of the conjugate base $[A^-]$ will be exactly equal to the concentration of the weak acid $[HA]$ at one-half of the equivalence point. When one-half of the acid taken is converted into its conjugate salt due to neutralization with the strong base, and one-half of the acid remain un-neutralized, it can be mathematically represented as,

At half equivalence point:

$$[HA] = [A^-]$$

Equation (9.10) becomes

$$pH = pk_a + \log\frac{[HA]}{[HA]}$$

$$\therefore pH = pk_a$$

pH vs Volume of NaOH added graph

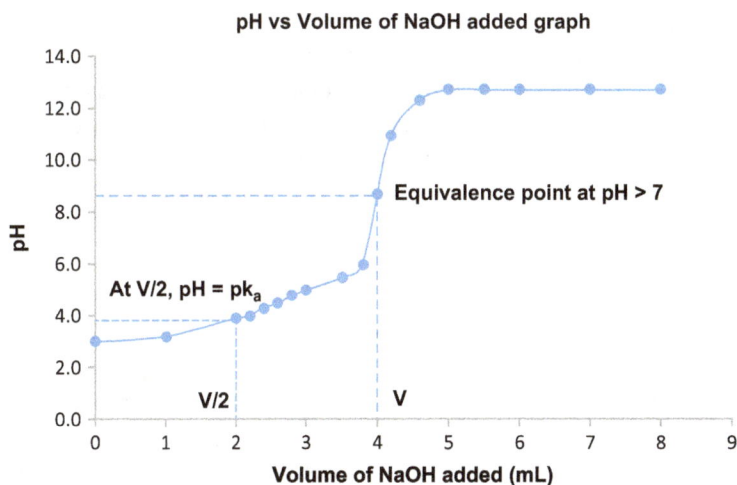

Figure 9.5: Determination of pk_a of weak acid.

pk_a can be located or calculated from the graph as shown in Figure 9.5. Once the equivalence point has been located, we can determine the volume of the alkali used (V). Now, locate the pH corresponding to V/2 volume, which is equal to pk_a.

As

$$pk_a = -\log k_a$$

$$k_a = -\text{antilog}\, pk_a$$

9.6.1.4 Titration of a strong acid against a weak base

The titration of a strong acid against a weak base may be best illustrated by the reaction of hydrochloric acid with ammonia,

$$HCl + NH_3 \rightarrow NH_4^+ + Cl^-$$

It is an analogous situation as NH_4^+ ions get hydrolyzed; it comes from a weak base and that is the reason the equivalence point in this case is below pH 7.0. The procedure for obtaining the curve is the same, and the curve obtained is shown in Figure 9.6. When NH_4OH is added to HCl, the formation of NH_4Cl takes place. The presence of both, NH_4OH and NH_4Cl, form the buffer, which further resists any change in the pH of the solution, and is responsible for pH > 7 at the equivalence point. The vertical portion of the graph will occur approximately between pH 3.0–8.0.

9.6.1.5 Titration of diprotic acid against strong base

For now, we have discussed the titration of a monoprotic acid with a strong and weak base. In this section, we will discuss pH metric titration of a dibasic/diprotic acid, for example, oxalic acid (COOH·COOH) or carbonic acid (H_2CO_3). Both dissociate in two steps; the loss of the first proton, H^+, is easier than the second proton, as the first one comes from a neutral species, while the second one comes from the negatively charged species. It can be shown as

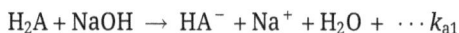

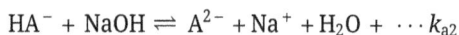

$$H_2A + NaOH \rightarrow HA^- + Na^+ + H_2O + \cdots k_{a1}$$

$$HA^- + NaOH \rightleftharpoons A^{2-} + Na^+ + H_2O + \cdots k_{a2}$$

The dissociation constant, pk_{a1}, for the first step is usually higher than the dissociation constant (pk_{a2}) for the second step. There will be two inflexion points observed in the graph; the first inflexion is not too sharp, and occurs between pH 3.0–4.0, while the second one is obtained in the basic medium at pH around 8.0. The total volume of alkali used for the complete neutralization is the volume corresponding to the second inflexion point. pk_{a1} and pk_{a2} can also be calculated from the graph. The pH corresponding to half the volume of the first equivalence point is equal to pk_{a1}, while the pH corresponding to half the volume of the second equivalence point is equal to pk_{a2}. The graph observed for the titration of oxalic acid against a strong base (NaOH) is shown in Figure 9.7.

pH vs Volume of NH₄OH graph

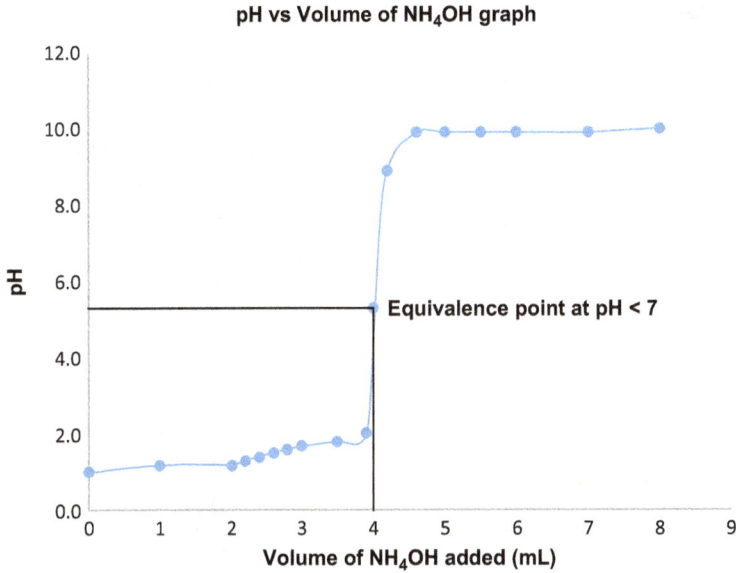

Figure 9.6: pH metric titration curve of strong acid versus weak base.

pH vs Volume of NaOH graph

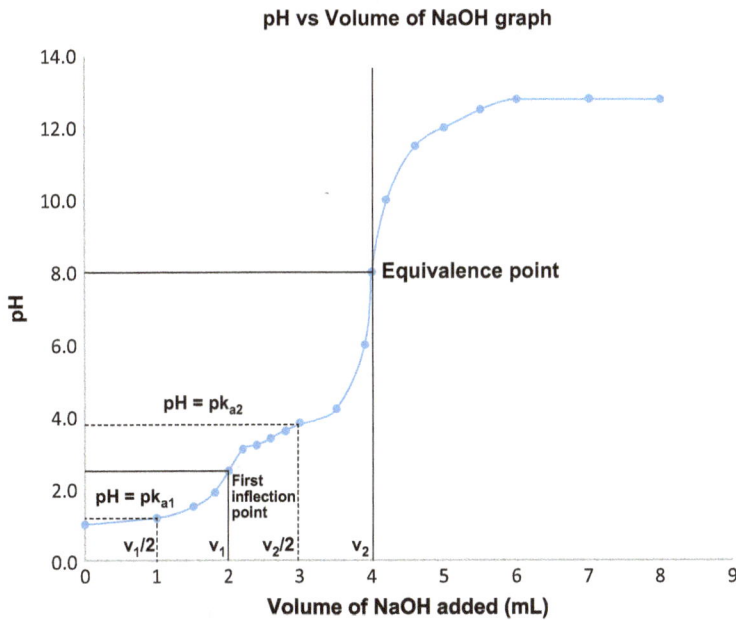

Figure 9.7: pH metric derivative curve of diprotic acid versus strong base.

9.7 Theory of acid–base indicators

The selection of an appropriate indicator for an acid–base titration should be done on the basis of the pH range of the indicator, which should fall within the pH change of the reaction. An indicator used should signal the end of the titration by changing the color of the solution. A random indicator cannot be used for any combination of acid–base titration. There is a unique indicator that shows a particular color change for a particular titration. The indicators used for acid–base titrations are usually hydrogen ion indicator, which may be defined as a substance that changes color with the variation in the pH. Every indicator has a different pH range within which it shows the color change. According to the Ostwald's theory of acid–base indicator, an H^+ ion indicator is either a weak acid or a weak base. To understand the working of an indicator, consider HIn as a weak acid whose dissociation can be represented as

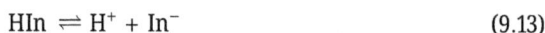

$$HIn \rightleftharpoons H^+ + In^-$$
(9.13)

where HIn is a weak acid and In^- is its conjugate base. Both, HIn and In^-, are different colored species. For example, phenolphthalein is a weak acid, and can lose two protons. Phenolphthalein molecule in non-ionized state is colorless, while doubly deprotonated species is pink in color. The equilibrium constant, K_{In}, for eq. (9.13) can be written as

$$k_{In} = \frac{[H^+]\,[In^-]}{[HIn]}$$
(9.14)

The color shown by the solution will be governed by the concentration of the $[H^+]$ ion. If the concentration of the $[H^+]$ ion is higher, the solution is acidic, the equilibrium in eq. (9.13) will be shifted toward the left (according to the Le Chatelier's principle), and the solution will show the color of the HIn species (colorless). If the concentration of the $[OH^-]$ ions is higher, the solution is basic, the equilibrium will be shifted toward the right, In^- species predominates in the solution, and the color is pink. Therefore, the color of the solution changes as the medium of the solution changes from acidic to alkaline. From eq. (9.14), the Henderson–Hasselbalch equation can be written as

$$pH = pk_{In} + \log\frac{[In^-]}{[HIn]}$$

If $\frac{[In^-]}{[HIn]} \leq 0.1$, the acidic color of the indicator predominates, and if $\frac{[In^-]}{[HIn]} \geq 10$, the base color predominates. The pH range of an indicator over which it changes color is given as

$$pH = pk_{In} \pm 1$$

As we have already discussed, the pH changes from 4.0 to 10.0 near the end point; therefore, phenolphthalein is the best choice as it has $pk_{In} = 9.4$, and hence, it works within the range of 8.4–10.4.

Another common and important indicator employed in acid–base titrations is methyl orange, which is a weak base, and can simply be written as MeOH. Its dissociation can be represented as

$$MeOH \rightleftharpoons Me^+ + OH^-$$

The base MeOH is yellow in color while its conjugate acid is red in color. In acidic medium, H^+ ions furnished by the acid will combine with OH^- ions, which shifts the equilibrium to the right and more of Me^+ species will be available; so the solution will be red in color. In the basic medium, due to the common ion effect of the OH^- ions, the equilibrium will shift toward the left, MeOH predominates, and the color of the solution will be yellow. So, the color change at the end point is from red to yellow. The pH range of methyl orange is 3.1–4.5.

References

[1]   Atkins, P., Atkins, P.W., and de Paula, J. Atkins' physical chemistry. Oxford university press; 2014.
[2]   Castellan, G.W. Physical chemistry, 4th edn, 2004.
[3]   Chang, R. Physical chemistry for the chemical and biological sciences. Viva Books; 2015.
[4]   Kapoor, K.L. A textbook of physical chemistry, 5th edn. Chapter 4, McGraw Hill; Vol. 1, pp. 231–405.

Meenakshi Gupta

Chapter 10
Conductance

Abstract: Conductance of a solution is the current-carrying capacity of a solution due to the movement of ions. This chapter gives an introduction to conductance and various related terms such as conductivity, resistance, resistivity, molar conductivity, equivalent conductivity, etc. along with Ohm's law and Kohlrausch's law. It also elucidates the principle of conductivity, its measurements, and how measurement of conductivity helps in determining the nutrient concentrations in liquid fertilizers, control quality of water, etc. Conductivity is affected by many factors like amount and type of salt in solution, temperature of the solution, etc. Conductometric titrations, their advantages over volumetric titrations, and calculation of dissociation constant (K_a) of weak acid have also been discussed at the end of this chapter.

10.1 Introduction

Solutions or metals conduct electricity due to flow of charged species. In electrolytes, there is a movement of ions, while in metals there is a movement of electrons due to which conduction occurs. Conductance is a very important technique, which is very helpful in our day-to-day life. Conductors not only allow current but also allow the transmission of heat and light from one source to another. Metals, earth, humans, and animals are all conductors. A conductor has many applications in our daily lives, for instance, mercury is an excellent liquid conductor, which is used in thermometers to check the temperature of the body. Conductivity has many uses; for example, the conductivity measurements help in determining the ionic content in a given solution, and thus used in water purification systems, soil salinity, etc. Pure water has no conductivity, but if impurities (salt) is added, conductivity arises. Thus, we can detect the presence of any salt or impurity in a water sample by measuring conductivity. This is also an important parameter while studying environmental monitoring processes. It can help in determining total dissolved solids (TDS, the amount of organic or inorganic solids dissolved in water) and salinity (i.e., the amount of salt in water) in a water sample. Measurement of salinity is an important parameter for sustenance of aquatic life in industries, i.e., aquaculture. This technique can be also used to determine pK_a value of an acid, equivalence point, and ionic product of solvents, etc.

https://doi.org/10.1515/9783110794816-010

10.2 Types of conductors

i. **Metallic conductors:** They are also known as electronic conductors. These conductors conduct electricity due to flow of free electrons, without producing any change in the chemical properties of the conductor. Examples in this category include PbS (lead sulfide), CdS (Cadmium sulfide), metals, alloys, etc.
ii. **Electrolytic conductors:** These conductors are also known as ionic conductors. These conductors conduct electricity due to migration of ions towards oppositely charged electrodes. In this type of conduction, new products are formed, for example, molten salts, aqueous solution of salts, etc.

10.3 Conductance of solution

When electricity passes through wire, current flows and potential difference arises; according to *Ohm's law*, potential difference (V) is directly proportional to current flow (I):

$$V \propto I$$

$$V = IR \tag{10.1}$$

where R is the resistance, i.e., a friction by the component to the flow of current. Its unit is ohm. Ohm is represented by the Greek letter "omega" (Ω):

$$R = \frac{V}{I} \tag{10.2}$$

$$1\,ohm = \frac{1\,Volt}{1\,Ampere}$$

Thus, 1 ohm is the resistance to flow of 1 ampere (A) of current when it passes through a potential difference of 1 volt (V).

Ohm's law is obeyed by ohmic materials like resistors, wires, etc., while nonohmic materials like battery, diodes, fluorescent lamps, etc. do not obey this law. The resistance is measured by an instrument known as ohmmeter.

Conductance is the term used to express the ease with which current flows through solids, metals and nonmetals, and solutions. Conductance is the reciprocal of resistance, i.e.,

$$R = \frac{1}{G} = \frac{V}{I} \tag{10.3}$$

It is symbolized by "G." The SI unit of conductance is mho or ohm^{-1} (Ω^{-1}) or Siemens (S).

Substances that show the property of conductance are conductors, for example, copper (Cu) and aluminium (Al), while insulators are substances that do not conduct electricity, like plastic.

$$G = \frac{I}{V} \tag{10.4}$$

If a current of 1 A (Ampere) passes through a component across which a potential difference of 1 V (Volt) exists, then the conductance of that component is said to be 1 S (Siemens):

$$1S = \frac{1A}{1V}$$

Thus, 1 S is equivalent to 1 ampere per unit volt.

The resistance (R) offered by the conductor to flow of current (I) is directly proportional to length (l) and inversely proportional to the area (A) of the cross section of a conductor:

$$R \propto \frac{l}{A}$$

$$R = \rho \frac{l}{A} \tag{10.5}$$

where ρ (rho) is a proportionality constant called *resistivity or specific resistance*. Its unit is ohm·meter (ohm·m). It is defined as resistance of the conductor having unit length and unit area of cross section, and l/A is called cell constant.

Therefore, resistance = resistivity × cell constant.

On inverting the equation,

$$\frac{1}{R} = G = \frac{1}{\rho} \times 1 \tag{10.6}$$

where G is conductance and $1/\rho$ is conductivity.

Hence,

$$Conductance = \frac{Conductivity}{cell\ constant} \tag{10.7}$$

Conductivity is the degree up to which any given material can conduct electricity. It is the ratio of material current density to the electric field that is generated when current flows. Conductivity is also known as specific conductance and conductivity of a material is its intrinsic property. It is denoted by κ (kappa). Its SI unit is S/m, and in CGS system, it is S/cm or ohm^{-1} cm^{-1}.

Resistance and conductance depend on the type of material used, size, shape of the object, and temperature; resistivity and conductivity are proportionality constants, and hence, they depend on material of wire (not on geometry or size) and temperature.

Note: Siemens is not a plural unit but this unit has been coined after the name of an electrical engineer named Sir William Siemens.

Equivalent conductivity is the conductivity of an electrolytic solution having 1 g (gram) equivalent of electrolyte dissolved in V cc (cubic centimeter) of solvent, which is placed between two electrodes that are 1 meter (m) apart from each other. It is denoted by Λ_{eq} and is equal to the product of specific conductance (κ) and volume in V (cc) containing 1 g equivalent of electrolyte:

$$\Lambda_{eq} = \kappa \times V_{eq} = \frac{1,000 \times \kappa}{C} \tag{10.8}$$

where C is g equivalent per litre, so volume (V) having 1 g equivalent is 1,000/C cm^3.

Its unit is
$$\Lambda_{eq} = \kappa \times V = \frac{1}{R} \times \frac{l}{A} \times V$$

$$= \frac{1}{ohm} \times \frac{cm}{cm^2} \times \frac{cm^3}{equiv}$$

$$= ohm^{-1} cm^2 equiv^{-1}$$

$$= S\,cm^2/equiv$$

Similarly, *molar conductivity* is the conductivity of an electrolytic solution having 1 mol of an electrolyte dissolved in V cc of solvent, which is placed between two electrodes that are 1 m apart from each other. It is denoted by Λ_m. It is equal to the product of specific conductance (κ) and volume, V (cc) containing 1 mol of electrolyte:

$$\Lambda_m = \kappa \times V = \frac{1,000 \times \kappa}{c} \tag{10.9}$$

where V is volume containing 1 mol of electrolyte in 1000/C cm^3. c is concentration, i.e., no. of moles/volume (L).

Its unit is
$$\Lambda_m = \kappa \times V = \frac{1}{R} \times \frac{l}{A} \times V$$

$$= \frac{1}{ohm} \times \frac{cm}{cm^2} \times \frac{cm^3}{mol}$$

$$= ohm^{-1} cm^2 mol^{-1}$$

$$= S\,cm^2/mol$$

Some examples have been given below for better understanding of these terms.

Example 10.1: The specific conductance of a decinormal solution of potassium chloride is 0.00112 ohm^{-1} cm^{-1} at 18 °C. If the resistance of the solution contained in the cell is 65 ohm, what is its cell constant?

Given:
$$\frac{N}{10} K_{KCl} = 0.00112 \, \text{ohm}^{-1} \, \text{cm}^{-1} \, \text{at} \, 18\,°C$$

Resistance = 65 ohm

Cell constant = ?

$$\text{Specific conductance} = \frac{\text{cell constant}}{\text{resistance}}$$

$$\text{Cell constant} = \text{specific conductance} \times \text{resistance}$$

$$= 0.00112 \times 65$$

$$= 0.0728 \, \text{cm}^{-1}$$

Example 10.2: Calculate the equivalent conductance of the solution, if the electrodes are 2 cm apart and area of cross section is 3.5 cm^2. The resistance of a decinormal electrolytic solution in conductometric cell is 245 ohm.

Given:
$$\frac{N}{10} R_{\text{electrolytic solution}} = 245 \, \text{ohm}$$

Concentration of solution (c) = 0.1 N

Length = 2 cm

Area = 3.5 sq cm

$$\text{Cell constant} = \frac{\text{length}}{\text{area}}$$

$$= \frac{2}{3.5} = 0.57 \, \text{cm}^{-1}$$

$$\Lambda_{eq} = \frac{1,000 \times k}{C}$$

$$\text{Conductance} = \frac{1}{\text{resistance}}$$

$$= \frac{1}{245} \, \text{ohm}$$

Specific conductance (k) = cell constant × conductance = $\frac{1}{245} \times 0.57 \, \text{ohm}^{-1} \, \text{cm}^{-1}$

$$\Lambda_{eq} = \frac{1,000 \times k}{C}$$

$$= \frac{1}{245} \times 0.57 \times \frac{1,000}{0.1}$$

$$= 23.27 \, \text{S} \, \text{cm}^2 / \text{eqiv}$$

Example 10.3: What is the resistance of the electrolytic solution placed in a conductometric cell having electrodes 6.5 cm apart and area of cross section is 13.0 cm^2? (Given resistivity of solution = 373.4 ohm • cm)

Given:
$$Length = 6.5 \text{ cm}$$
$$Area = 13.0 \text{ cm}^2$$

$$Resistivity \text{ of solution} = 373.4 \text{ ohm} \cdot \text{cm}$$

$$Resistance = Resistivity \times \frac{length}{Area}$$

$$= 373.4 \times \frac{6.5}{13}$$

$$= 186.7 \text{ ohm}$$

Example 10.4: The resistance of a cell containing N/50 potassium chloride was found to be 400 ohm. Determine the equivalent conductance of solution if cell constant is 1.10 cm^{-1}.

Given:
$$Resistance = 400 \text{ ohm}$$

$$Cell \text{ constant} = 1.10 \text{ cm}^{-1}$$

$$Concentration \text{ of KCl} = N/50$$

$$Conductance = \frac{1}{Resistance}$$

$$= \frac{1}{400} \text{ ohm}^{-1}$$

Specific conductance (k) = cell constant × conductance

$$= \frac{1}{400} \times 1.10 \text{ ohm}^{-1} \text{ cm}^{-1}$$

$$\Lambda_{eq} = \frac{1,000 \times k}{C}$$

$$= \frac{1}{400} \times 1.10 \times \frac{1,000}{1/50}$$

$$= 137.5 \text{ S cm}^2 / \text{ equiv}$$

Example 10.5: A conductometric cell having electrodes 3.0 cm apart and cross section area of 1.5 cm^2 contains 1 molar sodium chloride solution. If the molar conductivity of solution is 48.15 ohm^{-1} cm^2 mol^{-1} at 25 °C, what will be the current flow when potential difference of 5 V applied?

Given:
Length = 3.0 cm
Area of cross section = 1.5 cm^2
Concentration of NaCl = 1 M
Molar conductivity Λ_m = 48.15 ohm^{-1} cm^2 mol^{-1} at 25 °C
Voltage = 5 V

$$\Lambda_m = \frac{1{,}000 \times k}{C}$$

$$k = \frac{48.15 \times 1{,}000}{1}$$

$$\text{Conductance} = \frac{k}{\text{cell constant}} = \frac{48.15 \times 1{,}000}{1} \times \frac{1}{2}$$

$$= 2.4 \times 10^4 \, \text{ohm}^{-1}$$

$$\text{Resistance} = \frac{1}{\text{conductance}}$$

$$= \frac{1}{2.4 \times 10^4} \, \text{ohm}$$

$$= 4.2 \times 10^{-5} \, \text{ohm}$$

$$\text{Current} = \frac{\text{Voltage}}{\text{resistance}}$$

$$= \frac{5}{4.2 \times 10^{-5}} = 1.2 \times 10^{-5} \, \text{A}$$

10.4 Measurement of conductance

Conductance is the reciprocal of resistance, so we measure it by determining the resistance of an electrolytic solution using Wheatstone bridge in a laboratory.

Conductance cell: It is the cell in which an electrolytic solution is kept to determine its conductance.

Conductivity water: This water is free of ions, so it has zero conductance. This water is used in laboratory experiments.

Wheatstone bridge: It has four resistances out of which one resistance is a conductance cell which contains electrolytic solution whose resistance has to be determined (Figure 10.1).

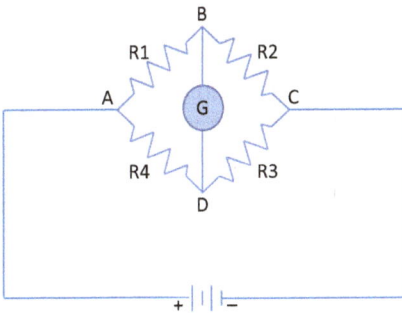

Figure 10.1: Wheatstone bridge.

Principle of Wheatstone bridge: If current flows through a circuit, there is no deflection in galvanometer, so ratio of resistances is equal, i.e.,

$$\frac{R1}{R4} = \frac{R2}{R3} \tag{10.10}$$

If no current flows, then bridge is balanced.

By using this, we can determine unknown resistance, if three other resistances are known.

A conductance cell having an electrolytic solution whose conductance is to be measured and two electrodes that are made up of platinum (Pt) plates coated with Pt black to avoid polarization are required. In this, we use alternating current (AC), not direct current (DC), because loss of energy during AC transmission is less and AC voltage can be stepped up and down, according to need. Conductance cell is also known as conductivity cell (Figure 10.2).

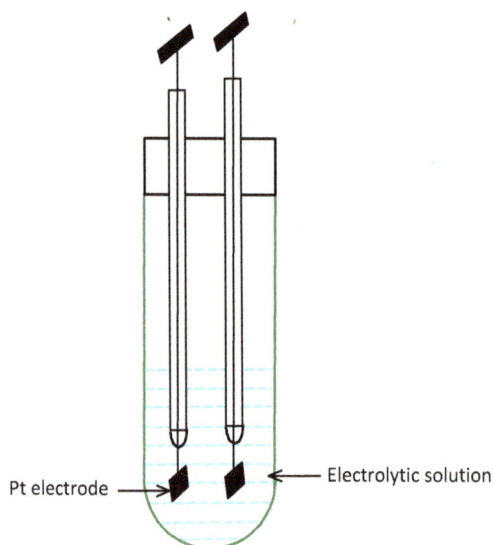

Figure 10.2: Conductivity cell.

Working: A Wheatstone bridge has a galvanometer, but in a conductance cell, galvanometer is replaced by an earphone, since on using AC, null point in galvanometer cannot be detected. AB is a wire along the bridge, and the bridge is connected to AC source, J is the jockey, and S is conductivity cell (Figure 10.3). Once resistance is measured, other factors like conductance, conductivity, resistivity, equivalent conductance, and molar conductance can be measured, as discussed earlier.

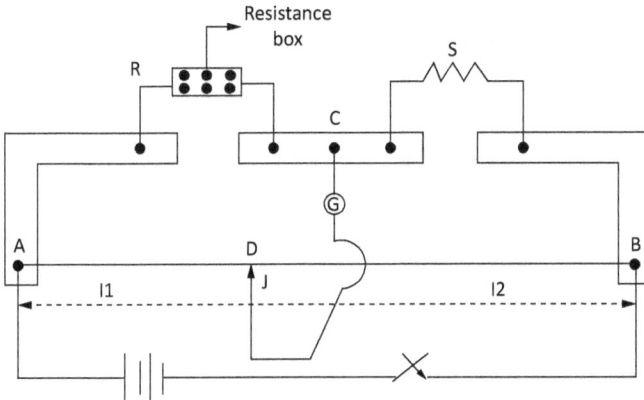

Figure 10.3: Wheatstone connected to a conductivity cell.

A suitable resistor, R is plugged out of the resistance box, so that jockey "J" is moved along wire AB. Let sound in earphone be minimum at point D. Then X, resistance of conductance cell is given by

$$\text{Wheatstone bridge} = \frac{\text{Resistance of } R}{\text{Resistance of } S} = \frac{\text{Resistance of } AD}{\text{Resistance of } BD} = \frac{\text{Length } AD}{\text{Length } BD}$$

$$\text{Resistance of } S = R \times \frac{\text{Length } BD}{\text{Length } AD}$$

$$\text{Conductance} = \frac{1}{R} \times \frac{\text{Length } AD}{\text{Length } BD} \tag{10.11}$$

10.4.1 Determination of cell constant

Cell constant is the distance between the electrodes (l) divided by area of cross section (A). Thus, if we determine these two parameters, cell constant can be measured; but direct measurement of these two is difficult. Thus, indirect method is employed to measure them, i.e.,

$$\text{Conductance} = \frac{\text{conductivity}}{\text{cell constant}}$$

$$\text{Cell constant} = \frac{\text{conductivity}}{\text{conductance}}$$

$$\text{Cell constant} = \frac{\text{specific conductance}}{\text{observed conductance}} \tag{10.12}$$

For a standard solution of KCl (potassium chloride), specific conductance at a given temperature is known, which can be used to determine the cell constant. A solution of known strength of KCl is taken in a conductance cell at a given temperature, and conductance value is measured. Plug these two values in the above equation and determine the cell constant.

10.5 Variation of conductivity with concentration

Conductivity depends on the concentration of solution, i.e., number of ions in a given solution. If concentration of an electrolyte increases, conductivity increases due to increase in number of ions. This occurs in strong as well as weak electrolytes.

10.5.1 Variation of molar conductance with concentration (or dilution)

The effect of dilution on molar conductance can be studied by plotting a graph of molar conductivity versus $\sqrt{c}$ (Figure 10.4). This graph shows that an increase in concentration decreases the molar conductivity for weak as well as strong electrolytes, though the variation is different for strong and weak electrolytes. There is a sharp increase in molar conductivity with dilution in strong electrolytes, while in weak electrolytes, molar conductivity gradually increases at high concentration, but for a very dilute solution, the increase is very sharp as observed in graph:

$$\Lambda_m = \kappa \times V = \frac{1{,}000 \times \kappa}{c}$$

This equation shows that molar conductivity (Λ_m) is directly proportional to κ, while it is inversely proportional to c (concentration). On dilution, number of ions decreases, so κ decreases but $1/c$ factor increases. Out of these two opposite factors, increase in $1/c$ factor is more powerful, so equivalent conductance increases with dilution, irrespective of the nature of electrolyte.

10.5.1.1 Strong electrolytes

For strong electrolytes, the graph is linear. This can be explained on the basis of Kohlrausch's equation. According to Kohlrausch, molar conductivity depends on the concentration by the following relation:

$$\Lambda_m^c = \Lambda_m^\infty - b\sqrt{c} \tag{10.13}$$

where Λ_m^c is molar conductivity at concentration c and Λ_m^∞ and b are constants; a straight-line graph with negative slope will be observed. In essence, in strong electrolytes, ion–ion interaction decreases on dilution.

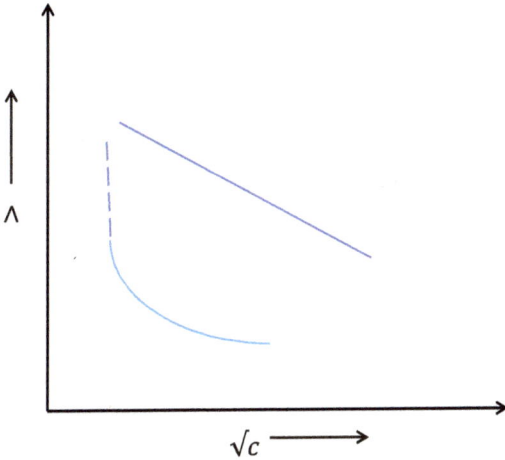

Figure 10.4: Graph of molar conductivity versus $\sqrt{c}$.

10.5.1.2 Weak electrolytes

They have low ionic concentration as they do not ionize completely, and interionic forces are infinitesimally small. The increase in molar conductivity with dilution is due to the increase in number of current-carrying ions. This can be explained on the basis of Ostwald's dilution law.

10.5.2 Ostwald's dilution law

Ostwald's dilution law states that as weak electrolyte solution is diluted, its degree of dissociation will be more.

10.5.2.1 Mathematically, Ostwald's dilution law states

$$\alpha = \sqrt{\frac{K}{c}} \tag{10.14}$$

where α is the degree of dissociation, K is the equilibrium constant of reaction $AB \rightleftharpoons A^+ + B^-$, and c is the concentration.

This law is applicable for weak electrolytes only. According to this law, as c decreases, α increases, so concentration of A^+ and B^- increases, and hence, molar conductivity increases due to increase in number of ions on dilution.

At infinite dilution, the degree of dissociation is

$$\alpha = \frac{\Lambda_m^c}{\Lambda_m^\infty} = \frac{\Lambda_{eq}^c}{\Lambda_{eq}^\infty} \tag{10.15}$$

Λ_m^c and Λ_m^∞ are molar conductivities at concentration c and at infinite dilution, respectively. Similarly Λ_{eq}^c and Λ_{eq}^∞ are molar conductivities at concentration c and at infinite dilution, respectively.

Here are a few examples:

Example 10.6: What is the degree of ionization of decinormal solution of benzoic acid, if its equilibrium constant is 6.46×10^{-5}?

Given: Concentration (c) of benzoic acid = N/10
                   Equilibrium constant (K) = 6.46×10^{-5}

$$\alpha = \sqrt{\frac{K}{c}}$$

$$\alpha = \sqrt{\frac{6.46 \times 10^{-5}}{N/10}}$$

$$= 0.025$$

Hence, the degree of ionization = 0.025.

Example 10.7: The specific conductance of centinormal solution of acetic acid is $0.000158 \ ohm^{-1} \ cm^{-1}$ at 18 °C. The equivalent conductance of acetic acid at infinite dilution at 18 °C is $387.0 \ ohm^{-1} \ cm^2 \ equiv^{-1}$. Calculate the degree of ionization of centinormal solution of acetic acid at 18 °C.

Given: $K_{CH_3COOH} = 0.000158 \ ohm^{-1} \ cm^{-1}$ at 18°C

Concentration of acetic acid (c) = N/100

$$\Lambda_{eq(HAc)}^\infty = 387.0 \ ohm^{-1} \ cm^2 \ equiv^{-1}$$

$$\alpha = \frac{\Lambda_m^c}{\Lambda_m^\infty}$$

$$\Lambda_c = \text{Specific conductance} \times \frac{1,000}{\text{eq. wt.}}$$

$$= \frac{0.000158}{N/100} \times 1,000$$

$$= 15.8 \ ohm^{-1} \ cm^2 \ equiv^{-1}$$

$$\alpha = \frac{15.8}{387}$$

$$= 0.04$$

10.5.3 Ionic mobility

Ionic mobility is defined as the speed of an ion under electric field, and it is denoted by μ:

$$\mu = \frac{\text{ionic velocity}}{\text{potential gradient}} = \frac{\text{m/s}}{\text{V/m}} = \text{m}^2/\text{Vs} \tag{10.16}$$

The unit of ionic mobility is $\text{m}^2/\text{V s}$, and it depends on the following factors:

i.   Temperature: An increase in temperature increases the ionic mobility, as ions start moving faster due to high kinetic energy.

ii. Nature of the electrolyte.

iii. Size of an ion: If the size of ion is bigger, then it would have difficulty in moving, and hence its ionic mobility will be less, and vice versa.

Example of ionic mobility:

Example 10.8: What is the ionic mobility of ammonium ion if it moves 2.05 m in 1 h when placed in a conductometric cell having potential of 5.60 V and the two electrodes are placed 9.8 cm apart?

Given: Distance covered = 2.05 m = 205 cm
                        Time taken = 1 h = 3600 s
                        Potential = 5.60 V
       Distance between electrodes = 9.8 cm

$$\text{Formula, } \mu = \frac{\text{ionic velocity}}{\text{potential gradient}}$$

$$\text{Ionic velocity} = \frac{\text{distance}}{\text{time}} = \frac{205}{3,600} = 0.057 \text{ cm/s}$$

$$\text{Potential gradient} = \frac{\text{potential}}{\text{distance between electrodes}} = \frac{5.60}{9.8} = 0.57 \text{ V/cm}$$

$$\mu = \frac{0.057 \text{ cm/s}}{0.57 \text{ V/cm}} = 0.1 \text{ cm}^2/\text{Vs}$$

10.5.4 Kohlrausch's law

According to this law, at infinite dilution, molar conductance of an electrolyte is the sum of conductances of each anion and cation present in it. For example,

$$\Lambda^\infty_{\text{HAc}} = \Lambda^\infty_{\text{H}^+} + \Lambda^\infty_{\text{Ac}^-}$$

where $\Lambda^\infty_{\text{HAc}}$ is molar conductivity of acetic acid at infinite dilution, $\Lambda^\infty_{\text{H}^+}$ is molar conductivity of acetic acid at infinite dilution, and $\Lambda^\infty_{\text{Ac}^-}$ is molar conductivity of acetic acid at infinite dilution:

$$\Lambda^\infty_{HAc} = \Lambda^\infty_{HCl} + \Lambda^\infty_{NaAc} - \Lambda^\infty_{NaCl}$$

where HCl, NaAc, and NaCl are hydrochloric acid, sodium acetate, and sodium chloride, respectively. They all are strong electrolytes, so their Λ^∞ can be determined easily in the laboratory by plotting Λ versus $\sqrt{c}$ graph. However, for weak electrolytes, determination of Λ^∞ is not possible in the laboratory, thus Λ^∞ of weak acid can also be determined by applying Kohlrausch's law.

For example:

Example 10.9: If the equivalent conductance at infinite dilution of HCl, CH_3COONa, and NaCl are 426.76, 91.0, and 126.45 $ohm^{-1}cm^2equiv^{-1}$ respectively, find the equivalent conductance at infinite dilution of acetic acid. The equivalent conductance of acetic acid at a given concentration is 48.15 ohm^{-1} cm^2 $equiv^{-1}$ at 25 °C; calculate the degree of ionization of acetic acid.

Given:
$$\Lambda^\infty_{eq(HCl)} = 426.76\ ohm^{-1}\ cm^2\ equiv^{-1}$$

$$\Lambda^\infty_{eq(CH_3COONa)} = 91.0\ ohm^{-1}cm^2equiv^{-1}$$

$$\Lambda^\infty_{eq(NaCl)} = 126.45\ ohm^{-1}\ cm^2\ equiv^{-1}$$

$$\Lambda^c_{eq(CH_3COOH)} = 48.15\ ohm^{-1}cm^2equiv^{-1}\ at\ 25\ °C$$

$$\Lambda^\infty_{eq(CH_3COOH)} = \Lambda^\infty_{eq(HCl)} + \Lambda^\infty_{eq(CH_3COONa)} - \Lambda^\infty_{eq(NaCl)}$$

$$= 426.76 + 91 - 126.45$$

$$= 391.31\ ohm^{-1}\ cm^2\ equiv^{-1}$$

$$\alpha = \frac{\Lambda^c_{eq}}{\Lambda^\infty_{eq}}$$

$$= \frac{48.15}{391.31}$$

$$= 0.123$$

Hence, equivalent conductance at infinite dilution of acetic acid is 391.31 ohm^{-1} cm^2 $equiv^{-1}$ and degree of ionization is 0.123.

10.6 Conductometric titrations

Conductometric titrations is one amongst various applications of conductance, and these titrations are employed to measure the strength of a given unknown solution. There are various methods to perform a titration, such as volumetrically, potentiometrically, pH metrically, etc. Conductometric titrations are the titrations in which conductance is measured to determine the strength of a given solution. Volumetric titrations are easy to perform but they have some drawbacks over conductometric titrations, which are as follows:

- Conductometric titrations involve equivalence point determination, which is more accurate than the end point that is observed in volumetric titrations.
- They can be performed for colored solutions.
- Indicators are not required for conductometric titrations.
- They can be used for the titration of a weak acid against a weak base.
- The stepwise dissociation constant of polyprotic acids can be obtained by conductometric titrations.

10.6.1 Principle of conductometric titration

In conductometric titration, during the course of titration, one ion is replaced by another ion, which differs in mobility, due to the difference in sizes. Thus, there is a variation of conductance during these titrations, which is recorded and studied to obtain the equivalence point.

In conductometric titrations, a measured volume of the solution whose strength has to be determined is taken in a conductance cell. Then, a known volume of titrant from a burette is added gradually in small lots (say, 0.2 mL), the contents are stirred, and value of conductance is recorded after each addition. The equivalence point is determined by plotting a graph of conductance versus volume of titrant added, and the strength of the given solution can be calculated by applying the law of equivalence.

10.6.2 Titration of strong acid (HCl, hydrochloric acid) versus strong base (NaOH, sodium hydroxide)

The procedure for the conductometric titration of a strong acid with a strong base is as follows:

i.   Take 30–40 mL of HCl in a conductance cell, and dip the electrodes in it.
ii. Add 0.2 mL of NaOH solution from the burette, and record the conductance after each addition.
iii. Plot a graph between obtained conductance value and volume of added NaOH (in mL). A V-shaped graph (Figure 10.5) will be obtained, and the line of intersection will give the equivalence point where the acid is completely neutralized.

Note: Generally, in conductometric titration, the strength of titrant taken is 10 times stronger than titrand, so equivalence point can be achieved in a few readings.

The obtained graph is V-shaped, as initially, only HCl, which is a strong electrolyte, is present.

Thus, $HCl \rightarrow H^+ + Cl^-$

Note: H^+ and OH^- have exceptionally high mobility due to "Grotthuss law." It states that proton jumps from one water molecule to another, which causes transfer of ions, and hence results in high mobility and high conductivity. This mechanism is also shown by NH_4^+ and NH_2^- ions, but NH_4^+ and NH_2^- have low conductance as compared to H^+ and OH^- due to their large sizes, which results in less mobility, and hence, low conductance.

Initially, highly conducting H^+ is present in the solution, so conductance is high. On gradual addition of NaOH, the following reaction occurs:

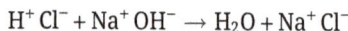

$$H^+ Cl^- + Na^+ OH^- \rightarrow H_2O + Na^+ Cl^-$$

Thus, highly conducting H^+ is replaced by slow-moving Na^+ and conductivity decreases by each addition of NaOH, until the whole of HCl is neutralized by NaOH. After equivalence point is achieved, only added NaOH will be left in the solution; therefore, conductance starts increasing again due to highly mobilized OH^- ions, and we get a V-shaped graph.

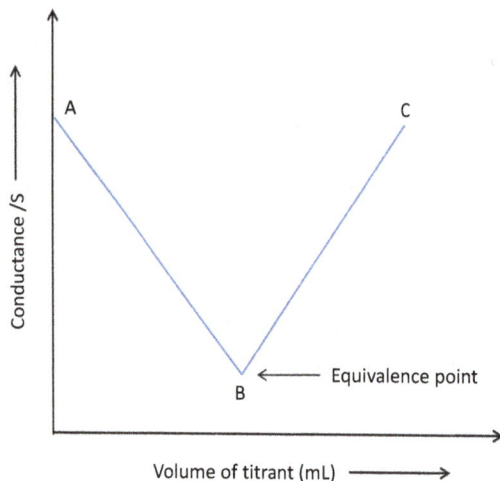

Figure 10.5: Conductometric titration curve of strong acid versus strong base.

10.6.3 Weak acid (CH₃COOH, acetic acid) versus strong base (NaOH, sodium hydroxide)

Acetic acid (CH_3COOH) is a weak electrolyte, so it is ionized weakly, and only a small amount of H^+ is present in the solution; initially, conductance of solution is very low:

$$CH_3COOH \rightleftharpoons CH_3COO^- + H^+ \qquad (10.17)$$

On addition of NaOH, the following reaction takes place:

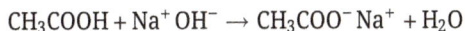

$$CH_3COOH + Na^+ OH^- \rightarrow CH_3COO^- Na^+ + H_2O$$

CH$_3$COONa also ionizes as

$$CH_3COONa \rightleftharpoons CH_3COO^- + Na^+ \tag{10.18}$$

Now, acetate ion (CH$_3$COO$^-$) is produced from two different reactions in system, i.e., reactions (10.17) and (10.18). Due to this, common ion effect takes place and suppression in ionization of acetic acid (reaction stops) happens, and point B (Figure 10.6, in curve) is obtained. Now, there is no free CH$_3$COO$^-$ and H$^+$, but when NaOH is added, following reaction takes place:

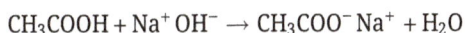

$$CH_3COOH + Na^+ OH^- \rightarrow CH_3COO^- Na^+ + H_2O$$

Due to CH$_3$COO$^-$ and Na$^+$, increase in conductance takes place, but it is not very sharp. When the whole of the acetic acid is neutralized by NaOH, equivalence point "C" is obtained. On further addition of NaOH, highly mobilized OH$^-$ ions increase, which enhances the conductance of the solution.

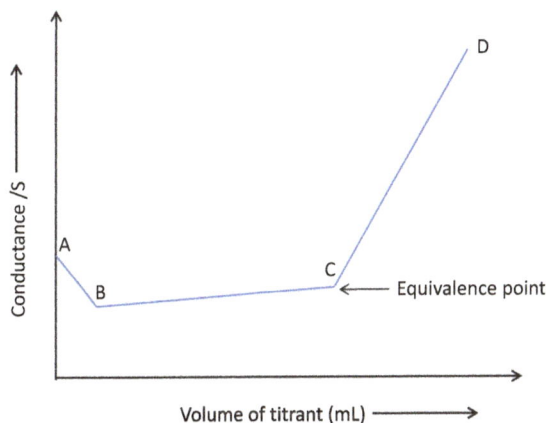

Figure 10.6: Conductometric titration curve of weak acid versus strong base.

10.6.4 Strong acid (HCl, hydrochloric acid) versus weak base (NH₄OH, ammonium hydroxide)

In the conductometric titration of a strong acid against a strong base, initially, the conductance of the solution is high due to highly mobilized H$^+$, as strong electrolyte (HCl) ionized completely. On addition of a weak base, NH$_4$OH, the following reaction takes place:

$$H^+ Cl^- + NH_4OH \longrightarrow NH_4^+ Cl^- + H_2O$$

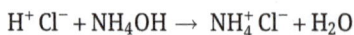

Now, the fast-moving H^+ ions are replaced by NH_4^+, so conductance of the solution decreases. Further addition of NH_4OH results in more and more decrease in conductance. On complete neutralization of HCl by NH_4OH, equivalence point (point B in Figure 10.7) is achieved. On further addition of NH_4OH, which is a weak electrolyte, no significant change in conductance is observed, and a straight line, BC, is obtained on adding more of NH_4OH.

[*Note:* In this case, graph before equivalence point is similar to strong acid versus strong base graph]

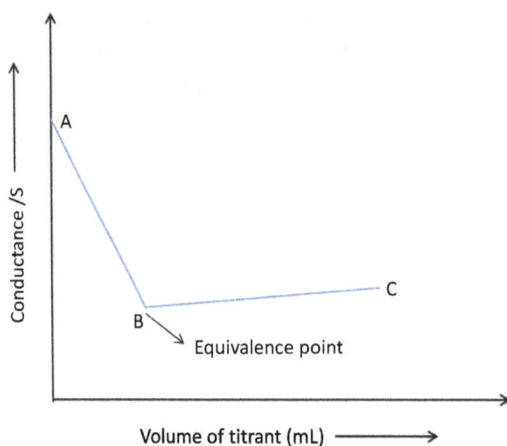

Figure 10.7: Conductometric titration curve of strong acid versus weak base.

10.6.5 Weak acid (CH$_3$COOH) versus weak base (NH$_4$OH)

Let us now discuss the conductometric titration of a weak acid (CH$_3$COOH) with a weak base (NH$_4$OH). In this case, the nature of the curve before equivalence point is the same as that observed in the case of acetic acid versus NaOH. But, after equivalence point is obtained, as NH$_4$OH ionizes feebly, conductance remains almost same virtually, on further addition of NH$_4$OH (Figure 10.8). Thus, we get an almost straight line after equivalence point.

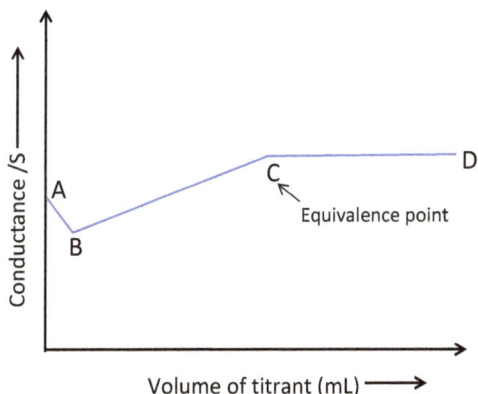

Figure 10.8: Conductometric titration curve of weak acid versus weak base.

10.6.6 Mixture of strong acid and weak acid versus strong base or weak base

After discussing conductometric titrations of different combination of acids and bases, now, let us consider the case of a mixture of strong acid and weak acid against strong base or weak base, for example, HCl and CH_3COOH against NaOH or NH_4OH.

In this graph, we get two break points, which represent two equivalence points. Firstly, the complete neutralization of strong acid takes place, so the first break point represents equivalence point of HCl. In the second stage, neutralization of weak acid starts, and the second break point represents the complete neutralization of acetic acid or equivalence point of weak acid. (Volume of titrant used to neutralize weak base is total volume of titrant till the second break point minus volume of titrant used up to the first break point.) Then, after two break points, if strong base is used, conductance increases due to presence of highly mobile OH^- ions; but if a weak base is used, conductance remains almost the same, due to weak ionization of NH_4OH, as shown in Figure 10.9.

10.6.7 Polyprotic acid versus strong or weak base

In case of polyprotic acid versus strong or weak base, the number of break points in the graph is equal to the basicity of acid. If we take diprotic acid, then stepwise dissociation of H^+ takes place, and each dissociation corresponds to one break point.

Suppose oxalic acid is titrated with a base. Since its basicity is 2, we will get two break points, and each break point corresponds to the neutralization of one proton (H^+):

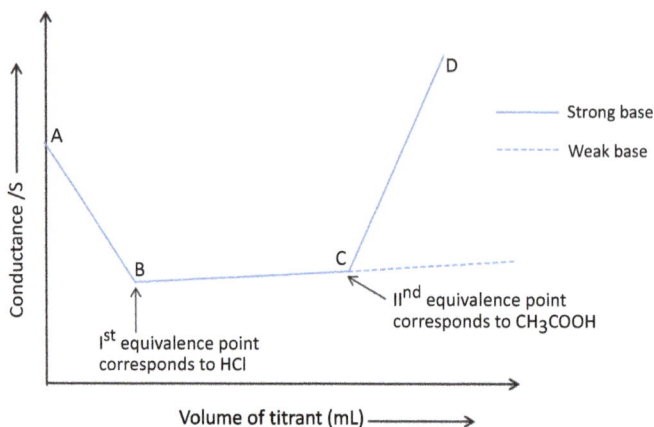

Figure 10.9: Conductometric titration curve of mixture of strong acid and weak acid versus strong base or weak base.

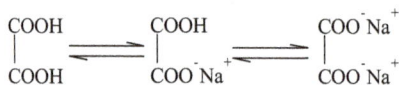

The ionization of first proton (H^+) gives the first break point (Point B in Figure 10.10), and second ionization (H^+) gives the second break point (Point C in Figure 10.10). After getting two break points, if strong base is used, then conductance increases due to highly mobile OH^- ions, but if a weak base is used, conductance remains almost the same due to its feeble ionization of weak base.

Thus, with the help of conductometric titrations, we can determine the stepwise ionization of each proton (H^+) as well as the basicity of an acid as

Basicity of an acid = No. of break points in graph

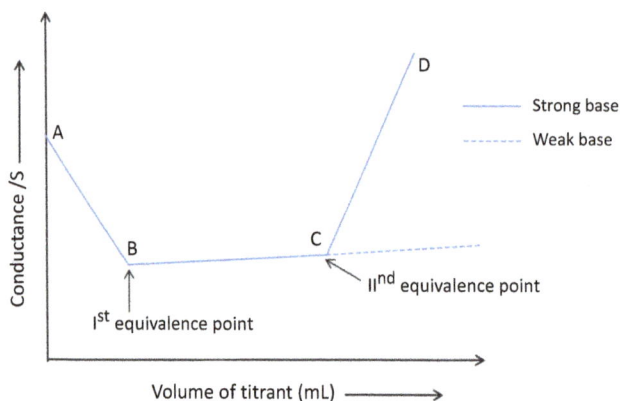

Figure 10.10: Conductometric titration curve of dibasic acid versus strong base or weak base.

10.6.8 Precipitation titration

To understand precipitation titrations, let us take the example of potassium chloride (KCl) and silver nitrate ($AgNO_3$). During this reaction, a precipitate of AgCl (silver chloride) is formed, which is insoluble in water. KCl and $AgNO_3$ both are strong electrolytes, so they dissociate completely in the solution:

$$KCl + AgNO_3 \longrightarrow AgCl + KNO_3$$

Initially, on addition of $AgNO_3$, there is virtually no change in the conductance, since Cl^- (chloride) ions are replaced by NO_3^- (nitrate ions) having almost same mobility. After the equivalence point, conductivity of the solution increases due to Ag^+ (silver) and NO_3^- (nitrate) ions (Figure 10.11).

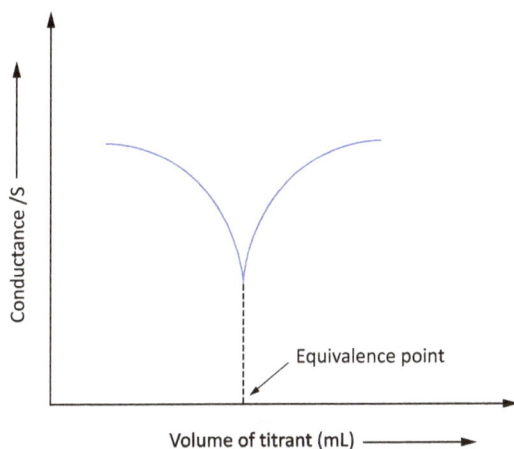

Figure 10.11: Precipitation titration.

10.6.9 Determination of the strength of a solution using equivalence point

The strength of an unknown solution can be determined by applying the law of equivalence, which is

$$N_{acid} \times V_{acid} = N_{base} \times V_{base}$$

where N_{base} is the normality of base of known strength, V_{acid} is the volume of acid whose strength is to be determined (i.e., volume of the titrand taken in conductance cell), N_{acid} is the normality of acid, which is to be determined, and V_{base} is volume of base, i.e., amount of titrant used till equivalence point is obtained from the graph.

So,

$$N_{\text{acid}} = \frac{N_{\text{base}} \times V_{\text{base}}}{V_{\text{acid}}}$$

(10.19)

Hence, strength of an unknown acid can be calculated using the relation:

$$\text{Strength} = N_{\text{acid}} \times \text{equivalent weight of an acid}$$

(10.20)

10.7 Determination of pK_a for a weak electrolyte

The pK_a value of a strong electrolyte can be easily determined as they ionize completely in the solution, while pk_a of a weak electrolyte is difficult to determine in the laboratory, as it does not ionize completely. However, the pK_a of a weak electrolyte can be determined using conductance as

$$pK_a = -\log K_a$$

(10.21)

where K_a is the dissociation constant of an acid.
    For example:

$$CH_3COOH \rightleftharpoons CH_3COO^- + H^+$$

$$K_a = \frac{[CH_3COO^-][H^+]}{CH_3COOH}$$

Let the initial concentration of acetic acid be "c", and α is the degree of dissociation after some time "t". Then, concentration of acetic acid, acetate ion, and hydrogen ion will be $c(1 - \alpha)$, $c\alpha$, and $c\alpha$, respectively:
    So

$$K_a = \frac{[c\alpha]\,[c\alpha]}{[c(1-\alpha]} = \frac{c\alpha^2}{(1-\alpha)}$$

But

$$\alpha = \frac{\Lambda_m^c}{\Lambda_m^\infty}$$

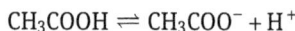

Plugging it in the above equation, we get:

$$K_a = \frac{c\Lambda^{c2}}{\Lambda^\infty\,(\Lambda^\infty - \Lambda^c)}$$

(10.22)

Hence, using the relation, pK_a = $-\log K_a$, pK_a can be determined.

For weak electrolytes, Λ^∞ can be determined by Kohlrausch's law. For example, the pK_a of acetic acid can be determined.

Example 10.10: Conductivity of acetic acid in aqueous solution at 25 °C was recorded as

c (mol/dm)	20	40	80	160	320
m (ohm^{-1} cm^2 equiv^{-1})	52.2	71.4	95.8	126.6	163.3

If equivalent conductance of acetic acid is 390.7 ohm^{-1} cm^2 equiv^{-1}, find its pK_a.

Solution:

$$K_a = \frac{c\Lambda^{c2}}{\Lambda^\infty (\Lambda^\infty - \Lambda^c)}$$

$$\frac{K_a}{c} = \frac{\Lambda^{c2}}{\Lambda^\infty (\Lambda^\infty - \Lambda^c)}$$

Hence, if we plot $\dfrac{\Lambda^{c2}}{\Lambda^\infty (\Lambda^\infty - \Lambda^c)}$ versus $1/c$ graph, slope will give value of K_a, and we can determine its pK_a value as

$$pK_a = -\log K_a$$

---

c (mol/dm)	m (ohm^{-1} cm^2 equiv^{-1})	$\dfrac{c\Lambda^{c2}}{\Lambda^\infty (\Lambda^\infty - \Lambda^c)}$
20	52.2	0.020603
40	71.4	0.040865
80	95.8	0.079655
160	126.6	0.15533
320	163.3	0.30015

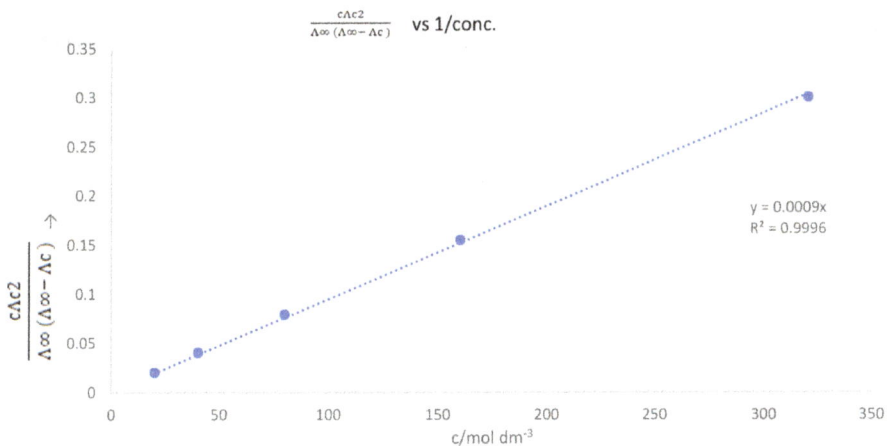

$\dfrac{c\Lambda c2}{\Lambda\infty\,(\Lambda\infty - \Lambda c)}$ vs 1/conc.

y = 0.0009x
R² = 0.9996

c/mol dm⁻³

$$\text{Slope from graph} = 0.0009 = K_a$$

$$pK_a = -\log K_a$$

$$= -\log(0.0009)$$

$$= 3.05$$

The symbols and units of important terms used in the chapter are summarized in Table 10.1.

Table 10.1: Important terms related to conductance, their symbols, and units.

Quantity	Symbol	Unit
Resistance	R	Ohm
Conductance	G	Siemens
Resistivity or specific resistance	ρ	Ohm cm
Conductivity or specific conductance	κ	S/cm
Equivalent conductance	Λ_{eq}	S/cm equiv
Molar conductance	Λ_m	S/cm mol
Ionic mobility	μ	m^2 /V s

References

[1] Kapoor, K.L. A textbook of physical chemistry, 5th edn. Chapter 5, McGrawHill; vol. 1, 406–473.
[2] Castellan, G.W. Physical chemistry. Narosa; 2004.
[3] Puri, B.R., Sharma, L.R., and Pathania, M.S., Principles of physical chemistry, Vishal Publishing Co.; 2020.

Meenakshi Gupta and Shramila Yadav

Chapter 11
Electrochemistry

Abstract: Electrochemistry, as the name implies, is the branch of chemistry which deals with the chemistry of electrical and chemical energy. A device which converts chemical energy into electrical energy or vice versa is an electrochemical cell, and it can be categorized in galvanic cell and electrolytic cell. This chapter deals with the construction of cells and chemistry underlying these cells, salt bridge, cell reaction, cell potential (oxidation and reduction potential) along with their examples. Standard hydrogen electrode, chosen as a standard for the determination of reduction potentials of all other cells and how a reactivity series has been formed along with application of standard potentials like prediction of oxidizing or reducing agent, determination of cell electromotive force, and spontaneity of the reaction, and Nernst equation for the half-cell and complete cell are also discussed. Various applications of electrochemistry such as determination of dissociation constant, potentiometric titrations, determination of thermodynamic quantities have also been taken up in this chapter.

11.1 Introduction

Electrochemistry, as the name implies, is the branch of chemistry which deals with the chemistry of electrical and chemical energy. Electrical energy can be generated using chemical reactions, or chemical reactions can be made possible with the help of electricity. In this chapter, we shall discuss the chemistry behind these two processes, their basic principles, sign conventions, various types of cells, and their applications.

A device which is used to convert chemical energy into electrical energy or vice versa is an electrochemical cell. So basically, an electric current is produced from a chemical reaction or with the help of current, a chemical reaction is made to occur. The chemical reaction which is involved in this is a redox reaction, a reaction that is the combination of reduction and oxidation. Redox reaction is defined as a reaction in which gain and loss of electron (e^-) takes place simultaneously. Oxidation is loss of electron by a species, whereas reduction is gain of electrons by another species. This can be illustrated with the following reactions:

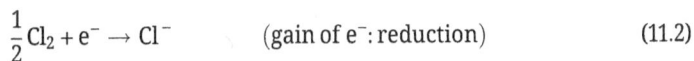

$$Na \rightarrow Na^+ + e^- \qquad \text{(loss of } e^- \text{: oxidation)} \qquad (11.1)$$

$$\frac{1}{2}Cl_2 + e^- \rightarrow Cl^- \qquad \text{(gain of } e^- \text{: reduction)} \qquad (11.2)$$

Adding eqs. (11.1) and (11.2), we get

https://doi.org/10.1515/9783110794816-011

$$Na + \frac{1}{2}Cl_2 \rightarrow Na^+ + Cl^- \quad \text{(loss and gain of } e^-: \text{redox reaction)} \qquad (11.3)$$

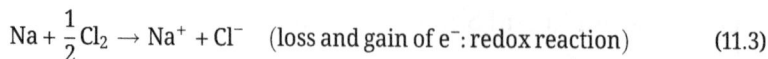

Oxidation and reduction are the two half components of a redox reaction; hence, they are also termed as "half reactions." Equation (11.1) is oxidation half reaction, eq. (11.2) is reduction half reaction, and eq. (11.3) is a redox reaction.

11.2 Faraday's laws of electrolysis

Faraday formulated two laws of electrolysis which are quantitative, and explain the phenomenon of electrolysis. These two laws were the findings of the electrochemical research of an English scientist Michael Faraday, which were published in 1833.

First law: According to the first law of electrolysis, the amount of substance deposited on electrode (W) is directly proportional to amount of electricity passed (Q):

$$W \propto Q \qquad (11.4)$$

Quantitatively, current (I) in amperes, time (t) in seconds, and Q (in coulomb) are related as follows:

$$I = \frac{Q}{t}$$

Or

$$Q = I \times t \qquad (11.5)$$

On putting the value of Q in eq. (11.4),

$$W \propto Q = z \times I \times t$$

where z is the proportionality constant and known as electrochemical equivalent.
If 1 ampere current flows for 1 s,

$$z = W \ (\text{in gram})$$

Thus, an electrochemical equivalent is the amount of substance deposited on electrode when 1 A current is passed for 1 s or simply, we can say, 1 C of charge is passed.

Second law: Faraday's second law states that "If the same amount of electricity is passed through the two different electrolytic cells, then the amount of substance deposited on the respective electrodes will be the ratio of equivalent weight of two electrolytes," which can be represented mathematically as follows:

$$\frac{\text{Amount of electrolyte 1 deposited}}{\text{Amount of electrolyte 2 deposited}} = \frac{\text{equivalent weight of electrolyte 1}}{\text{equivalent weight of electrolyte 2}}$$

Let us say, if we take two different electrolytes, H_2SO_4 (sulfuric acid) and $CuSO_4$ (copper sulfate), and the same amount of electricity is passed through them, then

$$\frac{\text{Amount of hydrogen deposited}}{\text{Amount of copper deposited}} = \frac{\text{equivalent weight of hydrogen}}{\text{equivalent weight of copper}}$$

$$\frac{W_1}{W_2} = \frac{E_1}{E_2}$$

where W_1 and W_2 are the amounts of electrolyte 1 and electrolyte 2, respectively, and E_1 and E_2 are the equivalent weights of electrolyte 1 and electrolyte 2, respectively. This law is helpful in calculating the equivalent weight of metals.

11.2.1 Faraday

Faraday (F) is the unit of electricity and can be defined as the quantity of electricity required to deposit 1 gram equivalent of a substance on an electrode after electrolysis.

One-gram equivalent mass requires 1 Faraday of electricity, so the amount of charge carried by 1 gram equivalent (g equiv) of any ion is 1 Faraday. In other words, we can say that 1 Faraday is the charge carried by 1 mol of electrons. Since 1 mol of electrons corresponds to one equivalent of the element; therefore, the units of the F can also be expressed as C/equiv.

Thus, $1\,\text{Faraday} = 1.6022 \times 10^{-19}\,C \times 6.022 \times 10^{23}\,\text{mol}^{-1} = 96,485\,C/\text{equiv}$

$$\approx 96,500\,C/\text{equiv}$$

If n is the valency of an ion, then 1 ion contains n g equiv, which therefore carries n F Coulomb (C) of electricity. These laws are independent of temperature, pressure, and nature of the solvent.

Now, some numerical on the above mentioned topic is given further:

Example 11.1 How many grams of lithium liberated when 250 C of charge passes through the cell having LiCl (lithium chloride) which is electrolyzed and Li (lithium) metal is liberated at cathode?

Solution: Given: charge = 250 C
Equivalent mass of Li metal = 7 g equiv
1 Faraday of charge liberates 7 g equiv of Li
$\therefore$ 96,500 C liberates 7 g equiv weight of Li
So, 250 C liberates $\frac{7}{96,500} \times 250 = 0.018$ g of Li
So, 0.018 g of Li will be liberated.

Example 11.2: What is the amount of electricity needed to liberate 710 g of chlorine (Cl_2) gas by electrolysis of sodium chloride (NaCl)?

Solution: 35.5 g of Cl_2 gas liberated by 1 Faraday of charge.
So, 710 g of gas liberated by $= \frac{1}{35.5} \times 710 \; F = 20 \; F$
So, 20 F of electricity is needed.

Example 11.3: What is the electrochemical equivalent of Ag (silver), if 0.198 g of Ag is deposited by passing current of 0.5 A for 20 min?

Solution: Given: time (t) = 20 min = 20×60 s = 1,200 s
Current (I) = 0.5 A
So, $Q = I \times t = 1{,}200 \times 0.5 = 600$ C.
Amount of Ag deposited by 600 C = 0.198 g.
Amount of Ag deposited by 1 C = $\frac{0.198}{600} \times 1 = 0.00033$ g
So, electrochemical equivalent of Ag = 0.00033 g.

Example 11.4: Two cells $CuSO_4$ (copper sulfate) and $AgNO_3$ (silver nitrate) are in series, and electric current is passed through them. What is the amount of Ag deposited if 2.5 g of Cu (copper) is deposited?

Solution: Weight of Cu = 0.25 g
Equivalent weight of Cu = 31.7 g equiv
Equivalent weight of Ag = 108 g equiv
As cells are in series and the same amount of current is passed

$$\frac{\text{Amount of copper deposited}}{\text{Amount of silver deposited}} = \frac{\text{equivalent weight of copper}}{\text{equivalent weight of silver}}$$

$$\frac{0.25}{\text{Weight of Ag}} = \frac{31.7}{108}$$

$$\text{Weight of Ag} = \frac{0.25 \times 108}{31.7} = 0.852 \text{g}$$

Hence, the weight of silver deposited will be 0.852 g.

11.3 Classification of electrochemical cells

As stated above, electrochemical cells convert chemical energy into electrical energy or vice versa. They can be broadly classified into two main categories:

i. Electrolytic cell – An electrolytic cell is an electrochemical cell in which electric current is used to carry out chemical reaction. For example, charging of lead storage battery.

ii. Galvanic cell – Galvanic cell is an electrolytic cell in which electric current is produced using spontaneous chemical reaction. For example, dry cell, lead storage battery.

11.3.1 Electrolytic cell

The chemical reactions involved in electrolytic cell are non-spontaneous therefore an external source (DC battery or AC current) is required to carry out the chemical change. This can be explained with the help of an example. Consider the electrolysis of molten sodium chloride (NaCl). The electrodes of platinum are dipped into NaCl solution and an external battery is connected to supply electric current. The negative terminal of the battery is connected to cathode (negative electrode), and cations (positively charged ions) will move toward it. The positive terminal of the battery is connected to anode (positive electrode) and negatively charged ions, anions are attracted toward it (Figure 11.1). The following reactions take place in the cell:

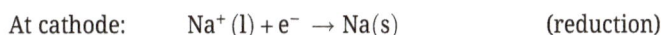

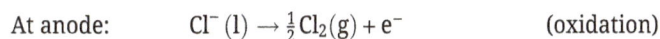

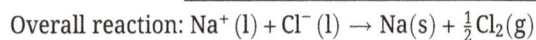

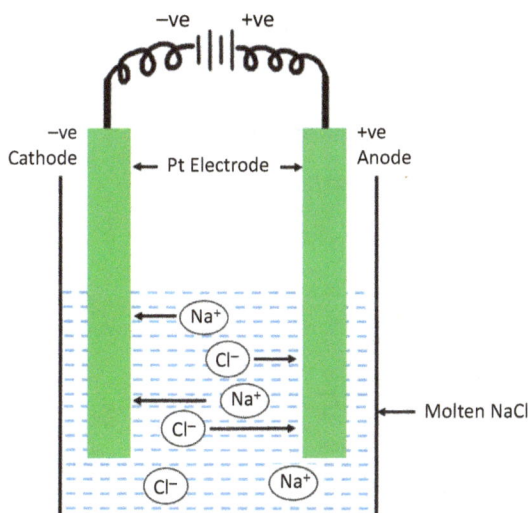

At cathode: $Na^+ (l) + e^- \rightarrow Na(s)$ (reduction)

At anode: $Cl^- (l) \rightarrow \frac{1}{2}Cl_2(g) + e^-$ (oxidation)

Overall reaction: $Na^+ (l) + Cl^- (l) \rightarrow Na(s) + \frac{1}{2}Cl_2(g)$

Figure 11.1: Electrolytic cell.

Sign convention of an electrolytic cell is summarized as follows:

Electrode	Position	Sign	Reaction
Cathode	Left	Negative	Reduction
Anode	Right	Positive	Oxidation

11.3.2 Galvanic cell

Galvanic cells, also known as voltaic cells, are electrochemical cells in which spontaneous chemical reactions are used to generate electrical energy. When a piece of zinc (Zn) metal is dropped in a solution of copper sulfate ($CuSO_4$,), what will happen? Zn metal starts dissolving by forming Zn^{2+} ions and the blue color of $CuSO_4$ solution starts fading as Cu^{2+} ions converting into Cu(s). This type of reaction is a spontaneous chemical reaction and can be represented as

$$Zn(s) + Cu^{2+}(aq) \rightarrow Zn^{2+}(aq) + Cu(s) \tag{11.6}$$

Such reaction is used to produce electric current by forming a device called cell. To formulate a cell, Zn electrode is dipped in a solution of $ZnSO_4$ (zinc sulfate) and Cu electrode is dipped in a solution of $CuSO_4$ in two separate compartments. Zn electrode dissolves and forms Zn^{2+} ions which goes into solution and releases two electrons which make the electrode negative and solution positive. The ionization process stops immediately due to these changes. Similarly, Cu^{2+} ions from solution get deposited on Cu electrode which makes the solution negative and electrode positive, and these changes stop the ionization process in this compartment too. These two processes can be written as follows:

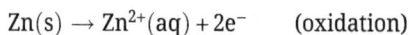

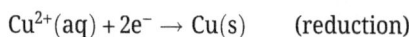

$$Zn(s) \rightarrow Zn^{2+}(aq) + 2e^- \quad \text{(oxidation)}$$

$$Cu^{2+}(aq) + 2e^- \rightarrow Cu(s) \quad \text{(reduction)}$$

So, sign conventions of a galvanic cell are:

Electrode	Sign	Charge on solution	Reaction
Cathode	Positive	Negative	Reduction
Anode	Negative	Positive	Oxidation

To resume the ionization process, oxidation at Zn electrode and reduction at Cu, these two electrodes should be connected with an external wire which facilitates the flow of electrons from Zn (where electron releases) to Cu or from anode to cathode and also brings these two solutions into contact without mixing, through a salt bridge as shown in Figure 11.2.

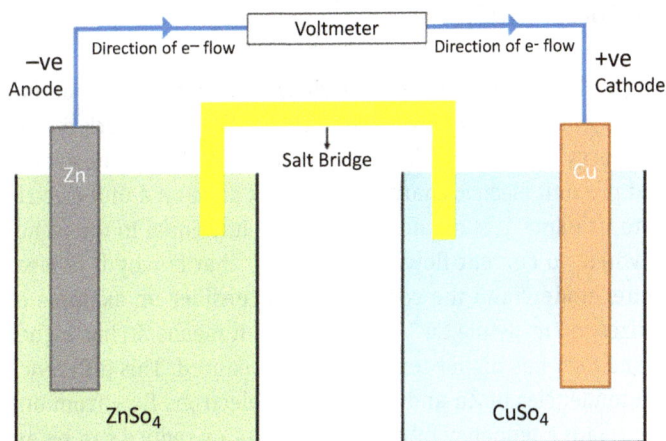

Figure 11.2: Galvanic cell (Daniel cell).

Salt bridge is a "U"-shaped tube that contains inert electrolytes such as KCl (potassium chloride) or NH_4NO_3 (ammonium nitrate) which is mixed with a gelatin material such as agar-agar that prevents the flow of electrolytes into the solution.

Functions of salt bridge are:

i.   It helps in neutralization of the charges of two solutions placed in separate compartment.

ii. It facilitates the movement of ions without intermixing of two solutions.

iii. It completes the electrical circuit.

11.4 Cell reaction

As mentioned earlier, the chemical reaction involved in electrochemical cells is the redox reaction which is a combination of oxidation and reduction reaction. Oxidation involves the loss of e^- taking place at the anode. These lost electrons flow from anode to cathode where they are gained by another species in the cathode compartment. The reactions at anode and cathode are half-cell reactions and by adding these two half reactions with equal number of electrons, net overall cell reaction is obtained. For example, in Daniel cell:

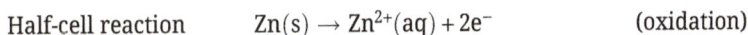

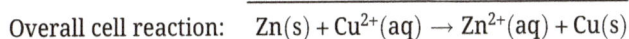

Half-cell reaction	$Zn(s) \rightarrow Zn^{2+}(aq) + 2e^-$	(oxidation)
Half-cell reaction	$Cu^{2+}(aq) + 2e^- \rightarrow Cu(s)$	(reduction)
Overall cell reaction:	$Zn(s) + Cu^{2+}(aq) \rightarrow Zn^{2+}(aq) + Cu(s)$	

11.4.1 Electromotive force (EMF)

Electromotive force (EMF) is the energy per unit electric charge, imparted by an energy source such as a battery or an electric generator. The device does work on the electric charge when the energy is converted from one form to another in the energy source. The energy gained per unit electric charge or the work done on a unit electric charge is EMF. Contrary to its name, it is not actually a force. It is equal to the terminal potential difference when no current flows in the circuit that is why it is measured in volts (V). To better understand the concept, let us consider an example of Zn–Cu cell. Zn (zinc) oxidizes to Zn^{2+}, while Cu^{2+} reduces to Cu. It means Zn has higher tendency to get oxidize, and Cu^{2+} has higher tendency to get reduced. This difference in oxidation or reduction tendencies of Zn and Cu make the electrons flow from one electrode to another. The relative tendency of atoms to oxidize or reduce can be expressed in terms of oxidation potential or reduction potential, respectively. Therefore, Zn has higher oxidation potential than Cu and electron flows from the electrode of higher oxidation potential to that of lower oxidation potential. The difference of potential between two electrodes which pushes electron to flow from one electrode to another is known as EMF of the cell, which is also represented as E_{cell}:

$$E_{cell} = \text{Higher oxidation potential} - \text{lower oxidation potential}$$

$$E_{cell} = \text{HOP} - \text{LOP}$$

Upon oxidation, an atom with highest oxidation potential will produce an ion with lowest reduction potential. The reduction potential is simply negative of the oxidation potential with the same magnitude. By convention, E_{cell} is written in terms of reduction potential and is also known as the standard potential:

$$E_{cell} = -(-\text{HOP}) + (-\text{LOP}) \text{ (LOP, lower oxidation potential)}$$

$$E_{cell} = -\text{LRP (lower reduction potential)} + \text{HRP (HRP, higher reduction potential)}$$

$$E_{cell} = \text{HRP} - \text{LRP}$$

11.4.2 Representation of a cell

An electrochemical cell can be represented in the form of cell diagram. A complete cell is a combination of two half-cells, and the chemical reactions that take place in these two half-cells are half-cell reactions or electrode reactions. A metal electrode dipped into metal ion solution constitutes a half-cell. In 1953, IUPAC (International Union of Pure and Applied Chemistry) recommended the following conventions for representing the cell as a cell diagram. Consider a Zn–Cu cell to illustrate the conventions:

i.   A single vertical line (|) represents a phase boundary. As the metal electrode and metal ion are in different phases, they are separated by a single vertical line:

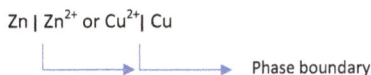

$$Zn \mid Zn^{2+} \text{ or } Cu^{2+}\mid Cu$$

Phase boundary

ii.   A double vertical line (‖) denotes the salt bridge. Two half-cells are separated by double vertical lines:

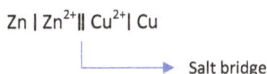

$$Zn \mid Zn^{2+}\| Cu^{2+}\mid Cu$$

Salt bridge

iii. Anode half-cell is written first on the left followed by the cathode half-cell on the right side.

iv. The concentrations of solution are written within the brackets () followed by their metal ions:

$$Zn \mid Zn^{2+}(0.1 \text{ M}) \| Cu^{2+}(0.1 \text{ M})\mid Cu$$

v.   If an inert electrode is involved, it is written with its gaseous component often enclosed in a bracket:

$$Zn \mid Zn^{2+}(0.1 \text{ M}) \| H^{+}(0.1 \text{ M}) \mid H_2(Pt)$$

vi. EMF value of a cell is written after cell diagram on the right side:

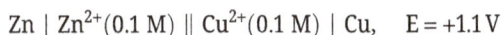

$$Zn \mid Zn^{2+}(0.1 \text{ M}) \| Cu^{2+}(0.1 \text{ M}) \mid Cu, \quad E = +1.1\,V$$

As reduction takes place at the cathode in a galvanic cell, the cathode will always have higher reduction potential and it is placed on the right-hand side. Oxidation takes place at the anode so it has the lower reduction potential. Therefore, E_{cell} can also be expressed as follows:

$$E_{cell} = E_{cathode} - E_{anode}$$

or

$$E_{cell} = E_{right} - E_{left}$$

Some of the important points regarding E_{cell} are worth remembering for better clarification:

i.   EMF is the potential difference between two half-cells.

ii.   An electrochemical cell is a combination of two half-cells.

iii. Half-cell kept on the left-hand side is anode half-cell and oxidation takes place on it.

iv. Half-cell kept on the right-hand side is cathode half-cell and reduction takes place on it.

v. $E_{cell} = E_{cathode} - E_{anode}$ or $E_{cell} = E_{reduction} - E_{oxidation}$ or $E_{cell} = E_R - E_L$

vi. If E_{cell} is positive, then half-cell on the right side has higher reduction potential than the half-cell on the left side.

vii. Half-cell having the positive reduction potential will have negative oxidation potential with the same magnitude.

viii. The negative (−ve) sign of E_{cell} or EMF indicates the feasibility of a cell reaction. If the reaction is reversed, the sign changes.

11.5 Reversible or irreversible cell

The thermodynamic principles are applicable to only those systems which are reversible in the thermodynamic sense. The thermodynamic properties such as free energy change, entropy change, and enthalpy change of an electrochemical cell can be determined, if the cell reaction is reversible. The reversible cell reactions are always in a state of equilibrium. There are two conditions which must be satisfied by the system to be reversible:

i. The driving and opposing force should be infinitesimally different from each other.

ii. By applying an infinitesimally greater force than the acting one, the chemical reaction can be reversed.

If these two conditions are satisfied by a cell, it is said to be reversible. Let us understand it by taking an example of Daniel cell. If the electrodes (Zn and Cu) are connected to an external source whose EMF is infinitesimally smaller than the E_{cell}, the net reaction is

$$Zn(s) + Cu^{2+}(aq) \rightarrow Zn^{2+}(aq) + Cu(s)$$

The direction of flow of current is from Zn to Cu electrode. If the EMF of the external source is increased such that it is infinitesimally greater than the cell EMF, the above reaction gets reversed:

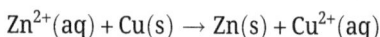

$$Zn^{2+}(aq) + Cu(s) \rightarrow Zn(s) + Cu^{2+}(aq)$$

When EMF of an external source is exactly equal to E_{cell}, no current will flow and the cell reaction will cease. Therefore, the Daniel cell is a reversible cell. If these conditions are not satisfied by changing the EMF of the external source, the cell reaction cannot be reversed, then the cell is said to be irreversible.

11.6 Determination of EMF

It is not possible to determine the potential of a half-cell or single electrode, and we can only measure the potential difference between two electrodes as a complete circuit which must contain two electrodes. The potential of an electrode to be measured can be combined with an electrode of known potential to make a complete cell, and the instrument used to measure the potential difference is called potentiometer. By convention, the potential of all electrodes is measured in reference to the standard hydrogen electrode (SHE). Arbitrarily, the potential of SHE at 298 K (1 bar pressure of H_2 and 1 M (molar) [H^+] concentration, which are the standard conditions) is fixed at zero. So, the potential of all other electrodes can be measured in reference to a common electrode (SHE). The half-cell reaction of SHE is

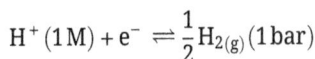

$$H^+ (1\,M) + e^- \rightleftharpoons \frac{1}{2} H_{2(g)} (1\,bar)$$

This electrode is reversible in nature and therefore can act as a cathode as well as an anode. In SHE, hydrogen gas at 1 atmospheric pressure is bubbled through 1 M solution of H^+ (hydrogen) ions maintained at 298 K. The Pt electrode is used to provide the surface of contact to H_2 gas as shown in Figure 11.3. The potential of the electrode under these conditions is referred to as standard potential, and represented as E^0.

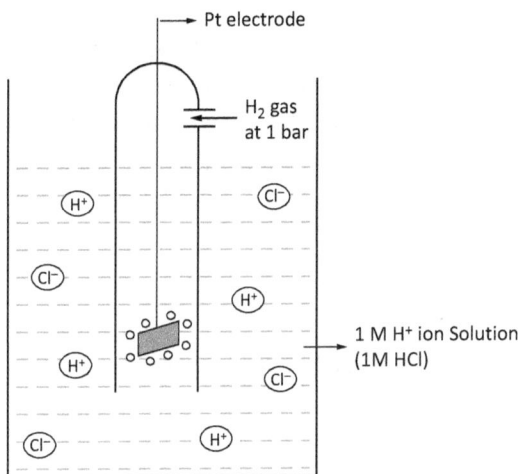

Figure 11.3: Standard hydrogen electrode (SHE).

To understand the process of measuring the potential of the electrodes in combination with a reference electrode, let us take an example of a zinc electrode. Zinc electrode dipped in the solution of $ZnSO_4$ ($Zn^{2+}|Zn$) is coupled with SHE and the resultant cell may be represented as follows:

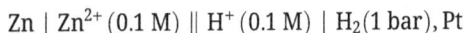

$$Zn \mid Zn^{2+} (0.1\,M) \parallel H^+ (0.1\,M) \mid H_2(1\,bar), Pt$$

where Zn electrode is placed on the left-hand side and SHE is placed on right-hand side. When $Zn^{2+}|Zn$ electrode is connected with SHE, whose electrode potential is zero, the potentiometer will give the potential of Zn electrode.

The EMF of the cell is

$$E^0 = E^0_R - E^0_L$$

$$E^0 = E^0_{SHE} - E^0_{Zn\,|Zn^{2+}}$$

$$= 0 - E^0_{Zn\,|Zn^{2+}}$$

$$= 0 - 0.76$$

$$= -0.76 \text{ V}$$

The electrode reactions of the above cell are represented as follows:

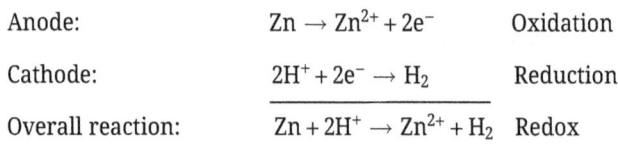

Anode:	$Zn \rightarrow Zn^{2+} + 2e^-$	Oxidation
Cathode:	$2H^+ + 2e^- \rightarrow H_2$	Reduction
Overall reaction:	$Zn + 2H^+ \rightarrow Zn^{2+} + H_2$	Redox

By convention, adopted by IUPAC, SHE is always placed on the left-hand side of the electrode under study. The reduction takes place on the electrode placed on the right-hand side, so the observed EMF of an electrochemical cell is the potential of half-cell on the right-hand side and is known as the standard reduction potential or simply standard potential:

$$E^0_{cell} = E^0_{unknown\ electrode} - E^0_{SHE}$$

$$E^0_{cell} = E^0_{unknown\ electrode} - 0$$

The standard potentials of various electrodes arranged in decreasing order form an electrochemical series.

11.7 Application of standard potentials

The various applications of standard potentials are:
i. **Predicting better reducing or oxidizing agent:** The positive value of standard potential indicates that the particular electrode has a tendency to get reduced, and therefore, can act as an oxidizing agent. More positive the standard potential of the half-cell, better is its oxidizing power. For example:

$$Au^{3+} + 3e^- \rightarrow Au \quad E^0 = +1.498$$
$$Ag^+ + e^- \rightarrow Ag \quad E^0 = +0.799$$
$$Cu^{2+} + 2e^- \rightarrow Cu \quad E^0 = +0.342$$
$$Pb^{2+} + 2e^- \rightarrow Pb \quad E^0 = -0.126$$
$$Zn^{2+} + 2e^- \rightarrow Zn \quad E^0 = -0.762$$

Strength as Oxidizing Agents

Strength as Reducing Agents

The negative value of a reduction potential indicates that it does not undergo reduction process, hence, it has a tendency to undergo oxidation. The half-cell which can undergo oxidation can reduce others; so, it is a better reducing agent. More negative the value of reduction potential, stronger the reducing agent. Zn has the most negative value of standard potential; hence, it is the best reducing agent among all the metals mentioned earlier and can reduce Au (gold), Ag (silver), Cu (copper), and Pb (lead), while Pb can reduce Au, Ag, and Cu. Au is the best oxidizing agent among all.

ii. **EMF of the cell:** The standard EMF (E^0) of a cell can be determined using the following relation:

$$E^0_{cell} = E^0_{Right} - E^0_{Left}$$

$$= E^0_R - E^0_L$$

iii. **Spontaneity of a reaction:** The positive value of E^0_{cell} means the reaction is feasible or spontaneous, while the negative value of E^0_{cell} indicates nonspontaneity of a reaction.

iv. **Displacement of a metal by another metal from its solution:** As already stated, using the magnitude of standard potential, one can predict the oxidizing or reducing power of a metal. Metals lying near the bottom of electrochemical series are strong reducing agents, while those lying higher up in the series are strong oxidizing agents. Whether a metal will displace another metal or not, can be predicted from the magnitude of standard potential; for example, Cu lying high up in the series with $E^0 = +0.34$ V and Zn lying down with $E^0 = -0.76$ V. According to their electrode potential values, Cu with positive reduction potential has a tendency to reduce Cu^{2+} to Cu, while Zn with negative reduction potential has a tendency to oxidize Zn to Zn^{2+}. Therefore, when zinc is placed in $CuSO_4$ solution, Cu metal will be displaced by Zn from its solution and Cu gets precipitated, while if Cu is placed in $ZnSO_4$ solution, no reaction will take place.

v. **Displacement of hydrogen by a metal from acidic solution:** As hydrogen has zero reduction potential, therefore, any metal lying up and having positive reduction potential is a weaker reducing agent than hydrogen. While a metal with a negative reduction potential or lying below hydrogen is a stronger reducing agent than hydrogen, and has the capability to convert H^+ to H_2. Therefore, Zn with the negative potential can react with acid and liberate H_2 from acid solution. This is the reason the zinc container cannot be used to store acid, while Cu with the positive reduction potential does not react with acid, and hence, can be used to store highly reactive acids.

11.8 The Nernst equation

Walter Nernst, in 1889, derived a mathematical relationship between electrode potential, temperature, and activities of the ions in solution. This equation, known as the Nernst equation, enables the determination of half-cell potential and cell potential under nonstandard conditions and can be stated as follows:

$$E = E^0 - \frac{RT}{nF} \ln K$$

where E is the half-cell potential, E^0 is the standard electrode potential, R is the gas constant, T is the temperature (in Kelvin, K), n is the number of electrons transferred in the half-cell reaction, F is Faraday of electricity, and K is the equilibrium constant for the reaction as given by equilibrium law. The various forms of Nernst equations can be given as

$$E = E^0 - \frac{2.303\,RT}{nF} \log K$$

Or at 298 K,
$$E = E^0 - \frac{0.0591}{n} \log K$$

It relates the measured cell potential to the reaction quotient (Q) and also allows the accurate determination of equilibrium constant.

Nernst equation for half-cell: The Nernst equation can be represented for the half-cell. Consider an oxidation half-cell reaction,

$$M \rightleftharpoons M^{n+} + ne^-$$

Equilibrium constant for the above reaction: $K = \dfrac{[M^{n+}]}{[M]}$

Therefore, the Nernst equation can be written as follows:

$$E = E^0 - \frac{2.303\,RT}{nF} \log \frac{[M^{n+}]}{[M]}$$

The concentration of a solid metal or liquid is equal to unity; therefore, $[M] = 1$.

$$\text{At 298 K, } E = E^0 - \frac{0.0591}{n} \log[M^{n+}]$$

Nernst equation for a cell: As we know that

$$E_{cell} = E_{right} - E_{left}$$

Nernst equation for a cell can be written by putting E_{right} and E_{left} in this equation or directly by writing the Nernst equation for the cell as

$$E_{cell} = E^0_{cell} - \frac{0.0591}{n} \log K$$

Consider a Zn–Ag cell which can be represented as

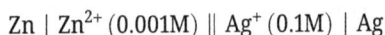

$$Zn \mid Zn^{2+}\ (0.001M) \parallel Ag^+\ (0.1M) \mid Ag$$

The cell reactions (oxidation and reduction) for the above cell can be written as

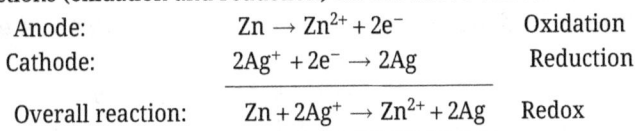

Anode:	$Zn \rightarrow Zn^{2+} + 2e^-$	Oxidation
Cathode:	$2Ag^+ + 2e^- \rightarrow 2Ag$	Reduction
Overall reaction:	$Zn + 2Ag^+ \rightarrow Zn^{2+} + 2Ag$	Redox

$$K = \frac{[Zn^{2+}][Ag]^2}{[Zn][Ag^+]^2} = \frac{[Zn^{2+}]}{[Ag^+]^2} \qquad \left([Ag]^2 \text{ and } [Zn] \text{ are unity}\right)$$

The E^0_{cell} can be calculated from the standard reduction potential values of the corresponding half–cells:

$$E^0_{cell} = E^0_R - E^0_L = E^0_{Ag^+|Ag} - E^0_{Zn|Zn^{2+}}$$

$$= 0.80 - (-0.76) = 1.56\,V$$

Nernst equation for the cell may be given as

$$E_{cell} = E^0_{cell} - \frac{0.0591}{2} \log \frac{[Zn^{2+}]}{[Ag^+]^2}$$

$$E_{cell} = 1.56 - \frac{0.0591}{2} \log \frac{[0.001]}{[0.1]^2}$$

$$E_{cell} = 1.56 - \frac{0.0591}{2} \log(0.1)$$

$$= 1.56 + 0.02955 = 1.5895\,V$$

Therefore, in this way we can calculate the potential of a cell.

11.9 Different types of reference electrodes

A reference electrode is an electrode whose potential is known and is used as a reference to measure the potential of other electrodes as determining the accurate potential of any electrode is difficult.

There are two types of reference electrodes:

i. *Primary reference electrode*: The Standard Hydrogen Electrode (SHE) is called a primary reference electrode.
ii. *Secondary reference electrode*: The electrode whose potential can be measured using primary reference electrode is termed as "secondary reference electrode."

11.9.1 Standard hydrogen electrode (primary reference electrode)

As we have already discussed, the hydrogen electrode is employed as a reference for measuring the electrode potential of a cell. It is considered as the primary reference electrode as it defines the zero point in the electrochemical scale. SHE is a reversible electrode which can act both as anode and as cathode in an electrochemical cell although SHE is a primary standard but it has many disadvantages which restrict its use in the laboratory.

Disadvantages of using hydrogen electrode as reference electrode in research purposes are:

i.   It requires pressure of hydrogen gas that should be maintained at 1 atm, which is difficult to maintain.
ii. Hydrogen electrode are sensitive to H_2S (hydrogen sulfide), dissolved oxygen and reducible substances or ions that have positive electrode potential like Cu, Ag, and Au, so cannot be used in the presence of these metals/gases.
iii. They are very sensitive to electrode poisoning by catalytic poisons.

11.9.2 Calomel electrode

Calomel electrode is a type of metal-sparingly soluble salt electrode. It is one of the most convenient secondary reference electrodes and can replace SHE. It can act as either anode or cathode depending on the other electrode connected to it. Calomel electrode, shown in Figure 11.4, is a mercury–mercurous chloride ($Hg–Hg_2Cl_2$) electrode. In this electrode, pure Hg is placed at the bottom of the glass tube which is covered with a paste of Hg_2Cl_2 and Hg (i.e., calomel). The remaining portion of the tube is with KCl (potassium chloride) solution. KCl of different concentration can be used, that is, normal, decinormal, or saturated solution. The potential of a calomel reference electrode varies with the variation in the concentration of KCl. A Pt (platinum) wire is sealed into a glass tube to make connection.

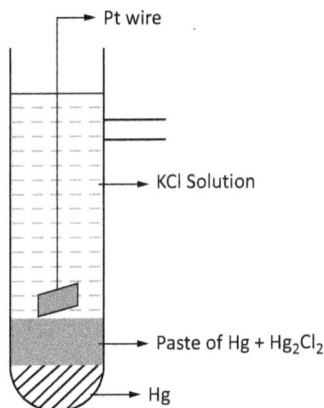

Figure 11.4: Calomel electrode.

It is represented by Hg|Hg$_2$Cl$_2$ (s), KCl (aq)

$$\frac{1}{2}Hg_2Cl_2(s) + e^- \rightarrow Hg(l) + Cl^-(aq)$$

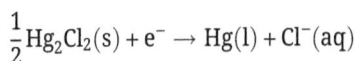

and Nernst equation for the calomel electrode is given as

$$E_{Cl^-|Hg_2Cl_2|Hg} = E^0_{Cl^-|Hg_2Cl_2|Hg} - \frac{RT}{F}\ln a_{Cl^-}$$

The value of an electrode potential for the different concentrations is as follows:

$$E_{Cl^-|Hg_2Cl_2|Hg} \text{ (0.1 N KCl)} = 0.336 \text{ V}$$

$$E_{Cl^-|Hg_2Cl_2|Hg} \text{ (1 N KCl)} = 0.283 \text{ V}$$

$$E_{Cl^-|Hg_2Cl_2|Hg} \text{ (saturated solution of KCl)} = 0.224 \text{ V at 298 K}$$

When calomel electrode is combined with the hydrogen electrode, the cell is represented as

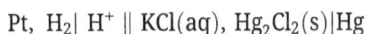

$$\text{Pt, } H_2| H^+ \| KCl(aq), Hg_2Cl_2(s)|Hg$$

Calomel electrode acts as a positive electrode when combined with the hydrogen electrode; therefore, the electron will flow toward the calomel electrode from the external circuit.

Let a Zn electrode be dipped in ZnSO$_4$ (0.1 N) solution and connected with (0.1 N KCl) calomel half-cell, and the cell formed is represented by

$$\text{Zn(0.1 N)}|ZnSO_4(aq) \| KCl(aq)(0.1 \text{ N}), Hg_2Cl_2(s)|Hg$$

So, EMF of a cell is measured by the potentiometer, that is, E_{obs}:

Thus, $E_{obs.} = E_{Zn} - (-0.336)$

$$E_{Zn} = E_{obs.} - 0.336$$

11.9.3 Silver–silver chloride electrode (Ag–AgCl electrode)

The silver–silver chloride is another example of reference electrode which is a widely used reference electrode as it is simple, very stable, inexpensive, and nontoxic. It is considered to be less toxic than calomel electrode (due to the presence of mercury). It is also a common choice of biological electrodes due to their low half-cell potential of about +222 mV (SHE) and also low impedance.

Ag–AgCl electrode consists of a strip of pure silver coated with AgCl (silver chloride). The electrolytic solution can be KCl or HCl. In laboratory, it is mainly used with saturated potassium chloride (KCl) electrolyte with an excess of AgCl crystals, but it can be used with lower concentrations such as 1 M KCl. The extra crystals of KCl dissolve into the electrolyte as the potassium and chloride ions present diffuse out through the liquid junction in normal use. Hence, extra buffer of KCl basically extends the time before the reference cell starts to drift due to the depletion of chloride ions in the electrolyte.

The electrode is represented as follows:

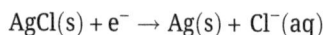

$$Ag|AgCl(s), KCl$$

$$AgCl(s) + e^- \rightarrow Ag(s) + Cl^-(aq)$$

$$E_{Cl^-|AgCl|Ag} = E^0_{Cl^-|AgCl|Ag} - \frac{RT}{F} \ln a_{Cl^-}$$

11.10 Applications of electrochemistry

11.10.1 Determination of pH of solution by using different electrodes

11.10.1.1 Hydrogen electrode

The unknown concentration of an acid can be determined by measuring the pH of the solution by combining the half-cell (unknown) to the another hydrogen electrode. Let us consider that a cell is made up of two hydrogen half-cells: one contains H^+ ion at unit activity and other contain H^+ ions of unknown activity.

The cell can be represented as

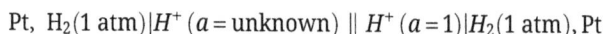

$$Pt, \ H_2(1 \ atm)|H^+ \ (a = unknown) \ || \ H^+ \ (a = 1)|H_2(1 \ atm), Pt$$

$$H^+ + e^- \rightarrow \frac{1}{2}H_2$$

H_2 gas is at 1 atm pressure in both half-cells.

$$\text{Now pH} = -\log[H^+] = \log\tfrac{1}{[H^+]}$$

By Nernst's equation

$$E = \frac{RT}{F}\ln a_{H^+} = \frac{2.303 \times 8.314 \times 298}{96,500}\log a_{H^+}$$

$$= 0.0591\log\frac{1}{a_{H^+}} = 0.0591\text{pH}$$

$$E_{cell} = E_{cathode} - E_{anode}$$

$$= E_c - 0.0591\,\text{pH}$$

So, at 298 K, $\quad pH = \dfrac{E_{cell}}{0.0591}.$

The hydrogen electrode has limitations so other electrodes are used generally in laboratories.

11.10.1.2 Quinhydrone electrode

Quinhydrone electrode contains equivalent amounts of *p*-benzoquinone and hydroqui-none. It is also used to determine hydrogen ion concentrations of unknown solutions. When the black-colored crystals of quinhydrone are dissolved in water, they decompose into quinone (Q) and hydroquinone (H_2Q) which establish the reversible reaction:

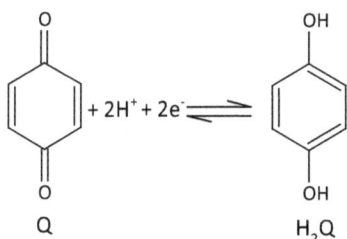

The quinhydrone electrode has Pt or Au which is dipped in solution whose pH is to be determined.

Then, Nernst's equation:

$$E_{Q, QH_2, H^+|Au} = E^0_{Q, QH_2, H^+|Au} = -\frac{RT}{2F}\ln\frac{a_{QH_2}}{a_Q \times a_{H_2^+}}$$

In solution, $a_Q = a_{QH_2}$

$$E_{Q, QH_2, H^+|Au} = E^0_{Q, QH_2, H^+|Au} - \frac{RT}{2F}\ln\frac{1}{a_{H_2^+}}$$

$$E_{Q, QH_2, H^+|Au} = E^0_{Q, QH_2, H^+|Au} + \frac{2.303 \times 298 \times 8.314 \times 2}{2 \times 96,500} \log a_{H^+}$$

$$E_{Q, QH_2, H^+|Au} = E^0_{Q, QH_2, H^+|Au} - 0.05913 \, pH$$

This electrode has limitation as it can only be used for pH less than 7. Above pH 7, hydroquinone suffers partially ionization and atmospheric oxidation. Hence, it is not useful when oxidizing or reducing substances are present.

Let quinhydrone electrode couple with the reference electrode to form a cell,

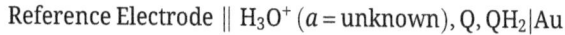

$$\text{Reference Electrode} \parallel H_3O^+ \, (a = unknown), Q, QH_2|Au$$

So experimentally, $E_{cell} = E_R - E_L$

$$E_{cell} = E^0_{Q, QH_2, H^+|Au} - 0.0591 \, pH - E_{ref}$$

So, $pH = \dfrac{E^0_{Q, QH_2, H^+|Au} - E_{ref} - E_{cell}}{0.0591}$

11.10.1.3 Glass electrode

Glass electrode is a type of ion-selective electrode which is very sensitive to hydrogen ions, that is why it is used to measure the pH of the solution. In 1909, F. Haber and Z. Klemensiewicz reported the results of their research on the glass electrode. In 1925, P.M. Tookey Kerridge developed the first glass electrode for analysis of blood samples which was the predecessor of the glass electrodes used in the present time. It was based on the principle that when two solutions of different pH are separated by glass membrane, potential difference between the two surfaces of the membrane develop. This value of potential difference varies with the type of glass and concentration of hydrogen ions. The potential of the glass electrode can be given as follows:

$$E_G = E^0_G - 0.0591 \log a_{H^+}$$

$$E_G = E^0_G + 0.0591 \log pH$$

where E_G is the potential of the glass electrode and E^0_G is the standard potential of the glass electrode.

Glass electrode is a glass tube having a thin-walled bulb in the end. The bulb contains a solution having hydrogen ion and electrode generally Ag–AgCl (silver–silver chloride) electrode or Pt electrode dipped in 0.1 N HCl solution (Figure 11.5). The glass electrode is dipped in solution whose pH has to be determined and then combined with the reference cell (like calomel electrode) using a salt bridge.

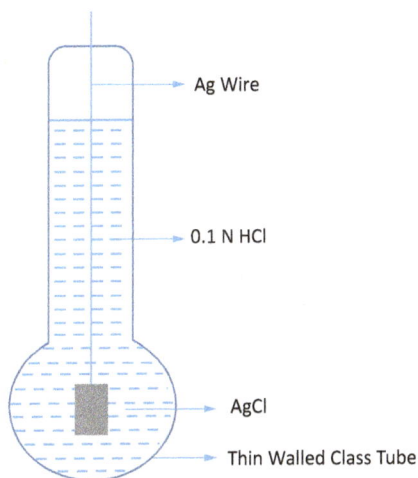

Figure 11.5: Glass electrode.

Cell representation of a glass electrode is

$$\text{Glass}|\text{pH(unknown)} \parallel \text{calomel, Pt, 0.1N HCl}$$

Glass electrode is made up of special glass having low melting point and high electrical conductivity. The EMF of a cell is determined by a potentiometer. The potential of a calomel electrode is known, hence, the potential of a glass electrode can be determined easily:

$$E_{\text{cell}} = E_{\text{SCE}} - E_G$$

$$= E_{\text{SCE}} - E_G^0 - 0.0591\,\text{pH}$$

$$\text{pH} = \frac{E_{\text{SCE}} - E_G^0 - E_{\text{cell}}}{0.0591}$$

E_{SCE} is known, E_{cell} can be determined experimentally, and if E_G or E_G^0 is known, then pH can be calculated.

E_G or E_G^0 can be calculated by using a solution of known pH. It has a limitation that E_G or E_G^0 depends on the type of glass, so it does not have a constant value.

11.10.1.4 Antimony electrode

Antimony electrode is a type of pH electrode. It is clinically useful because of its low cost and simple construction and it has no glass to break. It must be cleaned before use as the antimony surface should be pit free for better results. It consists of antimony and its trioxide and the electrode reaction may be represented as

$$2Sb(s) + 3H_2O \rightleftharpoons Sb_2O_3(s) + 6H^+ + 6e^-$$

$$E_{Sb, Sb_2O_3, H^+} = E^0_{Sb, Sb_2O_3, H^+} - \ln a_{H^+}$$

$$E_{Sb, Sb_2O_3, H^+} = E^0_{Sb, Sb_2O_3, H^+} + 0.0591\,pH$$

It has a limitation that it can be used for pH ranging from 2 to 7. It is set up for an approximate work and not precise work. They have an advantage over the glass electrode in case of solutions containing hydrofluoric acid, for which the glass electrode cannot be used. The glass reacts with solutions containing hydrogen fluoride.

11.10.2 Potentiometric titrations

The study of potential at various steps during an acid–base, redox, and precipitation titration is called potentiometry, and the titration carried out using this technique is known as potentiometric titration. It is one of the most widely used laboratory techniques to determine the concentration of an analyte in a sample, and employs a potentiometer. The potentiometric titration can be easily done in the laboratory by taking the known volume (usually 40.0/50.0 mL) of the acid solution (unknown strength) in a beaker. Dip the glass electrode in this solution, so that it should be completely immersed. After the potential of the solution is stable, start titrating it with the addition of a base solution of known concentration. The graph should be plotted with the obtained data between the potential and volume of the titrant added to obtain equivalence point. These titrations are better than volumetric titration as

i.   they do not require visual indicator,
ii. they can be performed with colorless solutions,
iii. they give more accurate equivalence point,
iv. they can be performed with precipitation reactions, and
v.   they also give stepwise dissociation constant of polyprotic acids.

11.10.2.1 Principle

Consider the reaction, Reactant (R) → Product (P).
 These titrations are based on Nernst's equation:

$$E = E^0 - \frac{2.303RT}{nF} \log \frac{[P]}{[R]}$$

$$E = E^0 - \frac{2.303RT}{nF} \log \frac{[reduced]}{[oxidized]}$$

According to this equation, as titration is performed, the ratio of activities of a product and a reactant or reduced and oxidized form changes, as their potential varies. The variation in potential is very sharp at an equivalence point which can be easily identified by plotting a graph.

Potentiometric titrations require three types of electrodes:

i. **Reference electrode**: The potential of these electrodes is known and remains constant. The potential of unknown solution is measured relative to reference electrodes.

ii. **Indicator electrode**: As the titrant will be taken in an electrolytic cell, the concentration of ions changes and hence EMF of an indicator electrode changes. Thus, any change in the potential of an indicator electrode, and the same change will be reflected in the cell potential.

iii. **Inert electrode**: These electrodes are used when both oxidized and reduced forms are present in the solution. The current is passed through inert electrodes. They help in connecting the redox systems with the external circuit, mostly Pt, Au, Hg, Mo, etc., which are used as an inert electrode.

Now, we will discuss different types of titrations that can be performed potentiometrically.

11.10.3 Acid–base titration

In this type of titration, two electrodes are used: (i) reference electrode, that is, calomel electrode and (ii) indicator electrode, that is, glass electrode. The following procedure can be followed to determine the concentration of an acid using a potentiometer:

i. Take 30–40 mL of an acid in electrolytic cell (beaker).

ii. Dip the glass electrode and standard calomel electrode (SCE) in it.

iii. Connect the cell to potentiometer. Measure the EMF of cell and record it.

iv. Then, add 0.2 mL of alkali to it and measure the EMF of the cell.

v. Keep on adding 0.2 mL of alkali to it, and measure the EMF of the cell after each addition till we get an "S"-shaped graph.

Basically, when alkali is added to the acid, the amount of H^+ changes, so, EMF of the cell changes, that is,

$$E_G = E_G^0 + \frac{2.303RT}{F} \, pH$$

Cell representation can be written as

Reference electrode ‖ Indicator electrode

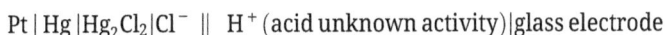

$Pt \,|\, Hg \,|Hg_2Cl_2|Cl^- \;\|\; H^+ \,(acid \, unknown \, activity)|glass \, electrode$

A graph is plotted between E_{glass} (EMF of indicator electrode) versus volume of titrant added. An "S"-shaped graph is obtained, which gives the equivalence point at the steepest point but this does not give a clear equivalence point; thus, first and second derivative graphs (i.e., dE/dV and d^2E/dV^2 resp.) of the potential are plotted against the volume of the titrant, and the equivalence point is determined (Figures 11.6, 11.7, and 11. 8).

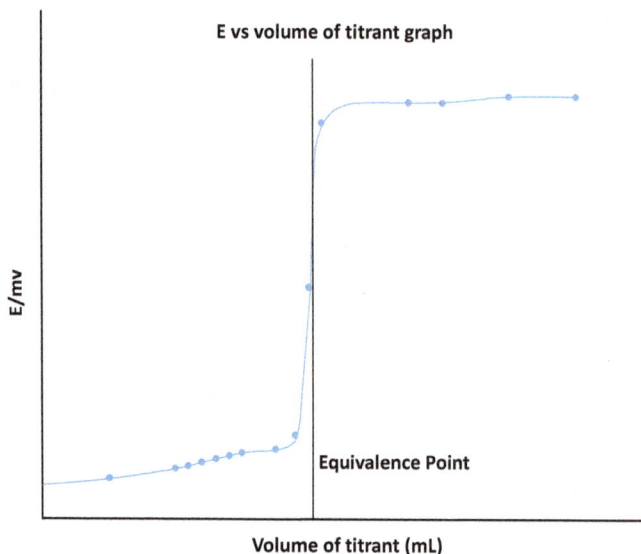

E vs volume of titrant graph

Figure 11.6: *E* versus volume of the titrant.

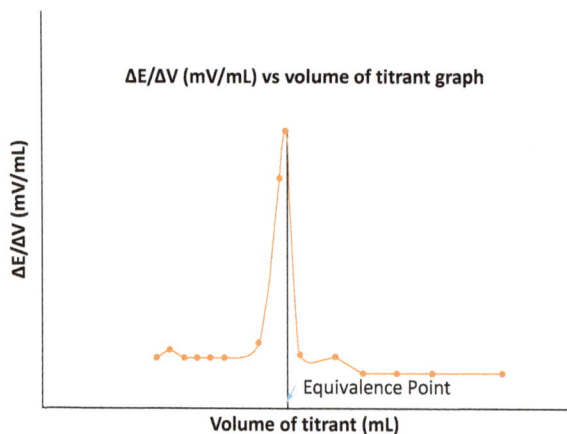

ΔE/ΔV (mV/mL) vs volume of titrant graph

Figure 11.7: Δ*E*/Δ*V* (mV/mL) vs volume of titrant graph.

Δ²E/ΔV² vs volume of titrant graph

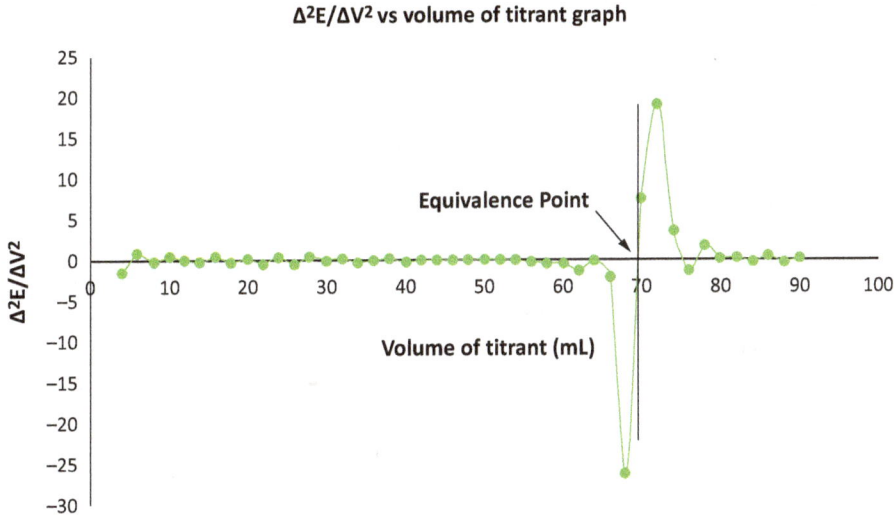

Figure 11.8: $\Delta^2E/\Delta V^2$ versus volume of the titrant.

11.10.3.1 Polyprotic acid and base

In case of a polyprotic acid, the number of steepness is equal to basicity. For example, oxalic acid versus NaOH (sodium hydroxide). Oxalic acid (COOH·COOH) is a diprotic acid and when it is neutralized with a strong base, proton dissociation occurs in two steps: first, H^+ dissociates, and then second H^+ dissociates. Thus, we get two steepness in the graph plotted between EMF versus volume of titrant as shown in Figure 11.9.

E vs Volume of titrant graph

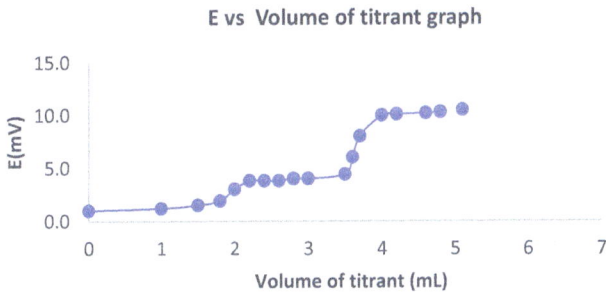

Figure 11.9: E versus volume of the titrant of diprotic acid.

Hence, no. of steepness = basicity of acid.

11.10.4 Redox titration

Redox titration involves reduction and oxidation reactions. In this titration, the ratio of activity of reduced and oxidized form changes on the addition of titrant, which reflects a change in cell potentials. At an equivalence point, the change in potential is sharp which can be easily determined.

Generally, SCE is used as a reference electrode, whereas Pt works as an indicator electrode in these titrations.

Let us consider a case of redox titration, say between Fe^{2+} (ferrous ion) and $KMnO_4$ (potassium permanganate). The procedure for the same is given:

i.   Take 30.0 mL of Fe^{2+} solution (Mohr's salt solution) in an electrolytic cell (beaker).
ii. Add 10 mL of acid (sulfuric acid) to provide acidic medium. Acidic medium is necessary to add, as it prevents the precipitation of MnO_2 (manganese oxide).
iii. Dip the reference and indicator electrode in the cell and leave it till the potential is stabilized.
iv. Titrate the solution by adding $KMnO_4$ solution and note down the potential after each addition.

The reaction taking place in the cell is given as

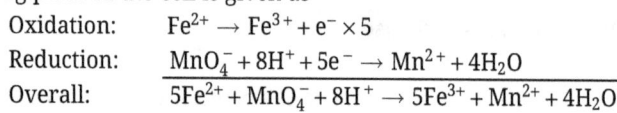

$$\text{Oxidation:} \quad Fe^{2+} \rightarrow Fe^{3+} + e^- \times 5$$
$$\text{Reduction:} \quad MnO_4^- + 8H^+ + 5e^- \rightarrow Mn^{2+} + 4H_2O$$
$$\text{Overall:} \quad \overline{5Fe^{2+} + MnO_4^- + 8H^+ \rightarrow 5Fe^{3+} + Mn^{2+} + 4H_2O}$$

The Nernst's equation for the cell can be written as

$$E = E^0 - \frac{RT}{nF} \ln \frac{[Fe^{3+}]^5 [Mn^{2+}]}{[Fe^{2+}]^5 [MnO_4^-][H^+]^8}$$

In this redox reaction, H^+ also affects the equivalence point.

At the equivalence point, $[Fe^{2+}] = 5[MnO_4^-]$ and $[Fe^{3+}] = 5[Mn^{2+}]$.

So,

$$\frac{[Fe^{3+}]}{[Fe^{2+}]} = \frac{[Mn^{2+}]}{[MnO_4^-]}$$

$$E_{eq} = E_{eq}^0 - \frac{RT}{nF} \ln \left[\frac{Fe^{3+}}{Fe^{2+}}\right]^6 + \frac{RT}{nF} \ln[H^+]^8$$

where

$$E_{eq}^0 = E_{MnO_4^-, Mn^{2+}}^0 - E_{Fe^{3+}, Fe^{2+}}^0$$

At 298 K,

$$E_{eq} = E_{eq}^0 - \frac{0.0591}{6} \log \left[\frac{Fe^{3+}}{Fe^{2+}}\right]^6 + \frac{0.0591}{6} \log[H^+]^8$$

or

$$E_{eq} = E_{eq}^0 - \frac{6 \times 0.0591}{6} \log \left[\frac{Fe^{3+}}{Fe^{2+}}\right] + \frac{8 \times 0.0591}{6} \log[H^+]$$

or
$$E_{eq} = E_{eq}^0 - 0.0591 \log \left[\frac{Fe^{3+}}{Fe^{2+}}\right] + \frac{8 \times 0.0591}{6} \log[H^+]$$

Thus, at an equivalence point, the ratio of activity of Fe^{3+} to Fe^{2+} sharply changes, so the EMF value shows a steep change with the volume of titrant. Potentiometric titration is useful in these types of colored solution as the indicator is not required.

11.10.5 Precipitation titration

Let us take an example of NaCl (sodium chloride) versus $AgNO_3$ (silver nitrate) to understand precipitation titration. Take NaCl solution (40.0 mL) of unknown strength in an electrolytic cell, and dip Ag electrode in it. Add $AgNO_3$ solution to it through burette. These two half-cells were connected through salt bridge. The reaction taking place is given as

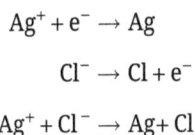

$$Ag^+ + e^- \rightarrow Ag$$

$$Cl^- \rightarrow Cl + e^-$$

$$Ag^+ + Cl^- \rightarrow Ag + Cl$$

The Nernst's equation for the cell can be written as

$$E = E^0 - \frac{RT}{nF} \ln \left[\frac{Ag}{Ag^+}\right]\left[\frac{Cl}{Cl^-}\right]$$

As, [Ag] = [Cl] = 1 (concentration of pure solid is unity)

$$\therefore \quad E = E^0 - \frac{RT}{nF} \ln \frac{1}{[Ag^+][Cl^-]}$$

$$K_{sp} = a_{Ag} + a_{Cl^-}$$

$$E = E^0 - \frac{RT}{nF} \ln \frac{1}{[Ag^+][Cl^-]}$$

$$E = E^0 - \frac{RT}{nF} \ln K_{sp}$$

At 298 K, $E = E^0 - 0.0591 \log K_{sp}$

Note: Calomel electrode was not taken in NaCl solution cell since in the presence of excess Ag^+, chloride ion, Cl^-, from the calomel electrode seeps out and reacts to form insoluble AgCl.

11.10.6 Determination of K_{eq}, equilibrium constant

The standard equilibrium constant of a reaction, K_{eq}, and EMF of a cell are related to each other as follows:

$$\Delta G^0 = -RT \ln K_{eq}$$

$$\Delta G^0 = -nFE^0_{cell}$$

Comparing the above two equations:

$$-RT \ln K_{eq} = -nFE^0_{cell}$$

or,

$$E^0_{cell} = \frac{RT}{nF} \ln K_{eq}$$

At 298 K,

$$E^0_{cell} = \frac{0.0591}{n} \log K_{eq}$$

On rearranging,

$$\log K_{eq} = \frac{nFE^0_{cell}}{2.303\,RT}$$

Let us take an example,

Example 11.5: Calculate the equilibrium constant of the following reaction at 298 K:

$$Sn + Pb^{2+} \longrightarrow Sn^{2+} + Pb$$

Solution: The reactions involved are:

(1) Oxidation at anode

$Sn \longrightarrow Sn^{2+} + 2e^-$

(2) Reduction at cathode

$Pb^{2+} + 2e^- \longrightarrow Pb$

The number of electrons involved in the given reaction is 2:

$\therefore n = 2$

$$E^0_{cell} = E^0_{Pb^{2+}|Pb} - E^0_{Sn^{2+}|Sn}$$

$$= -0.126 - (-0.136) \text{ (from table)}$$

$$= 0.010 \text{ V}$$

$$\log K_{eq} = \frac{nFE^0_{cell}}{2.303\,RT}$$

$$= \frac{2 \times 96,500 \times (0.01)}{2.303 \times 8.314 \times 298}$$

So, $K_{eq} = 2.18$

11.10.7 Dissociation constant (K_a) of a weak acid

Consider the titration of acetic acid (CH_3COOH, weak acid) with sodium hydroxide (NaOH, strong base):

$$CH_3COOH + NaOH \rightarrow CH_3COO^- Na^+ + H_2O$$

The cell formed is

$$Pt, H_2(1\,atm)|CH_3COOH(m_1), CH_3COO^- Na^+(m_2), NaCl(m_3)|AgCl(s), Ag$$

Two electrodes, hydrogen and Ag–AgCl, were used. The electrolytic solution contains $CH_3COO^-Na^+$ salt which is formed during the reaction and NaCl whose anion is common to the electrode anion:

$$E^0_{cell} = E^0_{Cl^-|AgCl|Ag} - E^0_{H^+|H_2|Pt}$$

$$E_{cell} = E^0_{Cl^-|AgCl|Ag} - \frac{RT}{nF}\ln a_{Cl^-} + \frac{RT}{nF}\ln\frac{1}{a_{H^+}}$$

$$E_{cell} = E^0_{Cl^-|AgCl|Ag} - \frac{RT}{F}\ln a_H + a_{Cl^-}$$

But the activity coefficient, denoted by $\gamma = \dfrac{\text{activity}}{\text{molality}} = \dfrac{a}{m}$.

Note: Activity is the actual amount or active concentration of a substance which will undergo a reaction, and molality is the amount of a substance taken for a reaction. Thus always $\gamma \leq 1$.

So, activity = $\gamma\, m$.
Now,
$$E_{cell} = E^0_{Cl^-|AgCl|Ag} - \frac{RT}{F}\ln(m_{H^+}\, m_{Cl^-})(\gamma_{H^+}\gamma_{Cl^-})$$

$$\frac{F}{RT}(E_{cell} - E^0_{Cl^-|AgCl|Ag}) = -\frac{RT}{F}\ln(m_{H^+}\, m_{Cl^-}) - \ln(\gamma_{H^+}\gamma_{Cl^-}) \qquad (11.7)$$

$$CH_3COOH \rightleftharpoons CH_3COO^- + H^+$$

So,
$$K_a = \frac{a_{CH_3COO^-}\, a_{H^+}}{a_{CH_3COOH}}$$

$$K_a = \frac{m_{CH_3COO^-}\, m_{H^+}}{m_{CH_3COOH}}\, \frac{\gamma_{CH_3COO^-}\, \gamma_{H^+}}{\gamma_{CH_3COOH}}$$

Taking natural logarithm,

$$\ln K_a = \ln\frac{m_{CH_3COO^-}\, m_{H^+}}{m_{CH_3COOH}} + \ln\frac{\gamma_{CH_3COO^-}\, \gamma_{H^+}}{\gamma_{CH_3COOH}} \qquad (11.8)$$

Substituting eq. (11.8) into eq. (11.7)

$$\frac{F}{RT}\left(E_{\text{cell}} - E^0_{\text{Cl}^-|AgCl|Ag}\right) = -\ln\frac{m_{CH_3COO^-}\,m_{H^+}}{m_{CH_3COOH}} - \ln\frac{\gamma_{CH_3COO^-}\,\gamma_{H^+}}{\gamma_{CH_3COOH}} - \ln K_a$$

$$\frac{F}{RT}\left(E_{\text{cell}} - E^0_{\text{Cl}^-|AgCl|Ag}\right) + \ln\frac{m_{CH_3COO^-}\,m_{H^+}}{m_{CH_3COOH}} = -\ln\frac{\gamma_{CH_3COO^-}\,\gamma_{H^+}}{\gamma_{CH_3COOH}} - \ln K_a \qquad (11.9)$$

The EMF of a cell is measured at different values of m_{CH_3COOH}, m_{H^+}, and $m_{CH_3COO^-}$. Then, it is plotted against the ionic strength. The graph is extrapolated to zero ionic strength. When all activity coefficients approach unity, the intercept in eq. (11.9) gives $-\ln K_a$ and hence K_a can be calculated.

11.10.8 Thermodynamic data

According to Gibbs–Helmholtz equation:

$$\Delta G = \Delta H + T\left[\frac{\partial(\Delta G)}{\partial T}\right]_P \qquad (11.10)$$

Free energy of a reaction and EMF of cell are related as

$$\Delta G = -nFE_{\text{cell}} \qquad (11.11)$$

Differentiate eq. (11.10) with respect to temperature, keeping pressure constant

$$\left[\frac{\partial(\Delta G)}{\partial T}\right]_P = -nF\left[\frac{\partial E}{\partial T}\right]_P \qquad (11.12)$$

where $[\partial E/\partial T]_P$ is known as temperature coefficient and its unit is V/K.

Putting relations (11.11) and (11.12) into eq. (11.10), we get

$$-nFE_{\text{cell}} = \Delta H - nF\left[\frac{\partial E}{\partial T}\right]_P$$

On rearranging the equation, we get

$$\Delta H = -nF\left[E_{\text{cell}} - T\left[\frac{\partial E}{\partial T}\right]_P\right] \qquad (11.13)$$

As entropy change is given by

$$\Delta S = \frac{\Delta H - \Delta G}{T} \qquad (11.14)$$

On substituting eqs. (11.13) and (11.11) into eq. (11.14), we get

$$\Delta S = nF \left[\frac{\partial(E)}{\partial T} \right]_P \qquad (11.15)$$

Thus, eqs. (11.11), (11.13), and (11.15) give thermodynamic data if E_{cell} is determined.

Example 11.6 Consider the following cell:

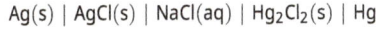

$$Ag(s) \mid AgCl(s) \mid NaCl(aq) \mid Hg_2Cl_2(s) \mid Hg$$

Write its half-cell reactions. The standard EMF of the cell at various temperatures are as follows:

T (K)	291	298	303	311
E (mV)	43.0	45.4	47.1	50.1

Calculate ΔG, ΔS, and ΔH for the reaction at 298 K.

Solution: The half-cell reactions of the given can be written as

$$Hg_2Cl_2 + 2e^- \rightarrow 2\,Hg(l) + 2\,Cl(aq) \qquad (11.16)$$

$$2AgCl(s) + 2e^- \rightarrow 2Ag(s) + 2\,Cl^-(aq) \qquad (11.17)$$

Subtracting eq. (11.17) from eq. (11.16), we will get

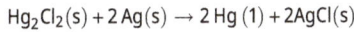

$$Hg_2Cl_2(s) + 2\,Ag(s) \rightarrow 2\,Hg\,(1) + 2AgCl(s)$$

Plot a graph between E (EMF) and T (temperature) to determine the temperature coefficient.

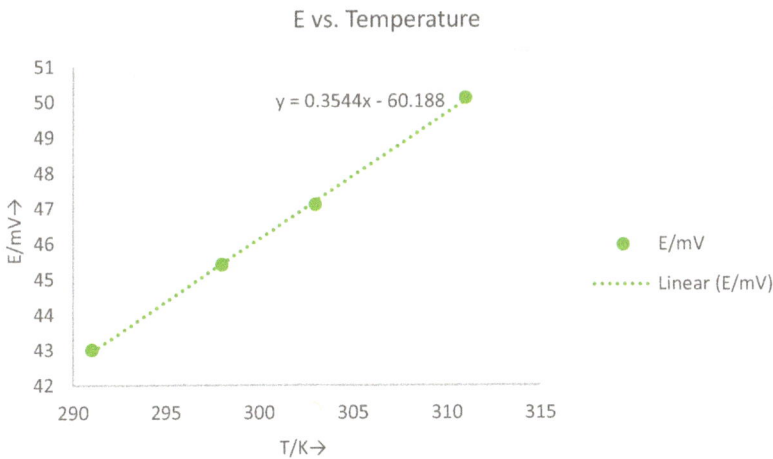

E vs. Temperature

From graph,

$$\text{Slope,} \quad \left[\frac{\partial E}{\partial T}\right]_P = 0.3544 = 0.0003544 \text{ V/K}$$

As, $\Delta G = -nFE_{cell}$

$$E_{cell} = 45.4\,mV\,(\text{given at 298 K})$$

$$= 0.0454\ V$$

$$\Delta G = -2 \times 96,500 \times 0.0454$$

$$= -8762.2\,J/mol$$

$$\Delta H = -nF\left[E_{cell} - T\left[\frac{\partial E}{\partial T}\right]_P\right]$$

$$= -2 \times 96,500[0.0454 - 298 \times 0.0003544]$$

$$= 11,620.76\,J/mol$$

$$\Delta S = nF\left[\frac{\partial(E)}{\partial T}\right]_P$$

$$= 2 \times 96,500 \times 0.0003544$$

$$= 68.3992\,J/K\,mol$$

References

[1] Kapoor, K.L. A textbook of physical chemistry, Edition IV. Chapter 6, Macmillan Publishers, India; 2001, Vol. 3, pp 403–558.
[2] Castellan, G.W. Physical Chemistry. Narosa; India 2004.
[3] Puri, B.R., Sharma, L.R., and Pathania, M.S. Principles of Physical Chemistry. Vishal Publishing Co; 2020 India.

Section VI: **Separation methods and chromatography**

Prashant Kumar, Ritika Nagpal, Jyoti Devi

Chapter 12
Separation methods in analysis

Abstract: Solvent extraction has emerged as one of the most significant methods for the separation, purification, and recovery of various metal ions from aqueous and non-aqueous media as well as diverse organic species. This analytical technique has gained utmost prominence in comparison to the other separation methodologies such as chemical precipitation, ion exchange method, and distillation, as it is faster, easier to operate, offers better separation, exhibits high degree of selectivity, consumes low energy, and shows prospects of ready scale up, thus, favoring real-time analysis. Owing to the superlative benefits, recently, significant achievements have been witnessed in the field of water purification as a consequence of implementation of solvent extraction processes. This chapter summarizes some of the most fundamental aspects of this technique, including its applicability in the recovery of metal ions from aqueous solutions and extraction of organic species from diverse media.

12.1 Introduction

Solvent extraction process has been widely used for the purification of water by eliminating organic pollutants, dyes, pesticides, metal ions, and for the quantitative analysis of several organic compounds present in water. This technique has also been found to play an important role in qualitative and quantitative estimation of oil and grease present in water. Solvent extraction includes a variety of techniques such as liquid–liquid extraction, liquid–solid extraction, supercritical fluid extractions, and many others. In this process, species transfers from one solvent to another, depending upon their differential solubility or distribution coefficient between two immiscible or partially miscible solvents. This process has been well-recognized as an important analytical separation method because of its ease, simplified equipment, wide applicability, and speed. Solvent extraction serves three purposes; pre-concentration of trace elements, elimination of matrix interference, and differentiation of chemical species.

The method relies on the use of the suitable extractant, which is a substance responsible for the transfer of a solute from one phase to another. For this, the extractant is dissolved in a suitable diluent, which collectively works as solvent. The diluent is immiscible with the other phase, which is usually water. The extractant reacts with solute by solvation/chelation/ion-pair formation to extract from the aqueous phase. The distribution equilibrium between two phases is governed by Gibbs' Phase Rule, given by

https://doi.org/10.1515/9783110794816-012

$$P + V = C + 2$$

where P is the number of phases, V is the variant of degree of freedom, and C is the number of components present in that phase.

In solvent extraction, $P = 2$ (two phases, namely, aqueous and organic) and component $C = 1$, which is a solute in solvent and water phase at constant temperature and pressure.

Recently, there has been an emphasis on reducing organic solvent consumption and rendering environmentally friendly processes, which has led to the development of innovative solvent extraction techniques. Some of the latest development techniques include miniaturization, use of hybrid techniques where solvent extraction combines with membrane separation, chromatography, distillation, nano- and micro-technological extraction, and pressurized liquid extraction (PLE). This chapter sheds light on basic principles and extraction mechanisms involved, along with novel technologies that have been developed in this area in recent years.

12.1 Solvent extraction: Principle, classification, and efficacy of the technique

Solvent extraction is based on the principle of Nernst distribution law, which states, "if a component A distributes itself between two immiscible solvents (aqueous and organic phase) at a constant temperature and pressure then ratio of concentration of A in organic phase and aqueous phase is constant" (Figure 12.1). This constant is termed as distribution coefficient or partition coefficient (K_D):

$$\frac{\text{Concentration of component in organic layer } (C_{org})}{\text{Concentration of component in aqueous layer } (C_{aq})} = K_D \qquad (12.1)$$

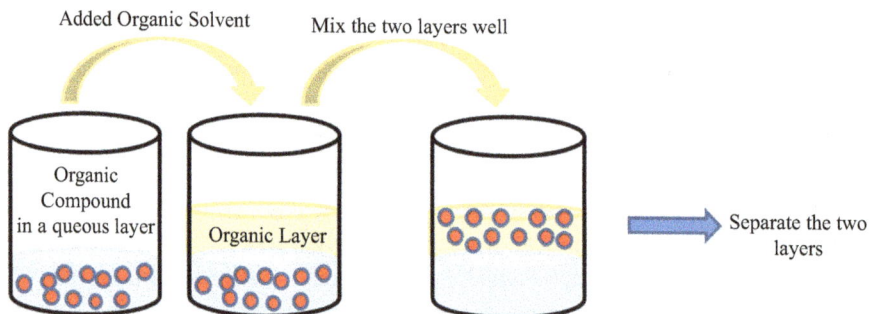

Figure 12.1: Principle of solvent extraction.

Most often, some of the substances are partially ionized in the aqueous layer as weak acids. In this case, the extraction process, now, depends on the pH of the solution [1]. Consider, for example, the extraction of benzoic acid from an aqueous solution into ether. Benzoic acid (HBz) is a weak acid in water with a particular ionization constant K_a (given by eq. (12.4)). Then, distribution coefficient (K_D) is given by

$$K_D = \frac{[HBz]_e}{[HBz]_a} \tag{12.2}$$

where e represents ether and a represents the aqueous solvent. However, depending on the extent of K_a and on the pH of the aqueous layer, a fraction of the benzoic acid in the aqueous layer will exist as Bz⁻ that does not transfer into ether; hence, when a significant fraction exists as Bz⁻, its quantitative separation may not be achieved.

The substances which undergo dissociation in aqueous layer, the distribution coefficient in modified form can be defined as distribution ratio (D), which is the ratio of the concentrations of all the species of the solute in each phase. It is given by

$$D = \frac{[HBz]_e}{[HBz]_a + [Bz^-]_a} \tag{12.3}$$

12.1.1 Relationship between K_D and D

The relationship between D and K_D can be derived readily from the equilibria involved. The acidity constant, K_a, for the ionization of the acid in the aqueous phase is given by:

$$K_a = \frac{[H^+]_a [B^-]_a}{[HBz]_a} \tag{12.4}$$

Hence,

$$[B^-]_a = \frac{K_a [HBz]_a}{[H^+]_a} \tag{12.5}$$

From eq. (12.2),

$$[HBz]_e = K_D [HBz]_a \tag{12.6}$$

After substitution of eqs. (12.5) and (12.6) into eq. (12.3),

$$D = \frac{K_D [HBz]_a}{[HBz]_a + K_a [HBz]_a / [H^+]_a} \tag{12.7}$$

$$D = \frac{K_D}{1 + K_a / [H^+]_a} \tag{12.8}$$

This equation implies that when $[H^+]_a \gg K_a$, D is nearly equal to K_D, and if K_D is large, the benzoic acid will be quantitatively extracted into the ether layer; hence, under these conditions D will be maximum. On the other hand, if $[H^+]_a \ll K_a$, then D will be minimum, and the benzoic acid will remain in the aqueous layer. In short, this predicts that the benzoic acid is ionized and cannot be extracted in alkaline solution, while in acid solution, it is largely undissociated. These conclusions are what we would possibly expect from the chemical equilibria. Moreover, like eq. (12.1), this equation also predicts that in solvent extraction, the extraction efficiency is usually independent of the original concentration of solute. This is one of the significant features of solvent extraction; it is applicable to tracer (e.g., radioactive) levels as well as to macro levels, a condition that applies only as long as the solubility of the solute in one of the phases is not exceeded and there are no side reactions, such as dimerization of the extracted solute.

12.1.2 Separation factor (β) and dissolution equilibrium

For better understanding of solvent extraction process, it is more meaningful to first understand these two terms: separation factor (β) and dissolution equilibrium.

Separation factor is the ratio of distribution coefficients of two solutes, A and B, and is denoted by β. If the solution to be extracted consists of two solutes, A and B, and we have to separate A from B, then we will use the extracting solvent, which will dissolve more quantity of A and a very little quantity of B. Under this condition, the efficacy of separation is expressed in terms of separation coefficient or separation factor (β), given by

$$\beta = \frac{(K_D)_A}{(K_D)_B}$$

For the clear separation of two solutes by solvent extraction, β should be very high.

The equilibrium in which the compound itself gets distributed in two immiscible liquids when placed in contact with these two liquids is known as dissolution equilibrium, which is governed by the temperature of the system. Suppose compound A is to be extracted from aqueous solution into organic liquid, then for compound A, dissolution equilibrium can be represented by the equation:

$$A_{(aq)} \rightleftharpoons A_{(org)}$$

The concentration of compound A in aqueous phase and in organic phase under dissolution equilibrium is definite.

Solvent extraction can be classified into various types, depending on the following variables:

12.1.2.1 Extraction mechanism

The mechanism chosen depends upon the component to be separated. An extractant is added to the organic phase, which interacts with the solute by forming chelates, solvates, or ion-pairs, resulting in extraction of solute from the aqueous phase. For example, if the components are metal ions, chelating extraction is chosen, preferably. The extraction mechanism can be further classified as solvating extraction, chelating extraction, cationic extraction, and anionic extraction, as displayed in Figure 12.2.

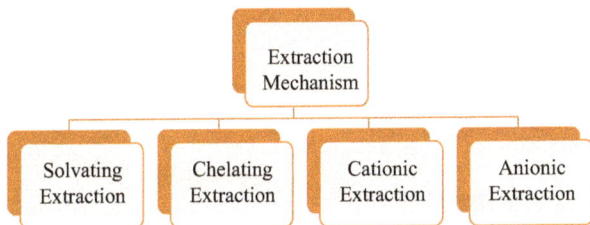

```
                    Extraction
                    Mechanism

Solvating      Chelating      Cationic      Anionic
Extraction     Extraction     Extraction    Extraction
```

Figure 12.2: Types of extractions based on extraction mechanism.

12.1.2.2 Nature of solvent

There are a variety of solvents that can be employed for the solvent extraction process. The organic solvent is chosen on the basis of several parameters such as high extraction capacity, selectivity, non or less toxicity, immiscibility in aqueous phase, favorable temperature coefficient, lower boiling point than solute, sufficient density difference with the aqueous phase, and cost effectiveness. Depending on the solute–solvent interactions, different solvent extraction processes can be used, which are shown in Figure 12.3.

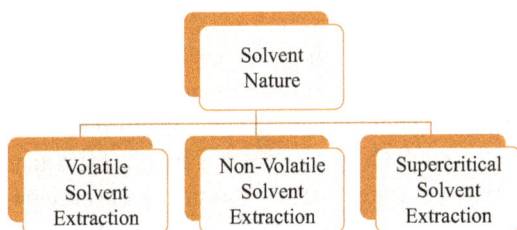

```
                    Solvent
                    Nature

Volatile       Non-Volatile   Supercritical
Solvent        Solvent        Solvent
Extraction     Extraction     Extraction
```

Figure 12.3: Types of extractions based on solvent nature.

12.1.2.3 Mode of operation

This classification is based on how the solute or sample is fed in to the system, i.e., whether the entire solute is introduced at one time (batch extraction) or continuously (continuous extraction) added to the equipment (Figure 12.4).

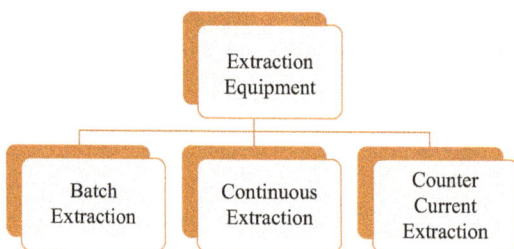

Figure 12.4: Types of extractions based on operation modes.

The solvent extraction process is more efficient than other separation techniques because of its easy functioning, low operation time, less energy consumption, almost-complete separation, and high selectivity. Therefore, different types of solvent extraction techniques are being used in various fields, such as in processing of perfumes, vegetable oil, or biodiesel, determination and recovering metal ions, such as plutonium (nuclear fuel), iron (alloys), lead (blood), copper (in alloys), separation and purification of organic compounds, in pharmaceutical industries, etc.

12.2 Mechanism of extraction: Extraction by chelation and solvation

The mechanism of extraction is based on the formation of neutral metal solvates or metal chelates. Metal salts, due to their ionic nature, coordinate with water forming neutral hydrates, since these are insoluble in organic solvents. During this process, the metal charge neutralization is performed with the help of a suitable organic extractant (ligand or solvent) by replacing the coordinated water molecules in the metal coordination sphere. As the organic extractant is soluble in the organic phase, it aids in the transfer of metal ions from the aqueous phase to the organic phase. All the multidentate ligands find useful applications in these extraction processes. There are three processes through which solvent extraction can be achieved: chelation, solvation, or supramolecular assembly formation.

12.2.1 Extraction by chelation

In extraction by chelation, the mode of operation proceeds via formation of a stable and neutral chelate complex (ring structure) between the added chelating ligand and the metal ion to be extracted. The organic ligands used in this mechanism are usually multi-dentate ligands, which can bind to the metal ion through more than one binding site. Since chelate ligands are organic in nature, they strongly interact with the organic solvent and become soluble in the organic phase. An ideal chelating ligand is that which is capable of forming stable complexes with metal ions easily, and the complex should be soluble in organic phase. There are some other important aspects that should be taken care while choosing this method of separation for the recovery of metals, which included rate of extraction and regeneration of chelating reagent, among others.

In 1955, Elving et al. reported the extraction of uranium using 8-hydroxyquinoline as chelating agent and chloroform ($CHCl_3$) as organic phase. The 8-hydroxyquinoline ligand binds to metal cations through an oxygen atom and a nitrogen atom, forming five-membered chelate rings. Likewise, many other research groups reported extraction of various metal ions such as iron and gallium using cupferron (ammonium salt of N-nitroso-N-phenylhydroxylamine, $NH_4[C_6H_5N(O)NO]$) as chelating agent and carbon tetrachloride (CCl_4) as organic phase. The ligand, cupferron, is a common reagent used for extraction of many metal ions. The anion binds to metal cations through two oxygen atoms, forming five-membered chelate rings. Moreover, ionic liquids are also being utilized in extraction and separation processes. Benabdallah et al. reported extraction studies of high amounts of copper(II) cations from different concentrated saline aqueous solutions with 2-((phenylimino)methyl)phenol extractant (HPIMP) in kerosene at 25 °C (Figure 12.5).

In 2015, Billard and coworkers extracted and separated lanthanides (from La to Gd) using 4-(4-fluorobenzoyl)-3-methyl-1-phenyl-pyrazol-5-one(HPMFBP) or 3-methyl-4-(4-methyl-benzoyl)-1-phenyl-pyrazol-5-one (HPMMBP) and 1-(2-thienyl)-4,4,4-trifluoro-1,3-butanedione (thenoyltrifluoroacetone, HTTA) as chelating agents in benzene and ionic liquid.

12.2.2 Extraction by solvation

Extraction by solvation is a mechanism in which metal ions get solvated and trapped inside the organic solvent used. Metal ions are extracted from aqueous phase to organic phase, as the chelating ligand used in the mechanism is soluble in organic phase. The examples of organic solvents involved in this extraction are carboxylic acids, tertiary amines, alkyl-substituted phosphoric acids, alcohols, ketones, ethers and esters, etc. These solvents have lone pairs of electrons on the oxygen atom, which assists in solvating metal ions to form complexes that can be extracted. Nowadays, ionic liquids have replaced organic solvents in the extraction process of various metal ions such as of Zn^{2+},

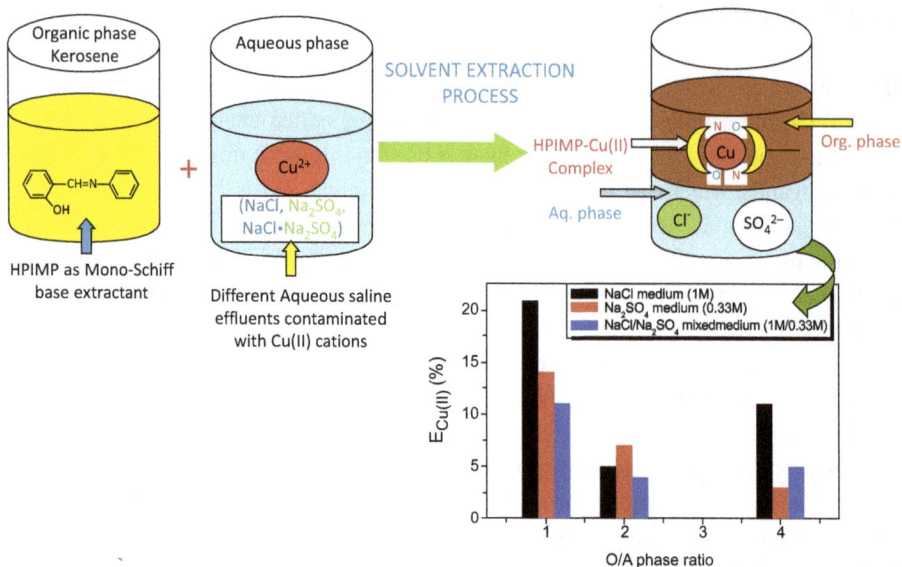

Figure 12.5: Formation of chelate between Cu^{2+} and the chelating ligand HPIMP with saline media, using kerosene as organic phase.

Cd^{2+}, and Fe^{3+} from their respective aqueous solutions to make the procedure more benign and greener in nature. Recently, nonaqueous solvents are also being employed in the process in which, instead of water, two organic phases that can behave as polar molecular organic solvents and ionic solvents are being employed.

12.3 Technique of extraction: Batch, continuous, and countercurrent extraction

12.3.1 Batch extraction

The most commonly and widely employed solvent extraction technique is batch extraction in which the entire solute that has to be extracted is loaded into the apparatus. This technique is commonly used in chemical laboratories employing a separating funnel as the equipment. The two immiscible phases are simply shaken up and then allowed to rest for attaining the equilibrium. Once the two layers are visible, they are collected out from the separating funnel. Batch extraction is applied to the systems in which the partition coefficient is relatively high. It is of two types: single and multiple batch extraction. In single extraction, the solvent is added in one lot whereas in multiple batch extraction, the solvent is added multiple times in succession, due to the moderate value of partition coefficient. As shown in Figure 12.6, the aqueous layer for the first extraction is added to 25 mL dichloromethane (DCM), mixed, and then separated. This aqueous layer still has the solute left, and hence, will again be extracted by DCM solution in the second extraction. This procedure will continue until almost all of the solute molecules are extracted in the organic layer. All organic layers are then combined to regenerate the solute back.

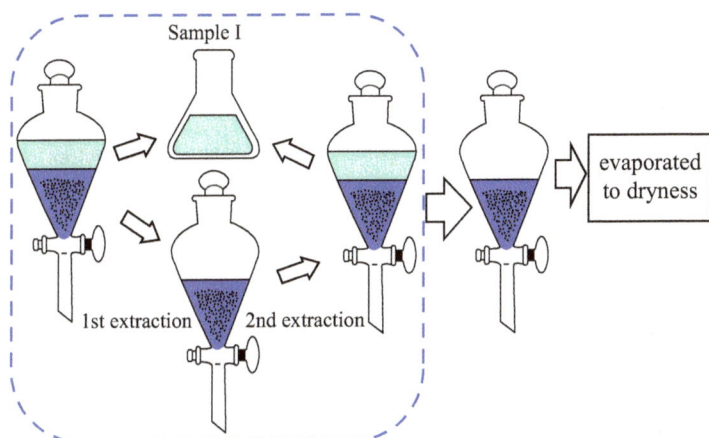

Figure 12.6: Multiple batch extraction (organic phase is denser than aqueous phase).

Some of the examples that employ batch extraction procedures are determination of gold in water by anion-exchange batch extraction and extraction of antioxidant compounds from seeds of *sterculia apetala* plant via batch extraction process.

12.3.2 Continuous extraction

In this type of extraction, continuous flow of the feed into the apparatus containing immiscible solvent, as shown in Figure 12.7, is done. These extractions are used when the partition coefficient is small. During the operation, the extracting solvent (organic layer – denser than water) is added to the flat-bottomed flask (reservoir) and boiled. The extracting solvent evaporates up to the top of the condenser and condenses through the water sample, extracting the solute to the condenser's bottom along with it. The extracting solvent then drops into the reservoir, where it is again boiled, while the extracted solute remains in the reservoir. The same kind of procedure can be followed with a lighter organic phase, too.

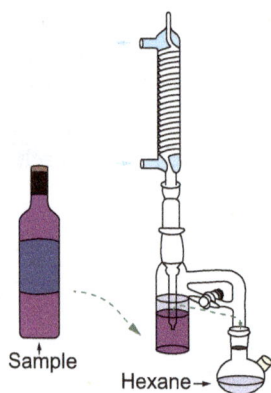

Sample

Hexane→

Figure 12.7: Continuous extraction using organic phase denser (light violet) than water (containing solute – dark violet).

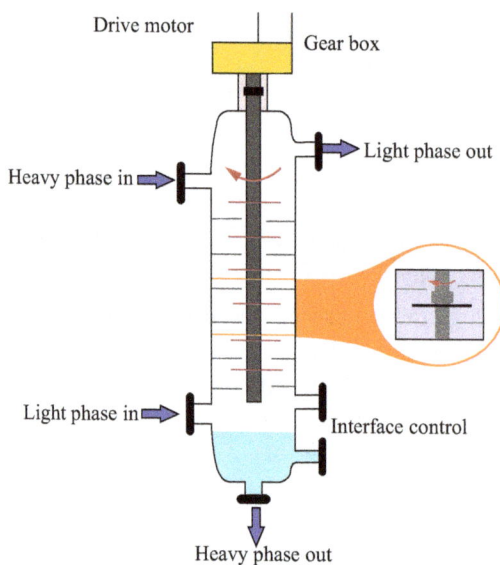

Drive motor

Gear box

Light phase out

Heavy phase in

Light phase in

Interface control

Heavy phase out

Figure 12.8: Countercurrent extraction.

12.3.3 Countercurrent extraction

This extraction is a type of continuous extraction, which is used for fractionation purposes using counter continuous flow. The principle lies in the fact that different fluids have different distribution coefficients. The apparatus used for this extraction is called Craig's apparatus. In this technique, both immiscible phases are continuously changed and as the extraction procedure progresses, they flow in opposite directions leading to their separation. This extraction is applicable to cases where the distribution coefficient is small. As this extraction employs large volumes of solvent, it is not advantageous to use it for large amounts of solute extraction. The apparatus consists of small tubes (20–25 in number); one such tube is shown in the Figure 12.8.

In the beginning, the first tube contains the mixture of solutes in the denser phase, and all the other tubes also contain the same phase. Now, the other immiscible phase is added (lighter one) to the first tube, shaken up, and left undisturbed for equilibrium to be reached. After that, the two phases separate, and the lighter phase of the tube is transferred to the next tube. Fresh lighter phase is added to the first tube for shaking and to attain equilibrium, again. Both the tubes (first and second) are left undisturbed for layers to get separated; the lighter layers of the first tube and the second tube are transferred to the second and third tube, respectively. This process will repeat itself through the other subsequent tubes. As the principle of the extraction states, the solute exhibiting higher partition coefficient moves faster than a lower partition coefficient solute. Therefore, the solute more soluble in lighter phase will move ahead with the lighter phase, and the other solute will be left more in denser phase. After travelling through a few tubes, the mixture will be separated. This extraction type can also be viewed as many multiple batchwise extractions taking place together.

12.4 Qualitative and quantitative aspects of solvent extraction

Qualitative and quantitative analysis are both indispensably important to understand any method or strategy to a greater extent and provide us the answer of two important questions: what is it? (qualitative) and how much is it? (quantitative). By definition, qualitative analysis gives the essential information about the presence or absence of different chemical components in a sample, whereas quantitative analysis deals with the amount of different chemical components present in a given sample. More simplistically, we can say that when we do qualitative analysis, we are measuring something by its quality rather than by quantity. On the other side, when we do quantitative analysis, we are exploring numbers, percentages, statistics, formulae, and data. In this section, we have tried to explain these two cornerstones of solvent extraction approach. Due to lower energy consumption, large production capacity, fast action, easy continuous operation, and ease of automation, the solvent extraction method is the most widely and favorably used separation method. Also, understanding the qualitative and quantitative aspects of solvent extraction becomes equally important.

During solvent extraction process, quantitative separation of a compound from aqueous phase into organic phase is not possible in single extraction. Thus, for quantitative extraction of a compound from aqueous phase into organic phase, aqueous phase must be extracted again or multiple times (Figure 12.9).

a) First extraction: **b) Second extraction:**   **c) Third extraction:**

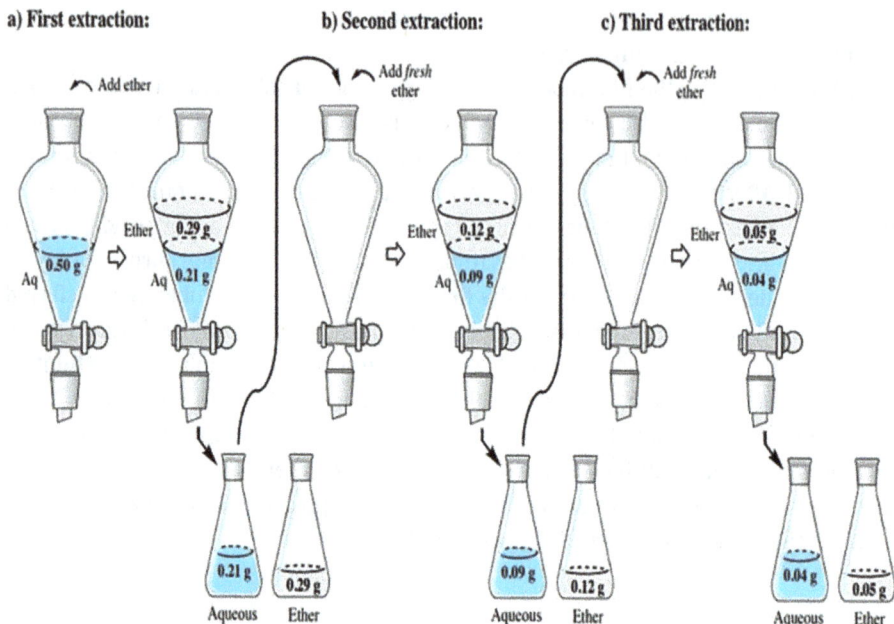

Figure 12.9: Quantitative extraction of a compound from aqueous phase into organic phase (ether).

Quantity of solute that remains in aqueous phase after n extractions can be determined by the equation

$$X_n = X_a \left[\frac{V_{aq}}{K_D \, V_o + V_{aq}} \right]$$

where n is the number of times solute is extracted from aqueous phase; X_n is the quantity of solute that remains in aqueous phase after n extractions; X_a represents millimoles of solute taken in aqueous phase; K_D is the distribution coefficient; V_{aq} is the volume of aqueous phase taken for extraction; and V_o is the volume of organic phase used for extraction at a given time in multiple extraction.

Percent extraction is another term used in quantitative solvent extraction to express efficiency of extraction. The fraction of solute extracted is equal to the millimoles of solute in the organic layer divided by the total number of millimoles of solute. Thus, the percent extracted or efficiency of extraction (%E) can be calculated as

$$\% E = \frac{m \text{ moles extracted}}{\text{Total } m \text{ moles in aqueous}} \times 100$$

12.5 Extraction of metal ions from aqueous solutions

Metals are abundant in nature; they are essential as well as toxic (in larger amount) for human health and also critical materials for modern industrialization and urbanization. For a cleaner environment and healthy human life, removal of heavy metal ions from water is of prime importance. Some metals like iron and zinc are necessary for human health, as they are involved in numerous metabolic processes. On the other hand, metals like arsenic, cadmium, mercury can cause harmful effects on skin, respiratory, cardiovascular, and nervous systems, and prolonged exposure to these metals can cause different types of cancer, and even death. Although these metals are present only in traces, they are still hazardous.

The Flint water crisis in 2014 affected about 100,000 people when lead from ageing pipes leaked into the water supply and polluted the drinking water [2]. This fatal incident emphasizes the significance of cautious monitoring of heavy metals in the water system and developing novel technologies to extract, recover, and examine them. According to WHO, the maximum amount of copper in drinking water is 2 mg/L, and no limiting values are provided for iron and zinc in drinking water, though high concentrations of these metals can affect human health. The retrieval of precious elements, including gold, platinum, and rare earth metals from natural and industrial wastes is also crucial for their economic, strategic, and national security values [3]. These elements have various applications in metallurgy, biomedical, and electronic industries. Various methods have been used for extraction and removal of metals from different sources of water, which includes microfiltration [4], chemical precipitation [5], coagulation and flocculation[6], electrochemical removal [6], liquid–liquid extraction [7], osmosis [8], crystallization and distillation [9], photocatalysis [10], and are adsorption membrane-based [11]. Nowadays, studies have been focused on developing more cost-effective, eco-friendly, and sustainable methods for effective removal of heavy metal ions from aqueous solutions.

Following the concept of "like dissolves like," the uncharged organic molecules tend to dissolve in the organic layer, while the charged anion from the ionized molecules remains in the polar aqueous layer. Metal ions do not tend to dissolve appreciably in the organic layer. To extract a metal ion into an organic solvent, its charge must be neutralized, and something must be added to make them organic-like. Thus, their extraction can be accomplished in three ways:

i.   Ion-association method
ii. Chelate formation method
iii. Synergic extraction method

12.5.1 Ion-association method

In this extraction method, metal ions form ionic complexes called ion-association complex with certain ions, by associating with another ion of the opposite charge to form an ion pair, or the metal ion associates with another ion of greater size (organic-like); for example, an ion pair formation of permanganate ion (MnO_4^-) with tetraphenylarsonium ion [$(C_6H_5)_4As^+$, MnO_4^-], which makes it organic-like, and it is extracted into methylene chloride.

12.5.2 Chelate formation method

The formation of a chelate molecule with an organic chelating agent is the most widely used method of extracting metal ions. The chelating agents are usually bidentate or multidentate organic ligands, which provide a hydrophobic pocket to the metal ion to form stable and neutral complexes. Such chelates are usually water-soluble as they readily ionize in water, giving an ionizable proton, which is then displaced by the metal ion to form a chelate, and the charge on the organic compound neutralizes the charge on the metal ion. The organic part of such ligands strongly interacts with the organic solvent, and thereby, the metal chelate becomes soluble into organic phase. For example, chelate formation of diphenylthiocarbazone (dithizone) with lead ion.

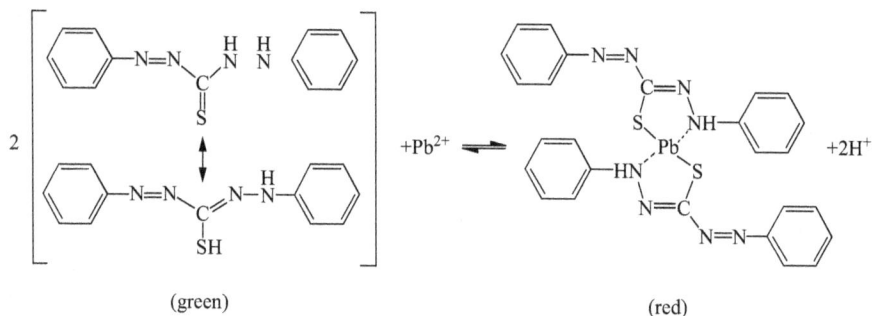

(green)   (red)

12.5.3 Synergic extraction method

In this method, addition of a mixture of extractants in organic phase is done to increase the efficiency of extraction of metal ions. The added extractants strongly interact with metal ions to give two kinds of interactions, namely mixed complex formation and reverse micelle aggregation [12] (Figure 12.10). These interactions lead to more solubility of metal ions into the organic phase and enhance extraction. This enhancement in extraction can be attributed to the increase in configurational entropy of an extracted ion; thus synergistic extraction can also be understood as a reciprocal effect of chelation [13].

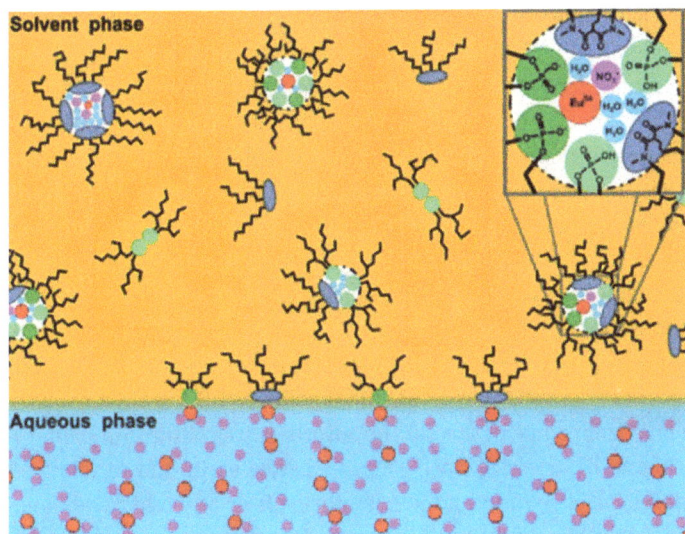

Figure 12.10: Schematic representation of the bulk model for the solvent extraction, using the mixture of the acidic and the neutral extractants. Various types of aggregates are present in the organic phase (solvent phase), and their probability at equilibrium is determined by the composition of their cores and extractant film.

12.6 Extraction of organic species from the aqueous and nonaqueous media

Efficient removal of harmful organics from aqueous and nonaqueous media is highly essential due to their severe effects on the ecological system as well as on human health. For analysis and removal consideration, of late, there has been much awareness about emerging organic contaminants (EOCs) [14]. Many strategies and methods are being used for the extraction process, which includes coagulation, flocculation, adsorption, ozonolysis, advanced oxidation, membrane system, and biological purifications from aqueous media [15], as also solvent extraction, spherical agglomeration, ionic liquid-assisted solvent extraction, adsorption and switchable hydrophilicity solvent extraction from nonaqueous media [16]. When suitable adsorbents are available, so far, the adsorptive removal technique has been found to be promising for the extraction (from both aqueous and nonaqueous media), in comparison to other available techniques. Moreover, its efficiency, operational ease under mild conditions, absence of poisonous by-products in the process, wide range of applications, and economical aspects also make it more favored than other approaches. However, the requirement of adsorbents with better properties including high adsorbent tendency, rapid adsorption, and mechanical strength are still underserved.

Numerous types of adsorbents have been developed for desulfurization, such as zeolites, metal oxides, metal organic frameworks, and boron-nitrite, but heteroatom-doped porous carbons from sucrose and phytic acid were being proposed as efficient adsorbents because of their reusability, well-developed porous structures, controllable chemical properties, and economic merits. An environmentally friendly route was adopted for synthesizing porous carbon from sucrose, in combination with phytic acid through hydrothermal carbonization (HTC) followed by calcination, without using additional chemical activation agents. To obtain the desired porous carbon with different textural and chemical properties, simple alteration in the ratios of sucrose and phytic acid was carried out.

Kim et al. developed a poly-HIPE alginate hydrogel scaffold coated with methylcellulose/tannic-acid complex particles for the removal of quinoline (QUI) from nonaqueous media and methylene blue from aqueous media. The adsorbent was named AAP and reported as biodegradable and bio-friendly, since it can be picked easily and regenerated for different cycles. Moreover, it required no temperature, pH or any other additional tedious separation step. This HIPE scaffold was proposed as the first supramolecular scaffold prepared for both aqueous and nonaqueous applications. In situ polymerization of polyacrylic acid in a solution of alginate (Alg) was done to prepare AAP. Methylcellulose-tannic acid (MC-TA) colloidal particles (which acted as both emulsifier and amphiphilic adsorbent) were used to stabilize the pickering emulsion (Figure 12.11). Amphiphilic surface wetting and highly exposed adsorptive sites were found as a result of MC-TA particle's localization on the framework of the scaffold. This distinctive property generated high adsorption tendency for the adsorptive removal of organics in both aqueous and nonaqueous media.

12.7 Conclusion

Solvent extraction, one of the most efficient separation technologies, has acquired immense significance in contributing towards productivity and global competitiveness of diverse industries including chemical, petroleum refining, and material processing. In recent times, this technique has also presented opportunities for minimizing waste and making proficient use of energy and raw materials. The chapter has summarized the basic mechanisms of extraction and has given due emphasis to some of the latest developments in this field in the extraction of organic as well as inorganic species. Ample examples have been presented to illustrate how various components are being extracted

(a)

(b)

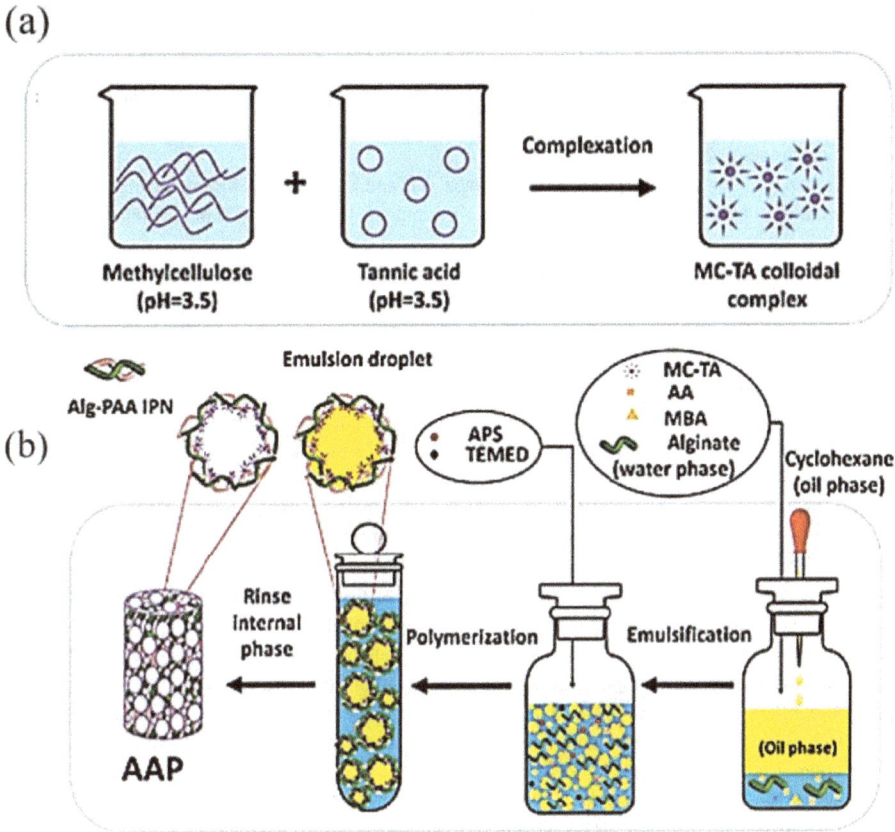

Figure 12.11: Preparation of (a) methylcellulose-tannic acid (MC-TA) and (b) AAP poly-HIPE (©Obtained with permission from [17]).

for water purification using fabricated nanocomposites, ultrafiltration membranes, polymer coated materials, etc.

References

[1]   Christian, G.D., Dasgupta, P.K., and Schug, K.A. Analytical chemistry. John Wiley & Sons; 2013 Hoboken, New Jersey, US.
[2]   Tariq, S.M., HanHui, Z., Yang, J., Ibrahim, M., Wang, L., and Tasleem, A. Extraction of zinc metal ions from aqueous solution using ionic liquids. Global Nest Journal. 2021; 23(1): 29–34.
[3]   Pieper, K.J., Tang, M., and Edwards, M.A. Flint water crisis caused by interrupted corrosion control: Investigating "ground zero" home. Environmental Science & Technology. 2017; 51(4): 2007–2014.

[4] Boudesocque, S., Mohamadou, A., Conreux, A., Marin, B., and Dupont, L. The recovery and selective extraction of gold and platinum by novel ionic liquids. Separation and Purification Technology. 2019; 210: 824–834.

[5] Du, R., Gao, B., and Men, J. Microfiltration membrane possessing chelation function and its adsorption and rejection properties towards heavy metal ions. Journal of Chemical Technology & Biotechnology. 2019; 94(5): 1441–1450.

[6] Carolin, C.F., Kumar, P.S., Saravanan, A., Joshiba, G.J., and Naushad, M. Efficient techniques for the removal of toxic heavy metals from aquatic environment: A review. Journal of Environmental Chemical Engineering. 2017; 5(3): 2782–2799.

[7] Charerntanyarak, L. Heavy metals removal by chemical coagulation and precipitation. Water Science and Technology. 1999; 39(10–11): 135–138.

[8] Duan, W., Chen, G., Chen, C., Sanghvi, R., Iddya, A., Walker, S., . . . Jassby, D. Electrochemical removal of hexavalent chromium using electrically conducting carbon nanotube/polymer composite ultrafiltration membranes. Journal of Membrane Science. 2017; 531: 160–171.

[9] Sahraeian, T., Sereshti, H., and Rohanifar, A. Simultaneous determination of bismuth, lead, and iron in water samples by optimization of USAEME and ICP–OES via experimental design. Journal of Analysis and Testing. 2018; 2(1): 98–105.

[10] Vital, B., Bartacek, J., Ortega-Bravo, J.C., and Jeison, D. Treatment of acid mine drainage by forward osmosis: Heavy metal rejection and reverse flux of draw solution constituents. Chemical Engineering Journal. 2018; 332: 85–91.

[11] Lu, H., Wang, J., Wang, T., Wang, N., Bao, Y., and Hao, H. Crystallization techniques in wastewater treatment: An overview of applications. Chemosphere. 2017; 173: 474–484.

[12] Wu, Q., Zhao, J., Qin, G., Wang, C., Tong, X., and Xue, S. Photocatalytic reduction of Cr (VI) with TiO2 film under visible light. Applied Catalysis B: Environmental. 2013; 142: 142–148.

[13] Abdullah, N., Yusof, N., Lau, W.J., Jaafar, J., and Ismail, A.F. Recent trends of heavy metal removal from water/wastewater by membrane technologies. Journal of Industrial and Engineering Chemistry. 2019; 76: 17–38.

[14] Liu, Y., Lee, M., and Senanayake, G. Potential connections between the interaction and extraction performance of mixed extractant systems: A short review. Journal of Molecular Liquids. 2018; 268: 667–676.

[15] Spadina, M., Bohinc, K., Zemb, T., and Dufrêche, J.F. Synergistic solvent extraction is driven by entropy. ACS Nano. 2019; 13(12): 13745–13758.

[16] Petrie, B., Barden, R., and Kasprzyk-Hordern, B. A review on emerging contaminants in wastewaters and the environment: current knowledge, understudied areas and recommendations for future monitoring. Water Research. 2015; 72: 3–27.

[17] Abebe, M.W., and Kim, H. Methylcellulose/tannic acid complex particles coated on alginate hydrogel scaffold *via* Pickering for removal of methylene blue from aqueous and quinoline from non-aqueous media. Chemosphere. 2022; 286: 131597.

Mohan Kumar

Chapter 13
Chromatographic methods

Abstract: Chromatography is an important biophysical tool employed for identifying, isolating, and separating the constituents of a mixture that are distributed between two phases: stationary phase and mobile phase. The components of the mixture have different affinities with the mobile phase and stationary phase, they travel at different rates, and get separated. When the mobile phase passes through the stationary phase, it (stationary phase) adsorbs and separates the components of a mixture. The molecular characteristics and types of interaction are considered when selecting separation methods, such as ion exchange, adsorption, partitioning, and size exclusion. While purifying a mixture, structure, charge, size, hydrophobic groups on the surface, and the capacity to bond with the stationary phase are some of the important factors that have to be taken into consideration. For biochemists, a chromatography method is a significant tool that is also relatively easy to utilize in clinical laboratory procedures. Paper chromatography, for example, is used for identifying sugars and amino acids in physiological fluids associated with hereditary metabolic disorders. In this chapter, the classification of the chromatographic methods, their mechanism of separation, efficiency, and applications are discussed.

13.1 Introduction

Today, chromatography has become the most widely used, efficient, and reliable technique that enables qualitative and quantitative analysis of mixtures by separating, identifying, and purifying their constituent elements. This invaluable technique was first developed by Russian-Italian botanist, Mikhail. S. Tsvet and the term "chromatography" is derived from its Greek roots *chroma* means "color," and *graphein* means "to write." IUPAC defines chromatography as the separation of sample components between two phases after they have been distributed between them. The stationary phase tends to be immobile (stationary), while the mobile phase usually permeates and surrounds the stationary phase [1]. Chroma is the Greek word for color, and chromatography was first used to separate colored compounds in plants, but its utility extends beyond these compounds. It is one of the most effective approaches for chemical purification. A wide range of analytical questions can be addressed by this versatile technique. Crystallization, distillation, sublimation, and similar processes are also used for the purification of solids, liquids, and gases, but they are not as efficient as chromatography. In a chemical laboratory, a frequently used separation technique for assessment, separation, and purification is the chromatography method. It is also used in the chemical production industry on

https://doi.org/10.1515/9783110794816-013

small as well as large-scale productions. A nanogram of material can be separated from smaller quantities of samples during analysis, while hundreds of kilograms of material can also be processed per hour into refined products. For many analytical purposes, such as structural analyses and in vitro biochemistry experiments, purified proteins are needed. The goal of the process of protein purification is to preserve the most functional protein while removing the minute impurities. Protein purification is influenced by various parameters, such as size, shape, charge density, surface chemical groups, and binding for the stationary phase [2]. The sample components are moved along with the mobile phase over the stationary phase in the general direction of migration, resulting in multiple sorption/desorption events that lead to the separation of components. In the process, several important criteria are involved or operate in concert, to achieve a satisfactory separation. Molecular interactions between different components in both phases, which are characterized by equilibrium constants and chemical thermodynamics, are what distinguish them.

Chromatography offers a wide range of applications in biological and chemical sciences. It is extensively used in biological studies to isolate and identify biologically derived chemical compounds. A variety of products are purified using chromatography, including sugar, pharmaceuticals, and rare earth elements. This methodology is being used in the petroleum sector to explore complex hydrocarbon mixture. As a result of the versatility, ease of use, and a very well-developed framework by which the various chromatography procedure's function, it is used widely in separation science.

13.2 Historical development

During his studies on plant pigments, Mikhail S. Tsvet (1872–1919), a Russian-Italian botanist, developed chromatography in 1906. He developed the process and coined the term "chromatography" in the first decades of the twentieth century, chiefly for the isolation of plant pigments like chlorophyll, carotenes, and xanthophylls. The plant pigments were separated by using a liquid-adsorption column with calcium carbonate [3]. A vertical glass column was filled with a lot of adsorbent materials, such as alumina, silica, or powdered sugar, and topped with a solution of plant pigments, which was then washed through the column with an organic solvent. The pigments are dispersed into a series of distinct colored bands on the column, separated by pigment-free zones. Since these components are separated into a range of colors (green, orange, and yellow, respectively), they are directly responsible for the name of the technique. In 1931, the German scientist Richard Kuhn and the French chemist Edgar Lederer revealed the use of chromatography in the resolution of several physiologically relevant compounds, bringing them out of relative obscurity. During the 1940s and 1950s, the chromatography technique was greatly influenced by the work of Archer John Porter Martin and Richard Laurence Millington Synge, for which they received the Nobel Prize in Chemistry in 1952. As a result of their research, partition chromatography became a key component of many chemical analyses; gas chromatography, paper chromatography, and high-performance liquid chromatography (known as HPLC). In their technique, they used a firmly bound liquid packed with finely granulated solids in a glass tube, while a second liquid percolates through it, mingling with the first liquid. The granular solid utilized was silica gel, whereas the mobile phase was chloroform. Their effectiveness with this strategy was astounding. This technique came to be known as partition chromatography. In their pioneering study, Martin and Synge also proposed that a carrier gas could be used as the mobile phase in partition chromatography. Martin and his coworkers came up with a new technique in which they used cellulose paper as a stationary medium. This technique was given the name of paper chromatography. Paper chromatography is widely used to analyze biologically relevant compounds, such as amino acids, steroids, and carbohydrates [4]. In the intervening years, technology has advanced rapidly, and with continued technological advancements, chromatography is becoming more and more effective in separating similar molecules.

13.3 Fundamentals of chromatography

Chromatography relies on the assumption that as soon as the molecules in a liquid are applied on a surface or encapsulated in solids, the liquid immobile or stationary phase separates from each other while moving under the influence of the liquid mobile phase. Adsorption (liquid–solid), partition (liquid–liquid), and molecular weight variations are useful characteristics of this separation method. As a consequence of these variances, particular components of the mixture stay in the stationary phase for a longer time and passage more slowly through to the chromatographic system, whereas others flow swiftly into the mobile phase and leave the system.

A mixture of miscible substances is separated into its constituent parts. Stationary phases (solid or liquid) adsorb the components of a mixture, while mobile phases (liquid or gas) move through into the immobile phase and gather up mixture components. The mixture's components are carried at different rates. The more highly attached constituents move very slowly with the moving phase, whereas the poorly adsorbed components move quickly. As a result, as the mobile phase travels through the immobile phase, the components of a mixture are isolated. The process of separation is depicted in Figure 13.1.

As previously stated, the stationary phase could be a solid or a liquid, whereas the moving phase can be a liquid or a gas. Selective adsorption is used for the separation of components with the stationary phase being a solid and the moving phase is a liquid; however, when the stationary phase is a liquid and the moving phase is a liquid or gas, the separation is dependent on the partitioning between the components.

The nature of the two phases in each type of chromatographic procedure is shown in Figure 13.2.

Separation Process

Mobile phases added when needed

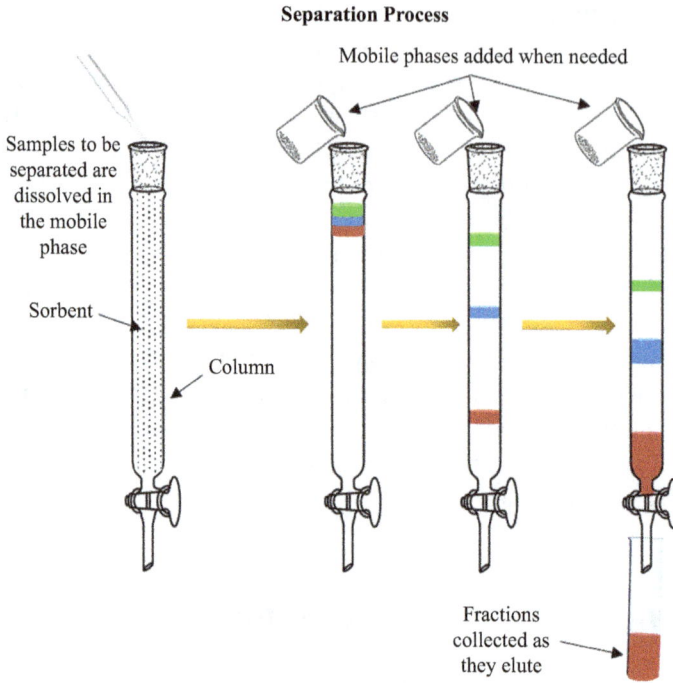

Samples to be separated are dissolved in the mobile phase

Sorbent

Column

Fractions collected as they elute

Figure 13.1: The process of separation in column chromatography.

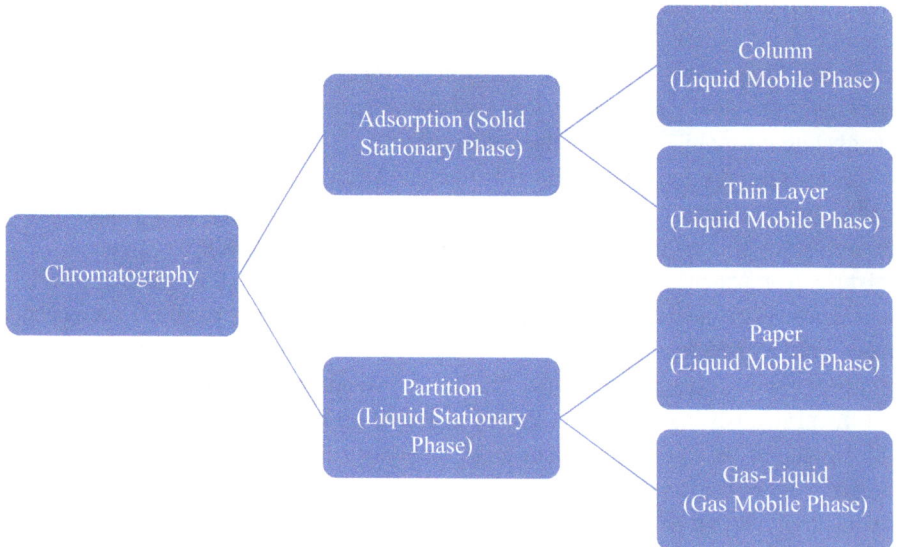

Chromatography

Adsorption (Solid Stationary Phase)

Column (Liquid Mobile Phase)

Thin Layer (Liquid Mobile Phase)

Partition (Liquid Stationary Phase)

Paper (Liquid Mobile Phase)

Gas-Liquid (Gas Mobile Phase)

Figure 13.2: Chromatographic methods.

The separation of molecules from each other depends on the type of association that occurs between the stationary phase, mobile phase, and substance in the mixture. Small molecules such as amino acids, carbohydrates, and fatty acids can be separated and identified using partition-based chromatography. As a result of affinity chromatography (also known as ion-exchange chromatography (IEC)), it is easier to separate macromolecules such as nucleic acids and proteins [5]. To separate proteins and examine protein production, paper chromatography is utilized; whereas for separating alcohol, ether, lipid, and amino groups, as well as for observing enzyme interactions, gas–liquid chromatography is employed [6, 7]. As a result of gas chromatography, it is possible to analyze gases, volatile liquid mixture, and solid materials. The liquid chromatography technique is ideally suited to nonvolatile and thermally unstable materials [8].

The purpose of utilizing chromatography as a methodology of quantitative analysis, other than separation, is to obtain an appropriate separation in a fair period of time. Many chromatographic methodologies have been introduced to achieve this purpose. Column chromatography, thin-layer chromatography (TLC), IEC, gel permeation chromatography, paper chromatography, gas chromatography, high-pressure liquid chromatography, and affinity chromatography are illustrations of these methods.

Separating molecules of different sizes using chromatographic separation is possible in addition to partitioning and adsorption (gel permeation chromatography and gel filtration). There is a gel-like phase commercially available in various porosities that is used as the stationary phase. Separation is performed by differential diffusion of molecules that are too small to be completely excluded from the pores.

13.4 Mechanism of separation

In chromatography, four types of sorption processes are used: adsorption, partitioning, ion exchange, and size exclusion.

13.4.1 Surface adsorption chromatography

The polarity differences between different feed components are used to separate the components. If a stationary phase is more polar, it will adsorb a molecule relatively firmly. In the same way, a molecule's non-polarity impacts how efficiently it is absorbed by the nonpolar stationary phase. The components to be separated and the mobile phase compete for stationary phase adsorption sites during a surface adsorption chromatography process. Low-polarity feed molecules spend proportionately more time in the mobile phase than highly polar feed molecules, which are retained for longer time. As a result, the components in a mixture are eluted in increasing polarity order. As a polar stationary phase, almost any polar solid may be used. The selection of the stationary phase is determined by the polarity of the feed components. It may be difficult to eliminate feed components that have been absorbed too heavily. Minutely polar compositions should be separated with very active absorbents; otherwise, separation will be minimal. The selection of the mobile phase is indeed crucial. The mobile phase's polarity should be chosen to complement the stationary phase's selection [9]. In general, using stationary phases with high polarity and lower polarity mobile phases such as hexane achieves effective separation. It should be emphasized that water is a fairly polar solvent. Two of the most popular chromatographic adsorbents are porous alumina and porous silica gel. Polar adsorbents like alumina are ideal for separating moderately polar components; with more polar components being kept by the adsorbent and eluted last. Alumina is also a basic adsorbent, which means it preferentially retains acidic components. Silica gel's low polarity and acidity make it a more effective adsorbent than alumina, thus, preferentially retaining basic components.

13.4.2 Partition chromatography

Originally developed in 1941 by Martin and Synge for the separation of acetylated amino acids, partition chromatography was subsequently used by Evans and Partridge to separate alkaloids in 1948. Partition coefficients are the characteristics that determine the separation of components in an aqueous and immiscible organic mixture. This chromatographic method separates solutes by dividing them among a liquid mobile phase and a stationary phase deposited on a solid substrate. It separates the components from the sample by partitioning the components into two phases, both of these components can be found in liquid form [10]. In this technique, the liquid surface on the stationary phase covers an immiscible solid surface in the mobile phase. The liquid surface is immobilized by the stationary phase, which finally becomes stationary. The constituents are separated as soon as the mobile phase leaves the stationary phase. The support in partition chromatography is typically silica, however, other materials have also been utilized [11]. Initially, partition chromatography was performed by covering the support with a liquid stationary phase that was incompatible with the mobile phase. Most contemporary partition chromatography columns, on the other hand, employ chemically linked stationary phases. Figure 13.3 depicts the pictorial representation of adsorption and partition chromatography.

13.4.3 Ion-exchange chromatography (IEC)

Proteins, peptides, amino acids, and nucleotides are the biological molecules which can be separated by IEC [12, 13]. Amino acids in proteins have both positively and negatively charged chemical groups, making them zwitterionic molecules. This strategy enables the separation of identical types of molecules that would have been difficult to separate

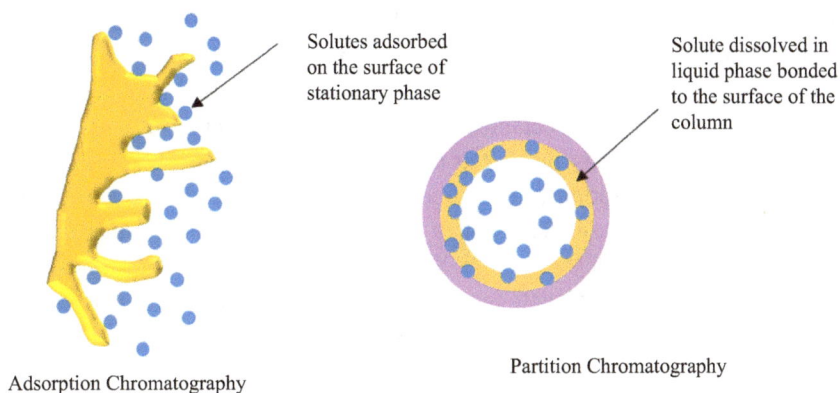

Solutes adsorbed
on the surface of
stationary phase

Solute dissolved in
liquid phase bonded
to the surface of the
column

Partition Chromatography

Adsorption Chromatography

Figure 13.3: Pictorial representation of adsorption and partition chromatography.

using other approaches since the charge held by the molecule of interest could be easily customized by changing buffer pH. The separation of ions in chromatography is achieved using ionic (or electrostatic) interactions between ions in a solution, ions in the eluent, and ions on the chromatographic support. During ion chromatography separation, two mechanisms are involved; ion exchange resulting from viable ionic binding (attraction) and ion exclusion resulting from repulsion among likewise charged analytes and ions fixed on the chromatographic support. Biomolecules interact variably with charged chromatographic materials due to their diverse charge properties. IEC, like other types of column-based liquid chromatography methods, has mobile and stationary phases. An aqueous buffer system is used in mobile phases to place the mixture which is to be resolved. The stationary phase is often composed of an innocuous organic substance altered with ionizable functional groups (fixed ions) that may contain displaceable oppositely charged ions. Stationary phases are often synthesized as copolymers based on monomer units like divinyl benzene (DVB) and styrene [14]. To control the porosity of resinous structures, the amount of DVB applied must be controlled carefully. Resins with a smaller extent of cross-linking possess large pores that allow large ions to permeate into the resin beads, allowing for fast ion exchange. Pores in a resin with a significant degree of cross-linking are comparable in size to those of tiny ions.

13.4.4 Size-exclusion chromatography

Among the most efficient, fast, and common methods for analyzing polymer molecular weight distributions, size-exclusion chromatography (SEC) is considered the most popular. This approach distinguishes molecules in a sample based on their hydrodynamic volumes rather than their molecular weight. In the SEC system, the sample solution is pumped into a stationary phase that comprises a porous packing material. Molecules larger than the pore size are eluted first since they cannot permeate the pores. However, smaller molecules will have a greater ability to penetrate pores, taking longer to elute. Molecules that can easily diffuse through the pores (in and out) cannot be separated since they are eluted at the complete permeation point. For aqueous or polar solvent separations, the stationary phase gels can be hydrophilic, while for nonpolar or weakly polar solvent separations, the stationary phase gels can be hydrophobic. In polar/hydrophilic mobile phases, Sephadex, a cross-linked polysaccharide substance in bead form, is used [15]. The degree of cross-linking may be varied to create beads with a diversity of pore sizes to fractionate materials over distinct molecular weight ranges. Hydrophobic gels are similar to ion exchange resins in that they are formed by cross-linking polystyrene with DVB, but they lack ionic groups. As a means of removing impurities from high-value products like peptides, proteins, and enzymes, SEC is widely used in the biochemical sector.

13.5 Development of chromatograms

13.5.1 Frontal chromatography

In this approach, the sample (liquid or gas) is continuously fed onto the chromatographic bed. In frontal chromatography, there is no need for an additional mobile phase [16]. Frontal chromatography is a quantitative separation method in which only the component with the least tendency of retention is separated from the rest. The separated mixture is constantly fed into the column under circumstances that encourage the binding of all but one of the components. Until the stationary phase's dynamic capacity is depleted and the other sample components breakthrough, this component is retrieved in pure form at the column outlet. Figure 13.4 depicts the pictorial representation of Frontal chromatography.

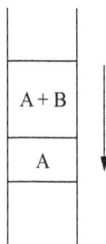

Figure 13.4: Frontal analysis: component B is more sorbed than component A.

13.5.2 Elution chromatography

Elution chromatography involves injecting the sample onto the column first, followed by the mobile phase. Each component of the sample moves at a different speed and elutes at different column heights. The migration rate of a component is affected by both its interactions with the mobile phase and the stationary phase. The retention of the substances to be separated can be adjusted in elution chromatography by altering the mobile phase's composition [17]. Elution chromatography usually achieves better resolution when a material has a strong affinity for the mobile phase and the bands are maintained properly. The mobile phase and the stationary phase serve the same purpose in elution chromatography since the driving force is the partition of the solutes between the two phases. Figure 13.5 depicts the pictorial representation of elution chromatography. A method of separation in which the composition of the mobile phase does not change during the separation is called isocratic elution.

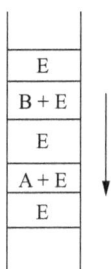

Figure 13.5: Elution development (E = eluent). B is more sorbed than A.

Throughout the isocratic elution process, the composition of the mobile phase is maintained [18]. With gradient elution, the composition of the mobile phase rises gradually throughout the elution process, while in step elution, the composition of the mobile phase varies regularly.

13.5.3 Displacement chromatography

Displacement chromatography is mostly used for preparative purposes. Any chromatographic technique in which a molecule termed as "displacer" is employed to displace other molecules adsorbed into the stationary phase, is known as displacement chromatography. Tiselius developed the displacement chromatography method along with the frontal, and elution modes of chromatography [19]. The material is initially maintained on a chromatographic column for separation (loading phase). Following the loading process, the "displacement" reagent dissolved in the mobile phase is passed through the column, which elutes the specific retained molecule. The displacement reagent is chosen because it can attach to the column more firmly than the analytes and hence, displace it. When analyte molecules are tightly or weakly coupled to the stationary phase, they are removed with greater difficulty or ease, respectively, and hence, elute at different retention durations. Figure 13.6 depicts the pictorial representation of displacement chromatography [20].

The Langmuir isotherm underpins the separation concept of displacement chromatography. There are a limited number of binding sites available on the stationary phase of the chromatographic support, and once an existing binding site is occupied by a molecule, it is inaccessible to other sample components. When molecules in the sample are at concentrations exceeding their dissociation constants, they become saturated, since, all binding sites have a finite number. Displacers have a strong affinity for the chromatographic substrate and will effectively compete with it for binding sites [21]. In this way, sample molecules are displaced and enriched in low-affinity solutes. The temperature affects the kinetics of this process; the displacement dynamics and mass transfer increases as the temperature rises. As a result, increasing the temperature lowers the overlapping of the sample component zones significantly. At higher temperatures, viscosity decreases, allowing more mass transfer and consequently, better separation. The sorbent's pore size is determined by the size of the sample molecules. For displacement chromatography of molecules less than 1 kDa (kilodaltons), pores smaller than 30 nm are suggested. For displacement chromatography of proteins, pores of 30 nm or bigger should be utilized.

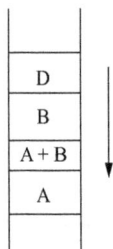

Figure 13.6: Displacement development (where D is displacer). D is more sorbed than B, which is more sorbed than A.

There are many advantages of displacement chromatography over other techniques. By using displacement instead of elution, we obtain higher concentrations of separated products, less solvent consumption, and less band tailing that resulted due to the self-sharpening boundary [22]. Although separations are often faster than elution, cycle times are sometimes longer due to the necessity to renew the column between injections.

13.6 Efficiency of technique

Chromatography is considered to be one of the most crucial analytical methods. The separation and subsequent qualitative and quantitative analysis of complicated mixtures are possible as long as the materials in the mixture are volatile or soluble in a suitable solvent. Gas, liquid, and supercritical fluid chromatography have differing mobile phases because chromatography relies on the partitioning of sample components between two phases one stationary, and one mobile phase. Chromatography is flexible, efficient, and can even be automated to a certain extent. A wide variety of analytical applications may be applied to chromatography. It is this adaptability that explains the variety of chromatographic techniques used today. By using this approach, not only previously difficult separation methods been made easier and more feasible, but previously impossible separations have also been made possible, specifically in organic chemistry, biochemistry, and pharmacology [23, 24]. Chromatographic separation is used to separate ionic species, inorganic or organic, and molecular species of all sizes, from lightest to smallest, such as hydrogen and helium, up to particulate matter, such as single cells. Using chromatography efficiently, hundreds of unknown identities and unknown-concentration components can be separated while keeping these components unaltered. For separating gases in laboratories possessing dispersive or London intermolecular interactions, gas–solid procedures or molecular sieves are used [25]. The application of gas–liquid chromatography can be used with species whose critical temperatures are high and whose boiling points are typically as high as 400 °C. A liquid–liquid or liquid–solid separation system is excellent for separating substances under typical laboratory conditions that are solids but have molecular weights under 1,000 g/mole.

References

[1]   Mallik, B., Chakravarti, B., and Chakravarti, D.N. Principles of chromatography. Current Protocols Essential Laboratory Techniques. 2016; 13(1): 6–1.
[2]   Coskun, O. Separation techniques: Chromatography. Northern Clinics of Istanbul. 2016; 3(2): 156.
[3]   Kumari, V.C., Patil, S.M., Ramu, R., Shirahatti, P.S., Kumar, N., Sowmya, B.P., . . . Patrick-Iwuanyanwu, K.C. Chromatographic techniques: types, principles, and applications. In Analytical Techniques in Biosciences. Academic Press; 2022, pp. 73–101.
[4]   Ebere EC, Obinna IB, Wirnkor VA. Applications of Column, Paper, Thin Layer and Ion Exchange Chromatography in Purifying Samples: Mini Review. SF J Pharm Anal Chem. 2019; 2(2): 1018.
[5]   Hage DS, Anguizola JA, Li R, Matsuda R, Papastavros E, Pfaunmiller E, Sobansky M, Zheng X (2017) Affinity chromatography. In: Fanali S, Haddad PR, Poole CF, Riekkola M-L (eds) Liquid chromatography. Elsevier, Amsterdam, pp 319–341.
[6]   Ahmmed, M.K,, Carne, A., Bunga, S., Tian, H.S., and Bekhit, A.E.D.A. Lipidomic signature of Pacific lean fish species head and skin using gas chromatography and nuclear magnetic resonance spectroscopy. Food Chemistry. 2021; 365: 130637.
[7]   Cassada, D.A., Zhang, Y., Snow, D.D., and Spalding, R.F. Trace analysis of ethanol, MTBE, and related oxygenate compounds in water using solid-phase microextraction and gas chromatography/mass spectrometry. Analytical Chemistry. 2000; 72(19): 4654–4658.
[8]   Burlet-Parendel, M. and Faure, K. Opportunities and challenges of liquid chromatography coupled to supercritical fluid chromatography. TrAC Trends in Analytical Chemistry. 2021; 144: 116422.
[9]   Poole, C.F., 2022. Applications of the solvation parameter model in thin-layer chromatography. JPC–Journal of Planar Chromatography–Modern TLC, 35(3), pp.207–227.

[10] Bezold, F. and Minceva, M. A water-free solvent system containing an L-menthol-based deep eutectic solvent for centrifugal partition chromatography applications. Journal of Chromatography A. 2019; 1587: 166–171.

[11] Song, L., Zhang, H., Chen, J., Li, Z., Guan, M., and Qiu, H. Imidazolium ionic liquids-derived carbon dots-modified silica stationary phase for hydrophilic interaction chromatography. Talanta. 2020; 209: 120518.

[12] Akhgar, C.K., Ebner, J., Spadiut, O., Schwaighofer, A., and Lendl, B. QCL–IR spectroscopy for in-line monitoring of proteins from preparative ion-exchange chromatography. Analytical Chemistry. 2022; 94(14): 5583–5590.

[13] Chakraborty, A., Puja, R., and Bose, K. Protein purification by ion exchange chromatography. In: Textbook on cloning, expression and purification of recombinant proteins, Singapore: Springer; pp. 173–198, 2022.

[14] Popov, A.S., Spiridonov, K.A., Uzhel, A.S., Smolenkov, A.D., Chernobrovkina, A.V., and Zatirakha, A.V. Prospects of using hyperbranched stationary phase based on poly (styrene-divinylbenzene) in mixed-mode chromatography. Journal of Chromatography A. 2021; 1642: 462010.

[15] Zhao, Y., Ouyang, X., Chen, J., Zhao, L., and Qiu, X. Separation of aromatic monomers from oxidatively depolymerized products of lignin by combining Sephadex and silica gel column chromatography. Separation and Purification Technology. 2018; 191: 250–256.

[16] Stone, M.T., Cotoni, K.A., and Stoner, J.L. Cation exchange frontal chromatography for the removal of monoclonal antibody aggregates. Journal of Chromatography A. 2019; 1599: 152–160.

[17] Qamar, S., Rehman, N., Carta, G., and Seidel-Morgenstern, A. Analysis of gradient elution chromatography using the transport model. Chemical Engineering Science. 2020; 225: 115809.

[18] Mukherjee, S. Isolation and purification of industrial enzymes: Advances in enzyme technology. In: Advances in enzyme technology, Elsevier; 2019, pp. 41–70

[19] Toraño, J.S., Ramautar, R., and de Jong, G. Advances in capillary electrophoresis for the life sciences. Journal of Chromatography B. 2019; 1118: 116–136.

[20] Nice, E.C. The separation sciences, the front end to proteomics: An historical perspective. Biomedical Chromatography. 2021; 35(1): e4995.

[21] Paladii, I.V., Vrabie, E.G., Sprinchan, K.G., and Bologa, M.K. Whey: Review. Part 2. Treatment processes and methods. Surface Engineering and Applied Electrochemistry. 2021; 57(6): 651–666.

[22] Ahmad Dar, A., Sangwan, P.L., and Kumar, A. Chromatography: An important tool for drug discovery. Journal of Separation Science. 2020; 43(1): 105–119.

[23] Gertsiuk, M. Chromatography in Ukraine: Development and achievements. Separations. 2022; 9(5): 114.

[24] Jie, M.S.L.K. The characterization of long-chain fatty acids and their derivatives by chromatography. In: Advances in chromatography, 2021, pp. 1–57.

[25] Villa, C.C., Galus, S., Nowacka, M., Magri, A., Petriccione, M., and Gutiérrez, T.J. Molecular sieves for food applications: A review. Trends in Food Science & Technology. 2020; 102: 102–122.

Prerna Singh

Chapter 14
Chromatographic techniques

Abstract: Chromatography is one of the most important and versatile analytical techniques that allow the separation of complex mixtures into their respective components and their subsequent quantitative and qualitative analysis. In this technique, two or more components of a mixture to be separated are distributed between two immiscible phases – stationary phase and mobile phase. The basic principle of separation is based on how the solute or components of a mixture equilibrate between these two phases.

This chapter will cover the classification of different chromatographic techniques based on mobile and stationary phases, and will mainly focus on the separation mechanism (adsorption, partition, ion-exchange, and pore-penetration) along with the analytical applications of these chromatographic methods.

14.1 Introduction

Chromatography is a versatile analytical technique used for the separation of different components of a mixture. In this technique, two or more components of a mixture to be separated are distributed between two immiscible phases: one phase is a stationary phase, which can be a solid or a layer of liquid adsorbed on the surface of a solid support, and the other phase is a mobile phase, which can either be a gas or a liquid [1].

Chromatography was first discovered by the Italian-born botanist, Mikhail Tsvet, in 1906, when he was trying to separate plant pigments, such as chlorophyll, carotenes, and xanthophylls, by passing a solution containing them through a column packed with calcium carbonate. The individual pigments passed down the column at different rates, depending on their repeated interactions with the mobile and stationary phase. They were separated from each other as different color bands; hence the name Chromatography: Chroma means "color" and graphy means "writing" [2].

14.2 Classification of chromatographic methods

Chromatographic methods can be classified according to their mobile phase into two categories:
(i) Liquid chromatography (LC)
(ii) Gas chromatography (GC)

https://doi.org/10.1515/9783110794816-014

These can be further classified into different types according to the nature of the stationary phase (Figure 14.1). The mobile phase can be a liquid or a gas, whereas the stationary phase can be a solid or a liquid (Table 14.1).

Table 14.1: Classification of chromatographic methods, based on stationary and mobile phase.

General classification	Stationary phase	Mobile phase	Type of equilibrium
Liquid–solid chromatography	Solid	Liquid	Adsorption
Gas–solid chromatography	Solid	Gas	Adsorption
Liquid–liquid chromatography	Liquid adsorbed or bonded to a solid surface	Liquid	The partition between two liquids
Gas–liquid chromatography	Liquid adsorbed or bonded to a solid surface	Gas	The partition between gas and liquid

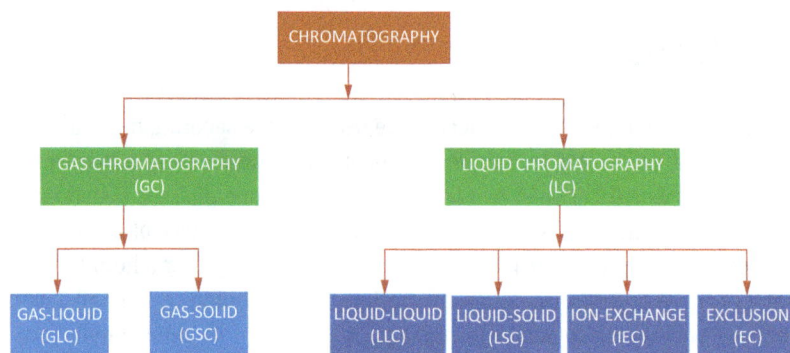

Figure 14.1: Classification of chromatographic methods.

14.3 Liquid chromatography

Liquid chromatography (LC) is distinguished by the predominant type of separation mechanism. In this technique, the mobile phase is a liquid and the stationary phase governs the separation mode. Various types of separation mechanisms are:
A. Adsorption
B. Partition
C. Ion-exchange
D. Size exclusion

14.3.1 Adsorption chromatography

Adsorption chromatography is categorized into two types, depending on the type of mobile phase: liquid–solid chromatography (LSC) and gas–solid chromatography (GSC), as shown in Figure 14.2. LSC is one where the stationary phase is solid and the mobile phase is liquid, whereas in GSC, the stationary phase is solid and the mobile phase is gas. The separation mechanism is based on the differential solute adsorption on an adsorbent's active sites (stationary phase). The adsorbent may be packed in a column (column chromatography) or spread on a plate (thin-layer chromatography, TLC) or more commonly on a porous paper (paper chromatography, PC). Some materials that are generally used as an adsorbent are silica gel, alumina, and charcoal. For example, silanol groups in silica gel act as an active site that interacts with the polar component of the compound to be separated, and the nonpolar component does not have much significance in the separation mechanism.

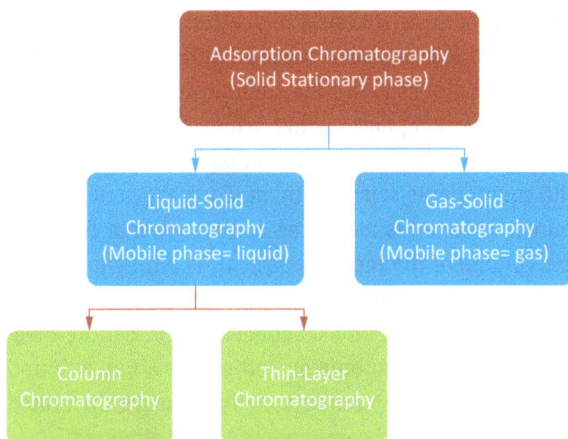

Figure 14.2: Types of adsorption chromatography.

14.3.1.1 Column chromatography

In column chromatography, the stationary phase is held in a glass tube, and the mobile phase is passed through the tube under pressure or by gravity. It separates substances based on the differential adsorption of compounds to the adsorbent, as the compounds move through the column at different rates, allowing them to get separated in fractions. This technique can be used on a small scale as well as on large scale to purify materials. Consider separation of two components sample in a closed chromatographic column. The column is a long, narrow tube that is packed with a stationary phase consisting of solid porous particles of smaller size (<150 μm in diameter). The sample is introduced at the top of the column, followed by the addition of a mobile solvent phase that moves the sample through the column packing (Figure 14.3). The individual component of a sample interacts with the stationary phase to a different degree, depending on its affinity for it, and undergoes sorption and desorption on the packing.

Each component is distributed between the stationary phase (s) and the mobile phase (m) as it passes down the column:

$$X_m \rightleftarrows X_s \tag{14.1}$$

The corresponding distribution coefficient for component X is given by

$$K_c = \frac{[X]_s}{[X]_m} \tag{14.2}$$

where K_c is the distribution coefficient/distribution constant/partition coefficient, $[X]_s$ is the concentration of component X on or in the stationary phase at equilibrium, and $[X]_m$ is the concentration of component X in the mobile phase at equilibrium.

A large value of K_c signifies that the component has a high affinity for the stationary phase and moves slowly through the column, whereas for small values of K_c, the component favors the mobile phase and moves along the column more rapidly. This is because a true equilibrium between the two phases is not achieved, which depends on the flow rate of the mobile phase and the degree of interaction with the stationary phase. The different speeds of the components separate them along the column, and if the concentration of the eluted components is measured as they emerge from the column, and is plotted against the volume of the mobile phase or as a function of the elution time to pass through the column, a chromatogram is obtained.

The individual component may be collected and its concentration can be measured externally by using methods like spectrophotometers, otherwise a flow cell and a detector are placed at the end of the column to automatically measure the eluted compounds, and a chromatogram of the separated components can be obtained.

Factors governing column chromatography
Capacity factor

The distribution coefficient is defined as a measure of the degree of retention for compound X. There is another criterion that is more practical and can be determined directly from the chromatogram i.e., capacity factor, represented by k'_x, and is defined as the ratio of the total moles of component (say X) in the stationary phase to the total moles of the component X in the mobile phase:

$$k'_x = \frac{\text{total moles of component X in the stationary phase}}{\text{total moles of component X in the mobile phase}} \tag{14.3}$$

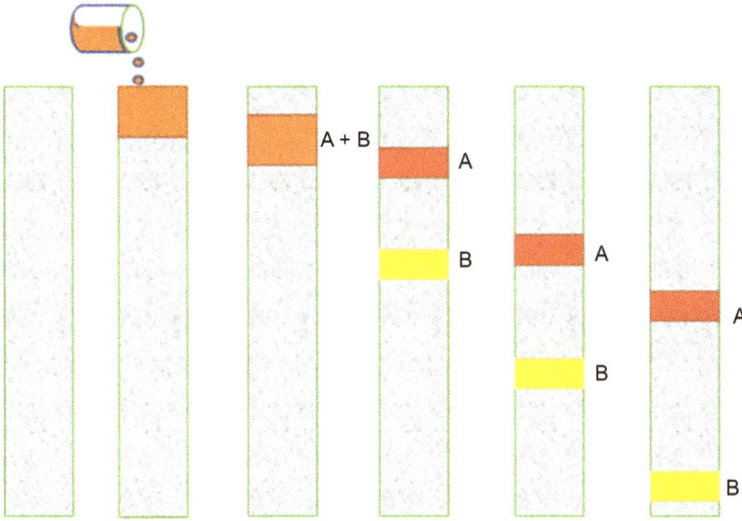

Figure 14.3: Progress of chromatographic separation of components A and B.

$$k'_x = \frac{V_s[X]_s}{V_m[X]_m} = \left(\frac{V_s}{V_m}\right)\left(\frac{[X_s]}{[X_m]}\right)$$

$$k'_x = \left(\frac{V_s}{V_m}\right)K_x \text{ (from equation 14.2)} \tag{14.4}$$

where V_s is the stationary phase within the column, V_m is the volume of mobile phase within the column and K_x is the distribution coefficient.

The volume of the mobile phase entering the column between the sample injection and the emergence of the peak max of the sample component is the retention volume, i.e. it is the volume of a mobile phase required to elute a solute to a maximum from a column during the chromatographic analysis.

Quantity eluted out = Quantity in mobile phase + Quantity in stationary phase

$$V_r C_m = V_m C_m + V_s C_s$$

Dividing by Cm

$$V_r = V_m + V_s K_x$$

V_r can also be related to capacity factor by substituting equation 14.4 in above equation and is given by

$$V_r = V_m (1 + k'_x) \tag{14.5}$$

V_r can also be obtained from chromatogram since

$$V_r = F \times t_r \tag{14.6}$$

where F is the flow rate (mL/min) and t_r is the peak retention time.

Retention times

Consider a simple chromatogram of a two component system, as displayed in Figure 14.4

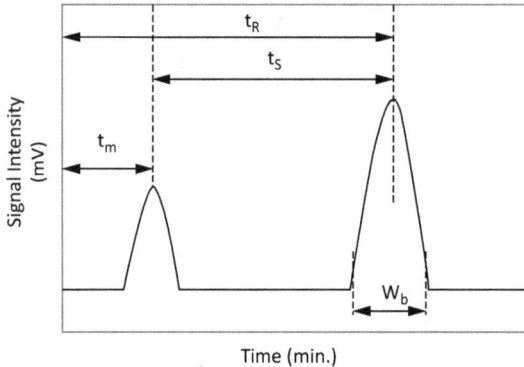

Figure 14.4: Chromatogram of a two-component system.

The small peak on the left represents a solute that is not retained by the stationary phase, and it traverses the column almost immediately. The time taken by the component that is not retained to traverse the chromatographic column is known as dead time or void time, represented by t_m. The dead time measures the average rate of the migration of the mobile phase, and is an important parameter in identifying the analyte peaks. All components spend at least this amount of time in the mobile phase. The larger peak on the right of the chromatogram is that of the analyte species. The time required to get this peak of an analyte after its injection into the column is called the retention time (t_r):

$$t_r = t_s + t_m \tag{14.7}$$

where t_s is the time spent by the analyte in the stationary phase.

The average rate of the analyte migration, represented by $\bar{v}$ (cm/s) is

$$\bar{v} = \frac{L}{t_r} \tag{14.8}$$

and average linear velocity, u of the mobile phase is

$$u = \frac{L}{t_m} \tag{14.9}$$

where L is the length of the column packing.

Relationship between migration rate and distribution constants

To relate the rate of migration of the solute to its distribution constant, we express the rate as a fraction of the velocity of a mobile phase:

$$\bar{v} = u \times \text{fraction of the time solute spends in the mobile phase}$$

This fraction, however, equals the average number of moles of solute in the mobile phase at any instant, divided by the total number of moles of the solute in the column:

$$\bar{v} = u \times \frac{\text{number of moles of solute in mobile phase}}{\text{total number of moles of solute}}$$

$$\bar{v} = u \times \frac{C_m V_m}{C_m V_m + C_s V_s} \tag{14.10}$$

where C_m is the concentration of the solute in the mobile phase, C_s is the concentration of the solute in the stationary phase, V_m is the volume of the mobile phase, and V_s is the volume of the stationary phase

$$\bar{v} = u \times \frac{1}{1 + \frac{C_s V_s}{C_m V_m}}$$

$$\bar{v} = u \times \frac{1}{1 + K_c \left(\frac{V_s}{V_m}\right)} \tag{14.11}$$

where K_c is the distribution constant.

Retention factor

The retention factor k_A for the solute A is related to the rate at which the solute A migrates through the column. It is the amount of time the solute spends in the stationary phase, relative to the time it spends in the mobile phase.

Thus, for solute A, retention factor k_A is defined as

$$k_A = \frac{K_A V_S}{V_M} \tag{14.12}$$

K_A is the distribution constant for solute A.

Substituting equation 14.12 in equation 14.11, we get

$$\bar{v} = u \times \frac{1}{1 + k_A} \tag{14.13}$$

To show how k_A can be calculated from a chromatogram, substitute eqs. (14.8) and (14.9) into eq. (14.13)

$$L/t_r = \frac{L}{t_m} \times \frac{1}{1+k_A}$$

$$1 + k_A = \frac{t_r}{t_m}$$

Rearranging and substituting equation 14.7 in above equation, we get

$$k_A = (t_r - t_m)/t_m = t_s/t_m \tag{14.14}$$

t_r and t_m can easily be obtained from the chromatogram.

Ideally, the retention factor for the solute in a sample is between 1 and 5.

If $k_A < 1$, the time taken by the solute to emerge from the column is near the void time.

If $k_A > 20$, the elution time becomes inordinately large.

Selectivity factor (α)

The selectivity factor of a column for two solutes A and B is defined as the ratio of a distribution constant for more strongly held species (B) to the distribution constant for less strongly held species (A)

$$\alpha = \frac{K_B}{K_A} \qquad \text{where } \alpha > 1$$

In terms of retention factor,

$$\alpha = \frac{k_A}{k_B} \tag{14.15}$$

Substituting equation 14.14 for two solutes in equation 14.15, we obtain an expression that permits the determination of selectivity factor from an experimental chromatogram

$$\alpha = \frac{(t_r)_B - t_m}{(t_r)_A - t_m} \tag{14.16}$$

Column efficiency

The column efficiency can be expressed in terms of the number of theoretical plates in the column. A theoretical plate is a hypothetical stage in which two phases such as the liquid and vapour phases of a substance establish an equilibrium with each other. The performance of many separation processes is enhanced by providing more of such equilibrium stages. A theoretical plate represents a single equilibrium step. The more the theoretical plates, the greater is the resolving power. The number of plates or efficiency can be obtained from a chromatogram:

$$N = 16 \left(\frac{t_r}{w_b} \right)^2$$

where N is the number of plates of a column, t_r is the retention time, and w_b is the baseline width, which is determined by extending the tangent lines from the inflection points on either side of the peak through the baseline.

14.3.1.2 Thin-layer chromatography (TLC)

TLC is a type of chromatography where the stationary phase is a thin layer of finely divided adsorbent spread on a glass or aluminum plate or plastic strip (20×20 cm) and a liquid mobile phase that consists of an organic solvent or a mixture of solvents.

Stationary phase for TLC
The most commonly used stationary phases used in TLC are silica gel, alumina, cellulose, and ion-exchange resins, coated to a depth of 0.15–0.5 mm on the plate. A substance called binders, such as plaster of paris, gypsum, or poly (vinyl alcohol) (5–10 wt%) is added to the stationary phase to improve the adhesion and mechanical strength of the TLC plate.

To prepare a TLC plate, aqueous slurry of the adsorbent powder is prepared and spread on the plate in a thin and uniform film. The slurry can be spread manually using a spatula or another plate or by using a spreading adapter (commercially available) to ensure uniform thickness. The solvent is evaporated off by drying in the air, and activated by heating in an oven at a temperature of 110 °C for a few hours since adsorbed water prevents other polar molecules from reaching the surface; thus, heating is required to remove the adsorbed water. Ready-made plates are available commercially though they are more expensive, but undoubtedly more convenient and offers good reproducibility. Silica gel contains silanol groups on its surface acting as active sites, forming hydrogen bonds with the polar molecules. Alumina, which contains –OH (hydroxyl) groups or oxygen atoms, is preferred for the separation of weakly polar compounds, whereas silica gel is preferred for strongly polar compounds such as amino acids.

Mobile phase for TLC
The mobile phase used in chromatography must be of high purity. A single solvent or mixture of two or three solvents is used as a mobile phase for TLC. The presence of a small amount of water or any other impurity can result in irreproducible chromatograms.

Development of chromatogram
Once the adsorbent plates are ready, a sample is spotted 2–3 times on the plate, approximately 2 cm above the lower edge of the plate. Spotting is done using a micropipette and the plate is placed in a jar containing the mobile phase, which is presaturated with the mobile phase vapors. The plate is kept in such a way that only the bottom of the

plate is dipped in the solvent, and not the sample spot. The solvent moves up the plate by capillary action, and the sample components move up at different rates, depending upon their solubility and degree of retention by the stationary phase.

After the development of the chromatogram, the individual component spots are noted if they are colored or are made visible by viewing the TLC plate in UV (ultraviolet) light or exposing the plate to iodine (I_2) vapors, or by spraying the color-developing reagent (e.g., ninhydrin for amino acids). A measure of the progress of each spot up the plate is given by the retention factor value (R_f), shown in Figure 14.5. The R_f value of a compound is equal to the distance travelled by the compound divided by the distance traveled by the solvent front, i.e.,

$$R_f = \frac{\text{Distance travelled by a solute}}{\text{Distance travelled by the solvent (solvent front)}}$$

The smaller the R_f value of a component, the more it is retained by the stationary phase. R_f values are always equal or less than unity because a solute cannot migrate farther than the solvent front.

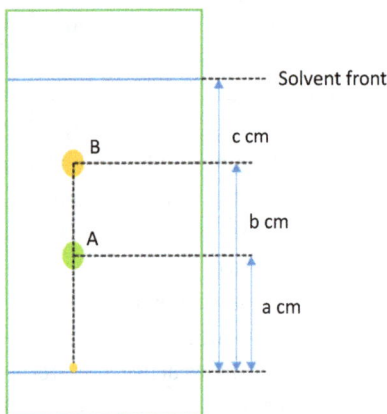

$$R_f(A) = \frac{a}{c} \qquad R_f(B) = \frac{b}{c}$$

Figure 14.5: Distance travelled by components A and B on a TLC.

14.3.1.3 High-performance thin-layer chromatography (HPTLC)

High-performance TLC (HPTLC) is a sophisticated and automated form of TLC with better and advanced separation efficiency and detection limit. There are three chief features of HPTLC that differentiate it from conventional TLC.

i. Quality of adsorbent layer:
The stationary phase of HPTLC is prepared by using purified silica gel with an average particle diameter of 5–15 μm and a narrow particle size distribution, instead of the

average 20 μm for conventional TLC. The use of very fine particles results in faster and more efficient separations.

ii. Method of sample application:
The introduction of the sample into the adsorbent layer is done by a mechanical applicator, which reduces the diameter of sample spots. The typical sample volume is 0.1–0.5 μL, which gives starting spots of 1–1.5 mm in diameter unlike the diameter of 3–6 mm in conventional TLC. The compact starting spots give a detection limit about 10 times better than the conventional TLC. Also, compact spots allow an increase in the number of samples applied onto the HPTLC plate.

iii. Mobile phase:
The movement of the mobile phase is slowed down in HPTLC because of the very fine particles of the adsorbent layer. To overcome this limitation, a forced flow technique has been employed using a pressurized chamber, which will pump up the mobile phase at a constant velocity [3].

HPTLC allows the processing of many samples in parallel, providing a low-cost analysis of a simple mixture, for which the sample workload is high. The greatest application of HPTLC is in the area of clinical biochemistry (e.g., analysis of drugs in the blood) and environmental analysis.

14.3.2 Paper chromatography/partition chromatography

PC was discovered by Synge and Martin in 1943. It is a type of planar chromatography where a sheet of cellulose filter paper, such as Whatman no. 1, serves as the separation medium. The water adsorbed in the pores of filter paper acts as a stationary phase, whereas the mobile phase can be a solvent or a mixture of solvents [4, 5]. The predominant mechanism in this type of chromatography is partition, in which the various components get distributed or partitioned between the liquid phases, depending on their affinity toward the stationary and the mobile phase.

The paper comes in various porosities (fine, medium, and coarse), which determines the rate of movement of the developing solvent. Low porosity paper gives slow movement to the solvent as also good resolution. The mobile phase that is used depends upon the nature of the substance to be separated. The sample should be sparingly soluble in the solvent; if it is too soluble, the distribution coefficient will strongly favor the mobile phase; the components will move with the solvent to the front, and poor resolution will result.

The sample is dissolved in a volatile solvent and applied to the paper above 2 cm from the lower edge of the filter paper, as a spot, using a syringe or a micropipette. The spot should be restricted to about 2 mm in diameter because larger spots result in poor separation. Once spotting is done, the paper is kept in a chamber containing a developing solvent, presaturated with the mobile phase vapors. The spot should not be dipped in the solvent, and the paper should not touch the boundary of the jar; it has to be straight.

Depending on the direction of the flow of the mobile phase, PC is categorized into three types:

14.3.2.1 Ascending paper chromatography

In ascending chromatography, the chromatogram is developed by placing the bottom edge of the suspended paper (not the spots) into the developing solvent, which ascends through the paper by capillary action. The paper is supported by a hook or a dip (Figure 14.6).

Figure 14.6: Apparatus for ascending paper chromatography.

14.3.2.2 Descending paper chromatography

In this type of chromatography, the development of a chromatogram is done by allowing the solvent to travel down the paper. The spotted edge of the paper strip is immersed in a trough at the top of the chamber containing the solvent. This descending chromatography is often preferred over the ascending chromatography since the downward flow of the solvent is governed by both capillary action and gravity, which helps the solvent to move farther, as compared to the ascending chromatography as shown in Figure 14.7.

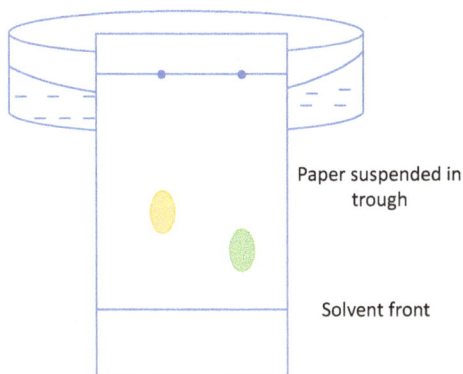

Figure 14.7: Apparatus for descending paper chromatography.

14.3.2.3 Radial chromatography

In radial or circular chromatography, a circular paper is used and the sample is applied at the center. After drying the spot, a small hole is made at the center of the spot, through which a thin cotton wick is introduced. The upper end of the wick is passed through it and the lower portion is dipped in the solvent taken in a petri dish. The solvent spreads out radially, and the component gets separated in the form of concentric circular zones as illustrated in Figure 14.8.

After the separation, the solvent front is marked and dried. If the separated components are colored, detecting the component is easy but if the components are colorless, they are sprayed with or dipped in a visualization reagent or detected under UV light.

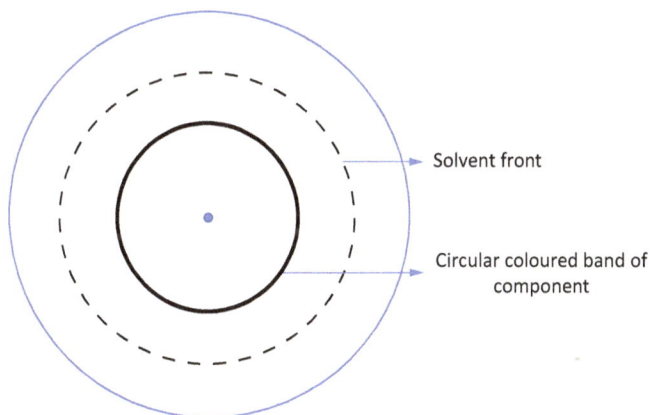

Figure 14.8: Chromatogram of radial/circular chromatography.

The movement of substances relative to the solvent is measured by the R_f value, which is a function of the partition coefficient, but the component has a specific R_f value, and an unknown compound can be identified by comparing their R_f values with the literature values.

14.3.2.4 High-performance liquid chromatography (HPLC)

High-performance liquid chromatography (HPLC) is an important qualitative and quantitative technique used to separate, identify, or quantify each component in a mixture. It is generally used for the estimation of pharmaceutical, biological, forensic, environmental, and polymer samples [6]. HPLC is basically a highly improved form of column LC. In column chromatography, solvent drips through a column under gravity, whereas in HPLC, it is forced through under high pressure of up to 400 atm [7]. HPLC has, over the past decades, become the method of choice for the separation of complex mixtures. Its main advantage over GC is that analytes do not have to be volatile; so macromolecules are suitable for HPLC analysis.

Conventional LC was done particularly in large columns with large particles under gravity, with a manual collection of eluents for measurement in the spectrophotometer, but in HPLC, one could use small particles of stationary phase, which increases the surface area under the increased flow pressure which, results in better separation and increased column efficiency.

HPLC is accomplished by the injection of a small amount of sample into a moving mobile phase, which passes through the column packed with the stationary phase. The separation of a mixture into its components depends on the degree of retention of each component in the column, which is determined by its partitioning of components between the liquid mobile phase and the stationary phase. This partitioning is affected by the relative solute/stationary phase and solute/mobile phase interaction.

The composition of the mobile phase has a huge impact on the separation of components unlike GC, where the mobile phase is just a carrier gas. Since the interaction of different components of a mixture with the mobile phase and stationary phase is different, they exit the column at different times, i.e., they have different retention times (time between injection and retention) [8].

HPLC consists of five principal parts.

Mobile-phase supply system

HPLC instrument is equipped with one or more glass reservoirs, each of which contains 500 mL or more of a solvent. The solvent must be pure and degassed to remove dissolved gases and gas bubbles from the mobile phase as they interfere with the performance of most detectors. The most common ways of degassing are sparging with helium, which removes 80% of the air from the mobile phase or vacuum degassing, which removes around 60% of air. Stainless steel or fritted glass filter removes particulate matter/dust from the mobile phase so that damage will not occur to the pumping system or column.

There are two types of elution in HPLC: isocratic elution and gradient elution. When elution takes place with a single solvent or mixture of the solvents of constant composition, it is termed isocratic elution. When two, and sometimes more, solvent systems that differ significantly in polarity and are varied in composition during the separation are used, it is termed as a gradient elution. It can be achieved in a preprogrammed way by changing the ratio of the solvents, either continuously or in a series of steps. Modern HPLC instruments have proportionating valves that introduce liquids from two or more reservoirs at ratios that can be varied continuously.

Solvents used for HPLC should be of HPLC grade, i.e., they should have been filtered through 0.2 μm filters, which reduces contamination from particulate matter, thereby extending the pump life.

Pumping system

A pumping system is used to provide accurate compositions, flows, and the pressure necessary to force the mobile phase through the tightly packed column, and often operates at a pressure ranging from 500–5,000 psi [9–11].

The most commonly used pump for HPLC is the reciprocating pump. This device consists of a small cylindrical piston chamber that is filled and emptied by the back-and-forth motion of the piston. This pumping motion produces a pulsed flow that must be damped; otherwise pulses appear as a baseline noise in the chromatogram. Reciprocating pumps have many advantages. They have small internal volume (35–400 μL), constant flow rates, high output pressure, and can be used for gradient elution, independent of the solvent viscosity or column backpressure. The other pumps used are motor-driven syringe pumps and pneumatic pumps (constant pressure pumps).

Sample injection system

The method used for the introduction of samples in LC is based on a sampling loop. These devices have interchangeable loops capable of providing a choice of sample size ranging from 1 to 100 μL or more. Modern HPLC instruments incorporate an autosampler with an automatic injector, which can introduce variable volumes of the sample from containers on the autosampler into the column. The major limitation of this injector is that the sample size is fixed, and the loop has to be changed to vary the sample injection size.

Column

A liquid chromatographic column is usually constructed from stainless steel tubing, which ranges from 5 to 25 cm in length and its inside diameter is between 3 and 5 mm. The most commonly used column is 10 or 15 cm long, with 4.6 mm inside diameter, and is packed with 5 μm particles.

Fast chromatography can be done with a wide bore and short columns, but better resolution can be obtained by using narrow columns.

A guard column or a pre-column, which is 3 to 10 cm in length, is placed between the injector and the analytical column. It has the same packing material as that of the analytical column. The sample is run through the guard column first, which retains highly sorbed compounds and particulate matter, and prevents them from reaching the analytical column; otherwise, it would contaminate the column by permanently adsorbing on it, and not eluting out. The guard column extends the life of an analytical column and it must be replaced or regenerated periodically.

Column temperature control

The temperature control of the column is usually not necessary for LSC, but is generally required for other forms of LC like liquid–liquid and ion-exchange. The most modern HPLC instruments are equipped with heaters that control the temperature from near room temperature to 150 °C. If the column is operated at a higher temperature than room temperature, a cooling jacket is placed between the ends of the column and the detector to bring the mobile phase to normal temperature.

Column packing

There are two types of packings used in HPLC; pellicular and porous particles. The pellicular particles are spherical, nonporous, glass or polymer beads, having a diameter of 30 to 40 μm, which are coated with a thin porous layer of silica, alumina, or ion exchange resin. Nowadays, small pellicular particles having ~5 μm diameter are used, which have completely replaced large pellicular particles for the separation of proteins and large macromolecules.

The porous particles have a diameter ranging from 3 to 10 μm, and are composed of silica, alumina, or an ion-exchange resin. The most commonly used packing material is silica, which is often coated with thin organic films that are chemically or physically bonded to the surface.

Detectors

The detector for HPLC should have high sensitivity, which ranges from micrograms to nanograms. The most widely used detectors are:

a) *Differential refractometer detector*: It is also called a universal detector, which measures the changes in the refractive index of the solvent caused by the analyte molecules. It detects concentrations between 10^{-5} and 10^{-6} g/mL (10 to 1 ppm).

b) *Ultraviolet (UV) detector*: It is the most commonly used detector for HPLC that has much better sensitivity than the refractometer detector (of about 10^{-8} g/mL i.e., 0.01 ppm). It cannot be used with solvents that have significant absorption in the UV or with sample components that do not absorb in the UV. During analysis, the sample goes through a glass cell, called a flow cell. When UV light is irradiated on the flow cell, the sample absorbs part of the UV light. Thus, the intensity of the UV light for the mobile phase and the eluent containing the sample will differ. By measuring the difference, the amount of the sample can be determined. A standard UV detector works in a wavelength range between 195 to 370 nm, and the most commonly used wavelength is 254 nm.

The most popular HPLC detector is the UV-Vis detector, which can measure nanogram amounts of UV absorbing analyte or those with suitable chromophores that absorb in the visible region.

c) *Fluorescence detector*: Fluorescent detectors are the most selective, sensitive, and specific than all other HPLC detectors. Fluorescence is 10–1,000 times more sensitive than UV detectors for strong UV-absorbing compounds. A specific wavelength is used to excite and then emit a light signal in the analyte atom. The intensity of the light is continuously monitored to quantify the concentration of the analyte. Compounds that do not have fluorescence absorbance are treated with fluorescence derivatives [12].

Types of HPLC

HPLC is used either in liquid–solid adsorption chromatography form or in liquid–liquid partition chromatography form. The polarity of a solute is an important criterion in the separation of mixtures in both partition and adsorption chromatography. The most common is partition chromatography, which is further divided into normal-phase chromatography and reverse-phase chromatography. Liquid–liquid partition processes are preferred for the separation of members of homologous series, whereas adsorption processes are more preferred for the separation of isomers having different steric configurations.

The highly polar materials are best separated using partition chromatography, while nonpolar materials are separated using adsorption chromatography. In adsorption chromatography, the adsorbent is usually kept constant and eluting solvent polarity is increased until elution is done. Polarity order of some commonly used solvents:

Petroleum solvents (hexane, petroleum ether) < cyclohexane < trichloroethane < toluene < dichloromethane < chloroform < diethyl ether < ethyl acetate < acetone < ethanol < water

In normal phase chromatography, the adsorbent is polar and the mobile phase is nonpolar. The stationary phase is a bonded siloxane with a polar functional group that retains polar compounds in preference to nonpolar compounds. In this process, a least polar component is eluted first, and increasing the polarity of a mobile phase decreases the elution time. In reverse phase chromatography, the stationary phase is nonpolar, often a hydrocarbon, and the mobile phase is a relatively polar solvent, such as water, methanol, tetrahydrofuran, etc. In this process, the most polar compound is eluted first, and increasing the mobile phase polarity increases the elution time. A wide range of organic compounds can dissolve in the mixed water-organic solvent phases and can be separated, so reverse phase chromatography is by far the most popular form of HPLC. If a sample is highly nonpolar and insoluble in a water mixture, normal phase chromatography is used; otherwise, reverse phase chromatography will be used.

The combination of LC and mass spectrometry (MS) would seem to be an ideal combination for separation and detection. A mass spectrometer could identify species as they elute out from the chromatographic column as is in GC/MS system. The coupling of these two techniques has some limitations. Firstly, a gas phase sample is needed for MS, while the output of the LC column is a solute dissolved in a solvent. Therefore, the solvent has to be removed before it is introduced into the mass spectrometer. Also, the separation of analytes that are nonvolatile and thermally labile is difficult. It took several years to develop such a technique. Today, there are several interfaces that make LC-MS a routine technique. The commonly used interfaces are electrospray ionization (ESI) source, thermospray ionization (TSI), atmospheric pressure chemical ionization (APCI), and particle beam ionization.

14.3.3 Size-exclusion chromatography

Size-exclusion chromatography is a powerful technique applicable to high molecular mass species [13–15]. It is also called molecular or gel chromatography. In this type, the stationary phase is a molecular sieve, i.e., the packing consists of a small (~10 μm) silica or polymer particles of uniform pores into which solute and solvent molecules can diffuse. The silica particles are more rigid, which permits easier packing and high pressure. The molecules that are larger than the average pore size of the packing are excluded, and pass straight through the column at the rate of the mobile phase. Molecules that are smaller than the pores can penetrate through the pores and entrap in them for a greater time, and thus are the last ones to elute out. The intermediate size molecules are retarded to a varying degree and will be eluted out in order of their decreasing molecular size. The size-exclusion separations differ from other chromatographic procedures in the respect that there are no chemical or physical interactions between the analytes and the stationary phase.

The term "gel filtration chromatography" is used when the mobile phase is water, and "gel permeation chromatography" is used when the mobile phase is an organic solvent. This technique can be used for the separation of polymers, proteins, enzymes, peptides, polysaccharides, and so on. The gels or the packing material must be equilibrated for a few hours to a day or more with the solvent to be used. Sephadex is the popular molecular sieve material for the separation of proteins. Other materials used are bio-gel, styragel, etc.

14.3.4 Ion-exchange chromatography (IEC)

This type of technique is particularly well suited for the separation of inorganic ions – both cations and anions – because separation is based on an exchange of ions in the stationary phase. The solid stationary phase consists of beads made up of a polystyrene polymer cross-linked with divinylbenzene. The cross-linked polymer (resin) has a free phenyl group attached to the chain, which can be treated to add functional groups. If a mixture of two or more different cations is passed through an ion-exchange column and their quantities are small as compared to the capacity of the column, then it might be possible to recover the absorbed ions by using a suitable eluting solution. If one cation is held more firmly by the resin in a mixture of two cations, then the cation that is less firmly held will be eluted first, followed by the cations that are more firmly adsorbed on the resin. This separation technique is ion-exchange chromatography (IEC) and gives spectacular results in the separation of a complex mixture of closely related compounds such as amino acids and lanthanides.

If a mixture of a solution of several ions of similar charge (say B, C, etc.) is being poured into the ion-exchange column, then different elution curves corresponding to each ion will be obtained by using an appropriate eluting agent. If elution curves are sufficiently apart (Figure 14.9), a quantitative separation is possible, but if the curves overlap, then an incomplete separation is obtained.

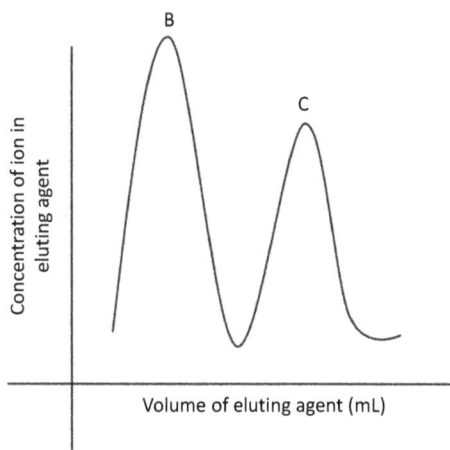

Figure 14.9: Chromatogram showing the elution of components B and C.

The rate at which two constituents separate in the column is determined by the ratio of their distribution coefficients:

$$\alpha = \frac{K_{d1}}{K_{d2}}$$

where α is the separation factor used as a measure of the chromatographic separation possible, and K_{d1} and K_{d2} are the distribution coefficients of the two constituents.
  Also,

$$K_d = \frac{\text{Amount of solute on resin}}{\text{Weight of the resin (g)}} \div \frac{\text{Amount of solute in solution}}{\text{Volume of solution(mL)}}$$

The greater the deviation of α from unity, the easier will be the separation.

  The principle of separation is the reversible exchange of ions between the ions present in the solution and those present in the ion-exchange resin. Ion-exchange separations are mainly carried out in columns packed with an ion exchanger (stationary phase). There are two types of ion exchangers:
i. Cationic exchangers
ii. Anionic exchangers

Cation-exchange resins: They possess negatively charged groups, and attract positively charged groups. These resins contain acidic functional groups added to the aromatic ring of the resin. The cation exchangers have sulfonic acid groups ($-SO_3H$), which are strong acids, much like sulfuric acid (H_2SO_4). The weak acid cation exchanger has carboxylic groups ($-COOH$), which are only partially ionized. The protons on these groups can exchange with other cations:

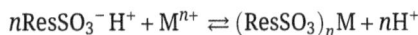

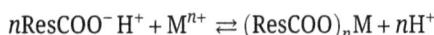

$$n\text{ResSO}_3{}^- \text{H}^+ + \text{M}^{n+} \rightleftharpoons (\text{ResSO}_3)_n\text{M} + n\text{H}^+$$

$$n\text{ResCOO}^- \text{H}^+ + \text{M}^{n+} \rightleftharpoons (\text{ResCOO})_n\text{M} + n\text{H}^+$$

where Res represents a resin and M^{n+} represents a cation.

  Cation-exchange resins are usually available in the hydrogen ion (H^+) form, which can easily be converted into sodium ion (Na^+) by treating it with sodium salt. The sodium ions then undergo exchange with other cations. When a cation exchanger comes in contact with the mobile phase, then the cation in the mobile phase replaces the cation in the organic resin till equilibrium is reached.

  The total number of equivalents of replaceable hydrogen per unit volume or per unit weight of the resin is called the exchange capacity of the resin, which affects the retention of the solute on the resin. Strong acid resins are used over the pH range of 1 to 14 and weak acid resins are used in a pH range of 5 to 14.

Anion-exchange resins: Anion-exchange resin has positively charged groups, which attracts negatively charged molecules. It is a polymer containing amines or quaternary ammonium groups as integral part of the polymer lattice in which basic groups on the resin, i.e., hydroxyl anion (OH^-) can be exchanged with other anions. These are strong base groups (e.g., quaternary ammonium groups) and weak base groups (amine group):

$$n\text{ResNR}_3{}^+\text{OH}^- + \text{M}^{n-} \rightleftharpoons (\text{ResNR}_3)_n\text{M} + \text{OH}^-$$

$$n\text{ResNH}_3{}^+\text{OH}^- + \text{M}^{n-} \rightleftharpoons (\text{ResNH}_3)_n\text{M} + \text{OH}^-$$

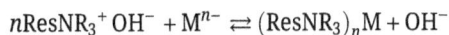

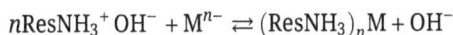

The strong base exchanges can be used over a pH range of 0–12, whereas weak base exchangers work in the pH range of 0–9.

Fundamental requirement of a resin

i. The resin must be sufficiently cross-linked to have negligible solubility.
ii. The resin must be sufficiently hydrophilic so that diffusion of ions through the structure can take place at a finite and usable rate.
iii. The resin must be chemically inert and stable, and should not undergo any kind of reaction with the eluting agent.
iv. The resin must have sufficient number of accessible ion-exchange groups.

Procedure for ion-exchange chromatography

i. Ion-exchange separations are carried out in a column packed with an ion exchanger, which contains charged groups covalently bonded to the surface of a highly cross-linked insoluble matrix.
ii. The choice of an ion-exchanger resin depends on the ions to be separated. To separate anions, an anion exchanger is used and to separate cations, a cation exchanger is used.
iii. Ion-exchange resin is firstly soaked in water and then the fully swollen resin is put into the column. The sample is then poured into the column, followed by the buffer (e.g., acetate buffer, citrate, and phosphate buffer).
iv. Based upon the affinity of ions toward the resin, the ions, like cation and anion, are separated. The ions that have less affinity toward the resin will elute first and the ions that have more affinity will elute later.

Application of IEC

i. IEC can be applied for the separation and purification of many molecules such as proteins, peptides, enzymes, antibiotics, and vitamins.
ii. Softening of hard water is best achieved using IEC. The hardness of water is due to the presence of Ca^{2+} (calcium), Mg^{2+} (Magnesium), and other divalent cations, which can be removed by passing the hard water through the cation exchanger containing Na^+ ions. The Ca^{2+} and Mg^{2+} ions from water are retained on the column, while Na^+ ions pass into the solution. These Na^+ ions are harmless for washing purposes.
iii. Deionized water can be obtained using this technique. It requires the complete removal of both cations and anions. For this, water is passed through an acidic cation exchanger where cations in water are replaced by H^+ ions on resins, and then water is passed through a basic anion exchanger where anions present in the water are

exchanged by OH$^-$. The H$^+$ and OH$^-$ ions that pass into the solution combine to form unionized water.
iv. Ion-exchange methods can be used to separate the complex mixture of amino acids obtained by the acid hydrolysis of proteins.
v. This method can also be used in the separation of a mixture of lanthanides that are quite similar in their properties.

14.3.4.1 Ion Chromatography

Ion chromatography (IC) was introduced in 1975 by Small, Stevens, and Baumann as a new analytical method. This technique is the high-performance form of IEC, which combined the separating power of ion exchange with the universal conductivity detector. Conductivity detectors were considered to be ideal detectors because they are highly sensitive, universal for charged species, and respond in a predictable way to concentration changes. Furthermore, such detectors are simple to operate, inexpensive to construct and maintain, easy to miniaturize, and usually give prolonged, trouble-free service. In normal IEC, a conductivity detector is limited in use because of the high conductivity of the eluting agent. In 1970, William Bauman at Dow Chemical Company suggested that this background conductivity of eluting agent can be removed by using a second ion-exchange column, which is called a suppressor column. This column is packed with a second ion-exchange resin that efficiently converts the elution solvent ions into molecular species with limited ionization, without affecting the conductivity caused by the analyte ions. For example, when cations are separated and determined, the eluent chosen is hydrochloric acid (HCl), and the suppressor column is an anion exchange resin in the hydroxide form (OH$^-$). The product of the reaction in the suppressor is water, i.e.,

$$H^+ (aq) + Cl^- (aq) + ResOH^- \rightleftharpoons ResCl^- + H_2O$$

The analyte cations are not retained by suppressor column.

For anion separations, the suppressor column is packed with the acid form of a cation-exchange resin, and sodium bicarbonate (NaHCO$_3$) or carbonate (Na$_2$CO$_3$) is the eluting agent:

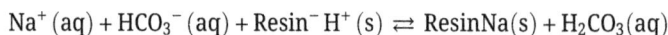

$$Na^+ (aq) + HCO_3^- (aq) + Resin^- H^+ (s) \rightleftharpoons ResinNa(s) + H_2CO_3(aq)$$

The carbonic acid (H$_2$CO$_3$) does not contribute much to conductivity measurement.

The major limitation with the original suppressor columns was the need to regenerate them periodically (typically every 8–10 h) in order to convert the packing back to the original acid or base form. To overcome this limitation, micromembrane suppressors were introduced in the 1980s [16].

14.4 Gas chromatography (GC)

GC is one of the most widely used sensitive analytical techniques for the effective separation of thermally stable and volatile organic and inorganic compounds. This technique was developed by Martin and Synge in 1941. The separation mechanism depends on the distribution of components of the vaporized sample into the mobile gaseous phase and a solid or a liquid (supported on a capillary wall) stationary phase [17, 18].

In this technique, a sample is vaporized, if it is not already a gas, by injecting it into a heating port, and introduced into the column through the head of the chromatographic column. Elution is done by the flow of an inert gaseous mobile phase, which does not interact with the molecules of the analyte, unlike other types of chromatographic techniques. The only function of the mobile phase is to transport the analyte through the column and separate the components of a sample, which are recorded as a sequence of the peak as they leave the column. The different components of the sample are separated and eluted at different times, which is called the retention time. The number of peaks determines the number of components present in the given sample. The identity of the components is determined by their characteristics, and the retention time and quantity are determined by the area under the peaks [19].

There are two types of GC: GSC (adsorption) and gas–liquid chromatography (GLC) (partition). In GSC, the mobile phase is a gas and the stationary phase is a solid that retains the analyte by physical adsorption and because of this reason, it has limited applications. GLC is the most important of the two, where the mobile phase is a

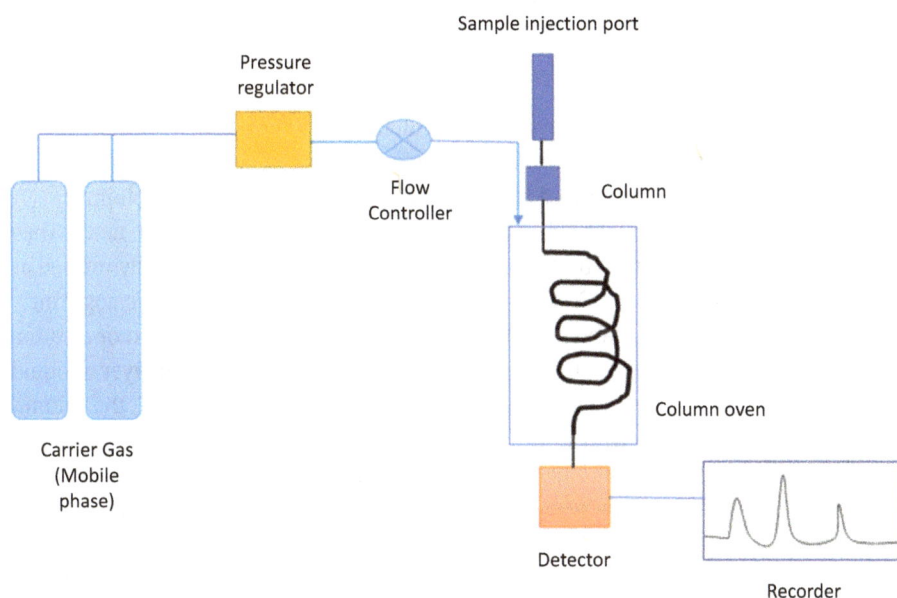

Figure 14.10: Schematic diagram of gas chromatography.

gas and the stationary phase is liquid retained on the surface of an inert solid by adsorption or chemical bonding.

The most important parameter in GC is the selection of the stationary phase, since the nature of the solid or liquid phase determines the exchange equilibrium with the component of the sample.

Generally, all GSC or GLC consist of six basic components as shown in Figure 14.10.

i. *Sample injection system*: The sample is introduced at the head of a chromatographic column through a silicone rubber septum, calibrated using a hypodermic micro syringe. The sample injection port, vaporization chamber, and detector are heated 50 °C above the lowest boiling point of the sample. The injection port and the detector are usually kept warmer than the column to promote rapid vaporization of the injected sample and prevent sample condensation in the detector. For packed columns, sample size ranges from 0.1 to 10 µL for liquid sample, and 1 to 10 mL for gaseous samples. Capillary columns require samples that are smaller by a factor of 100 or more because of the lower capacity of a column and often, sample splitter is needed to deliver such a small fraction of the injected sample, with the remainder going to waste.

Nowadays, gas chromatographs use auto injectors and auto samplers for the most reproducible sample injection.

ii. *Carrier gas*: The mobile phase gas in GC is called carrier gas, which is chemically inert and available in a pure form such as argon, helium, nitrogen, or hydrogen. A carrier gas is supplied at high pressure and passed to the instrument at a rapid and reproducible rate. The pressure regulator, flow meter, and gauzes are required to control the flow rate of the gas. The best efficiency is shown by high-dense gas because of lower diffusivity, but a low-density gas gives higher speed.

iii. *Separation column*: There are two types of columns used in GC; (a) open tubular columns or capillary columns, and (b) packed columns. For most current applications, packed columns have been replaced by more effluent and faster capillary columns.

The length of the capillary column varies from less than 2 to 60 m or more. They are made up of stainless steel, fused silica, glass, and Teflon. They are usually formed as coils having a diameter of 10–30 cm so that they can fit into an oven for thermosetting.

a) Capillary columns: The capillary column can be of two types: wall-coated open tubular (WCOT) and support-coated open tubular (SCOT). WCOT has a thin layer of liquid stationary phase coated along the wall of the column, whereas in SCOT, the column walls are first coated with a thin layer of adsorbent solid such as diatomaceous earth on which liquid stationary phase is adsorbed. SCOT columns are capable of holding a greater volume of stationary phase due to the greater sample capacity, than a WCOT column. Generally, the efficiency of the SCOT column is less than that of the WCOT column but significantly greater than that of the packed column. The most widely used capillary column is the coated fused silica open tubular column with an inside diameter of 0.25 and 0.32 mm. These columns are made of thin fused silica (SiO_2), coated on the outside with polyimide polymer for support and protection of the silica capillary. The resulting columns are quite flexible and can be bent into coils with a diameter of a few inches. Capillary columns offer the advantage of high resolution with narrow peaks, short analysis time, and high sensitivity.

b) Packed columns: Modern packed columns are typically 2–3 m long and have an inside diameter of 2–4 mm. These columns are fabricated from glass or metal tubing, and packed with uniform and finely divided packing material, which is coated with a thin layer (0.05–1 µm) of the stationary liquid phase. The column is packed with small particles that serve as stationary phase or solid support, coated with a nonvolatile liquid phase of varying polarity. GSC is useful for the separation of small gaseous species, such as hydrogen, nitrogen, CO_2 (carbon dioxide), and CO (carbon monoxide) using inorganic packings as solid support, such as alumina (Al_2O_3) or porous polymers, e.g., chromosorb (polyaromatic cross-linked resin). The solid support for a liquid phase consists of small, uniform spherical particles with good mechanical strength and high specific surface area, which should be chemically inert but wettable in a liquid phase.

    The most widely used packing for GC is prepared from diatomaceous earth, which is sold under different names such as Chromosorb W and Chromosorb P. The polarity of these materials is reduced by treating them chemically with dimethylchlorosilane, which gives a surface layer of methyl groups.

c) Stationary phase: Column packing material is mixed with the correct amount of liquid phase dissolved in a low boiling solvent such as acetone. After coating, the solvent is evaporated by heating, and kept in a vacuum to remove the last traces of solvent. The column is conditioned at elevated temperature by passing a carrier gas through it for several hours. The efficiency of the gas chromatographic column increases rapidly with decreasing particle size. The usual support particles are 60–80 mesh (250 or 170 µm) or 80–100 mesh (170–149 µm).

A proper choice of the stationary phase is a crucial step for successful separation. Liquid stationary phases are selected based on polarity, keeping in mind that "like dissolves like," which means that a polar stationary phase will interact more with the polar compounds, and vice-versa. To have a reasonable residence time on the column, an analyte must show some degree of solubility with the stationary phase. Nonpolar liquid phases are generally nonselective because there are few forces between the solute and the solvent; hence, the separation tends to follow the order of the boiling points of the solutes, with the low boiling one eluting first. Polar liquid phases exhibit various interactions, like dipole interaction and hydrogen bonds and thus there is not necessarily the same elution correlation with volatility.

    The most common phases for fused silica columns are polysiloxanes and polyethylene glycols (Carbowax). Commercial columns are available, having stationary phases that vary in thickness from 0.1 to 5 µm. Thick films are used with highly volatile analytes because such films can retain solutes for a longer time, thus providing greater time for separation, whereas thin films are used for separating low volatile analytes in a reasonable length of time. A film thickness of 0.25 mm or 0.32 mm is recommended for most applications.

iv. *Column oven or thermostat chambers*: The thermostat oven controls the temperature of the column, which is operated in two manners: isothermal programming (temperature of the column is held constant throughout the separation) and temperature programming (column temperature is increased continuously as the separation progresses).

v. *Detectors*: A detector is placed at the end of the column and gives the quantitative measurement of the components of the mixture as they elute out along with the carrier gas. The ideal detector for GC should have adequate sensitivity, good stability and reproducibility, short response time, high reliability, ease of use, nondestruction of the sample, and a temperature range from room temperature to at least 400 °C. There is no ideal detector that has all the above-mentioned characteristics. The most widely used detectors are:

a) Flow ionization detector (FID): This is the most widely used and extremely sensitive detector employed in GC. In FID, the eluted sample is directed at an air/hydrogen flame where it undergoes pyrolysis or chemical decomposition through intense heating. Pyrolyzed hydrocarbons release ions and electrons that are collected at an electrode and create a current (measured by a high impedance picoammeter), allowing the solute to be detected. This detector permits the measurement of components in the ppb (parts per billion) concentration range. Helium, nitrogen, and argon are most frequently used as carrier gas. FID is insensitive to inorganic compounds, including water, and so an aqueous solution can be injected. FID has many advantages, like the detector is unaffected by flow rate, noncombustible gases, or water. However, this technique requires flammable gas, and also destroys the sample.

b) Flame-thermionic detector: This detector is similar in construction to FID. It is widely used for compounds containing nitrogen and phosphorous, such as organo-phosphorous pesticides and pharmaceutical compounds.

   It has a two-stage flame ionization detector, which is divided by an alkali salt coated wire mesh screen. The second detector is mounted over the first one so that gases from the first flame pass into the second flame. The eluted components enter the lower flame of this device, which acts as a conventional FID, whose response may be recorded. A small current normally flows into the second flame due to evaporation and ionization of sodium from the screen. If a substance containing nitrogen or phosphorous is burned in a lower flame, the ions from these greatly increase the volatilization of alkali metal from the screen. This produces an increased current (at least 100 times more), which is recorded by a detector, and a chromatogram is obtained.

c) Thermal conductivity detector (TCD): TCD was one of the earliest detectors used for GC. This device consists of a heated filament wire made up of platinum, gold, or tungsten whose temperature depends on the thermal conductivity of the carrier gas. As gas is passed through a heated filament wire, the temperature and thus the resistance of the wire will vary according to the thermal conductivity of the gas. It usually employs two detectors; one of which is used as a reference for carrier gas, and the other monitors the thermal conductivity of the carrier gas and the sample mixture. These filaments are in opposite arms of a Wheatstone bridge circuit that measures the difference in their resistance. When there is no sample gas in the effluent, the resistance of the wires will be the same. When a sample component is eluted with the carrier gas, a small resistance change will occur in the effluent arm. The change in resistance is proportional to the concentration of the sample component in the carrier gas, and is recorded on a recorder. Helium and hydrogen gas are preferred as carrier gas as they have high thermal conductivity, and even a small addition of the sample largely decreases the thermal conductivity of the column effluent, which is readily detected by it.

The advantages of TCD are that it is simple to use, has large dynamic range, responds to both organic and inorganic species, and exhibits nondestructive character. The greatest drawback of TCD is the low sensitivity of the instrument to other detection methods, in addition to flow rate and concentration dependency.

d) Electron capture detector (ECD): It is an extremely sensitive detector that responds selectively to the electronegative analyte compounds (capable of capturing electrons), such as polychlorinated organometallics and nitro-containing compounds. In this device, an eluted sample is passed over a radioactive β-emitter (nickel-63 or tritium):

$$Ni^{63} \rightarrow \beta$$

These negatively charged particles collide with carrier gas molecules (often nitrogen, because of their low excitation energy) and ionize them, which results in the production of electrons:

$$\beta^- + N_2 \rightarrow N^{2+} + 2e^-$$

The free electron results in the generation of a standing current. When an electronegative analyte is eluted out, it captures some of the electrons due to which the current decreases markedly. The detector is insensitive to functional groups such as amines, alcohols, and hydrocarbons. ECD has the advantage of not altering the sample significantly (in contrast to FID, which consumes the sample), but the linear response of the detector is limited.

e) MS detector: This detector is one of the most powerful detectors for GC. The combination of GC and MS is known as GC/MS [20, 21].

A mass spectrometer is a sophisticated analytical technique that measures the mass-to-charge ratio (m/z) of the charged particles and therefore can be used to determine the molecular weight and elemental composition. The sample mixture is first separated by GC and then the eluted component is directed into the MS for detection [22]. They are transported by a carrier gas, which continuously flows through the GC and into the MS, where it is evacuated by the vacuum system. The neutral molecules in the mass spectrometer are first ionized, most commonly by electron impact (EI) or chemical ionization. The gaseous molecules are bombarded by a high-energy beam of electrons (usually 70 eV) generated from the tungsten filament. An electron that collides with a neutral molecule may impact sufficient energy to remove an electron from the molecule, resulting in a singly charged ion M^+ (molecular ion/parent ion). This high-energy ionization can result in an unstable molecular ion, and excess energy can be lost through fragmentation reaction, producing fragments of lower mass that are themselves ionized or converted for ions for further electron bombardment.

The next step is to separate the ions of different masses, which is achieved on their m/z ratio by a mass analyzer. The most common mass analyzers are quadrupole and ion-trap analyzers, where ions with very small differences in m/z ratio are separated. Once the separation is done, they reach the ion detector, where the signal is amplified by an electron multiplier. The signal is recorded by the software on the computer to produce a chromatogram and a mass spectrum for each data point. With MS, we can not only determine that a peak is due to more than one component, but we can also identify the various unresolved species.

f) Other detectors: Electrolytic conductivity detector is used for compounds containing halogens, sulfur, and nitrogen, which are mixed with a reaction gas in a small reactor tube. The products are then dissolved in a liquid, which produces a conductive solution. The change in conductivity due to the presence of the compound is then measured and a chromatogram is obtained.

In the photoionization detector, eluted compounds are photoionized by UV radiation due to which ions and electrons are produced, which are collected by a pair of electrodes, and current is produced. This device is generally used for ionic compounds.

vi. *Amplification and recorder system*: They use special electronic circuits to process and amplify the signals coming from the detector so as to display it in an understandable graphical format that represents several peaks of the constituents of the sample under analysis.

Applications of GC:
i.   Environmental monitoring: GC plays a significant role in the identification and quantification of environmental pollutants like volatile organic compounds, polyaromatic hydrocarbons, pesticides, and halogenated compounds [23].
ii. It is used in food, beverages, flavors, and fragrance analysis [24].
iii. It has medicinal and pharmaceutical applications.
iv. Quantification of drugs and their metabolites in blood and urine for both pharmacological and forensic applications.
v.   Analysis of industrial products for quality control.

References

[1]   Ettre, L.S. Nomenclature for chromatography. Pure and Applied Chemistry 1993; 65: 819.

[2]   Cazes, J. (ed.) Encyclopedia of chromatography. New York: Marcel Dekker; 2001.

[3]   Journal of Chromatographic Science, 18 (1980) 324.

[4]   Eigsti, N.W. Paper chromatography for student research. The American Biology Teacher, pp. 123–134, 1967.

[5]   Coskun, O. "Separation technique: Chromatography", Northen. Clinics of Istanbul. 2016; 3(2): 156.

[6]   Martin, M. and Guiochon, G. Effects of high pressure in liquid chromatography". Journal of Chromatography A 2005; (1-2)(7): 16–38.

[7]   Calvin Gidding, J. Composition of the theoretical limit of separation. Analytical Chemistry 1890; 36: 1964.

[8]   Liu, Y. and Lee, M.L. Ultrahigh pressure liquid chromatography using elevated temperature. Journal of Chromatography. 2006; 1104(1-2): 198–202.

[9]   Isenberg, H.D. (ed) Clinical microbiology procedures handbook. Chapter 12.19 and chapter 14. 1994 updates. Washington, D. C: American Society for Microbiology; 1992.

[10] Duffey, P.S., Guthertz, L.S., and Evans, G.C. Improved rapid detection of mycobacteria by combining solid-phase extraction with high-performance liquid chromatography analysis of BACTEC cultures. J. Clin. Microbiol. 1996; 34: 1939–1943.

[11] Durst, H.D., Milano, M., Kikta, E.J., Jr., Connelly, S.A., and Grushka, E. Phenacyl esters of fatty acids via crown ether catalysts for enhanced ultraviolet detection in liquid chromatography. Anal. Chem. 1975; 47: 1797–1801.

[12] G. R. Chatwal, S. K. Anand, "Instrumental methods of Chemical Analysis (5th ed), Himalaya publishing house, Mumbai, 2624–2639 (2007).

[13] Striegel, A., Yau, W.W., Kirkland, J.J., and Bly, D.D. Modern size-exclusion chromatography. In: Practice of gel permeation and gel filtration chromatography, 2nd edn, Hoboken, NJ: Wiley; 2009.

[14] Wu, C.S. ed. Handbook of size exclusion chromatography, 2nd edn, New York: Dekker; 2004.

[15] Wu, C.S. ed. Column handbook for size exclusion chromatography. San Diego: Academic Press; 1999.

[16] Fritz, J.S. and Gjerde, D.T. Ion chromatography, 4th edn, Chs. 6–7 Weinheim, Germany: Wiley-VCH; 2009.

[17] McNair, H.M. and Miller, J.M. Basic gas chromatography, 2nd edn. Wiley: Hoboken, NJ; 2009.

[18] Grob, R.L. and Barry, E.F. (eds). Modern practice of gas chromatography, 4th edn, Hoboken, NJ: Wiley Interscience; 2004.

[19] http://chem.libretexts.

[20] Sparkman, O.D., Penton, Z.E., and Ketson, F.G. Gas chromatography and Mass spectrometry. 2nd edn, Amsterdam: Elsevier; 2011.

[21] McMaster, M.C.. GC/MS: A practical user's guide. 2nd edn, Wiley: New York; 2008.

[22] Turner, D.C., Schafer, M., Lancaster, S., JanMohamed, I., Gachanja, A., and Creasy, J., Gas chromatography mass spectrometry, Royal society of Chemistry; 2020, ISBN-10; 1782629289.

[23] Santos, F. and Galceran, M. The application of gas chromatography to environmental analysis. Trends in Analytical Chemistry. 2002; 21(9-10): 672–685.

[24] Lehotay, S.J. and Hajsolva, J. Application of gas chromatography in food analysis. Trends in Analytical Chemistry. 2002; 21(9-10): 686–697.

Index

https://doi.org/10.1515/9783110794816-015

www.ingramcontent.com/pod-product-compliance
Lightning Source LLC
Chambersburg PA
CBHW080918220326

41598CB00034B/5610